Thin Films

VOLUME 20

ORGANIC THIN FILMS AND SURFACES: DIRECTIONS FOR THE NINETIES

Thin Films

ORGANIC THIN FILMS AND SURFACES: DIRECTIONS FOR THE NINETIES

Edited by

Abraham Ulman

Alstadt-Lord-Mark Professor
Department of Chemistry and
Polymer Research Institute
Polytechnic University
Brooklyn, New York

Volume 20

ACADEMIC PRESS

San Diego New York Boston
London Sydney Tokyo Toronto

This book is printed on acid-free paper. ∞

Academic Press, Inc.
A Division of Harcourt Brace & Company
525 B Street, Suite 1900, San Diego, California 92101-4495

United Kingdom Edition published by
Academic Press Limited
24-28 Oval Road, London NW1 7DX

International Standard Serial Number: 1079-4050

International Standard Book Number: 0-12-523485-6

PRINTED IN THE UNITED STATES OF AMERICA
95 96 97 98 99 00 EB 9 8 7 6 5 4 3 2 1

The Physics of Thin Films, consisting of 19 volumes since 1963, is one of the longest running continuing series in thin film science. The series contains some of the highest quality studies of the properties of various thin film materials and systems.

Since the origination of the series, the thin film field has expanded many times, becoming the basis for a number of major industries, such as semiconductors, hard coatings, optoelectronics, and magnetic storage. Simultaneously with the development of inorganic thin films, a special field of organic thin films has emerged and is developing rapidly at present. This latter field is also becoming one of increasing importance in materials science research and applications (e.g., polymer science and biomaterials).

In order to reflect the development of today's science and to cover all modern aspects of thin films, this series, beginning with Volume 20, will move beyond the basic physics of thin films. It will address the most important issues of both inorganic and organic thin films, in their theoretical as well as their technological aspects. Correspondingly, the series title has been slightly modified to *Thin Films* to better reflect modern technology-oriented problems.

Starting with Volume 20, the series will be led by two new editors: Dr. Stephen M. Rossnagel of IBM at Yorktown, who will be responsible for the field of inorganic films and applications; and Dr. Abraham Ulman, Alstadt-Lord-Mark Professor at the Polytechnic University of Brooklyn, New York, whose responsibility will include organic films, their chemistry, technology, and applications.

The *Thin Films* series will continue to be one of Academic Press's premier offerings, with high-quality editions in the science and technology field.

Publisher and Editors

Contents

Preface xv
Contributors xvii

Introduction

Supramolecular Assemblies: Vision and Strategy

Hans Kuhn and Abraham Ulman

1.1. Introduction 1
1.2. Efforts to Achieve Well-Defined Organized Molecular Assemblies 4
1.3. Perspective in Supramolecular Engineering 6
References 7

Advanced Materials

Oriented Growth of Nanocrystalline Particulate Films at Monolayers: A Colloid Chemical Approach to Advanced Materials

Janos H. Fendler

2.1. Introduction 11
2.2. Experimental Methodologies and Characterizations 12
2.3. Growth of Nanocrystalline Particulate Films 19
2.4. Size Quantization and Bandgap Engineering 21
2.5. Semiconductor Particulate Films 23
2.6. Silver Particulate Films 26
2.7. Organizational Imperative: Molecular Recognition and Epitaxial Growth 29
2.8. Conclusions and Future Directions 37
Acknowledgments 37
References 38

Third-Level Self-Assembly and Beyond: Polar Hybrid Superlattices via Postassembly Intercalation into Noncentrosymmetric Multilayer Matrices of Hydrogen Bonded Silanes

Rivka Maoz, Ruth Yam, Garry Berkovic, and Jacob Sagiv

3.1. Introduction 41
3.2. The Multilayer Matrix 45
3.3. Intercalated Superlattices 48
3.4. Conclusion 62
Acknowledgments 65
References and Notes 65

Building Two-Dimensional Polymers by the Langmuir-Blodgett Technique

Serge Palacin, Florence Porteu, and Annie Ruaudel-Teixier

4.1. Introduction 69
4.2. General Properties of Amphiphilic Porphyrins in LB Films 69
4.3. Ionic Network 72
4.4. Two-Dimensional Covalent Polymer 76
4.5. Conclusion 81
Acknowledgments 82
References 82

Surface Modification

Polymer Surface Modification

Thomas J. McCarthy, Timothy G. Bee, Joan V. Brennan, Elisa M. Cross, Anthony J. Dias, Nicole L. Franchina, Kang-Wook Lee, and Molly S. Shoichet

5.1. Introduction: Polymer Surfaces and Dynamic Interfaces 85
5.2. Polymer Surface Modification 87
5.3. Chemical Modification 89
5.4. Hydroxylation of Polypropylene 91
5.5. Functionalization of Poly(chlorotrifluoroethylene) 95
5.6. Dehydrofluorination of Poly(vinylidene fluoride) 99
5.7. Indirect Syntheses of Carboxylic Acid-Functionalized Fluoropolymers 100

5.8. Gas Phase Chlorination of Polyethylene 102
5.9. Carbonyl Surface Chemistry of Poly(ether ether ketone) 103
5.10. Conclusion 104
Acknowledgments 104
References 104

Lithographically Patterned Self-Assembled Films

Jeffrey M. Calvert

6.1. Introduction 109
6.2. Lithographic Considerations for Ultrathin Films as Imaging and Resist Layers 110
6.3. Deep-UV Photochemistry of Aromatic Hydrocarbon Organosilane Films 112
6.4. Selective Electroless Metallization of Patterned SA Films 117
6.5. Pattern Transfer and Microelectronic Device Fabrication by Means of Selectively Metallized SA Films 125
6.6. Deep-UV Photochemistry of Thiol-Terminal Organosilane Films and Patterned Proteins 128
6.7. Mixed Monolayers, Coplanar Molecular Assemblies, and Controlled Wetting 130
6.8. Patterned Amine Surfaces as Reactivity Templates 133
6.9. Selective Cell Adhesion and Outgrowth on Patterned SA Films. 135
6.10. Conclusions and Future Areas of Research 138
Acknowledgments 139
References 139

Recognition at Surfaces

Langmuir Films of Amphiphilic Alcohols and Surfaces of Polar Crystals as Templates for Ice Nucleation

Ronit Popovitz-Biro, Jaroslaw Majewski, Jinn-Lung Wang, Kristian Kjaer, Jens Als-Nielsen, Leslie Leiserowitz, and Meir Lahav

7.1. Introduction 145
7.2. Self-Aggregates of Amphiphilic Alcohols as Templates for Ice Nucleation 146
7.3. Nucleation of Ice Under Monolayers of Aliphatic Alcohols . . . 149
7.4. Nucleation of Ice Under Mixed Monolayers of Aliphatic Alcohols 160

7.5. Nucleation of Ice Under Alcohol Monolayers Bearing Amide or Ester Groups in the Hydrocarbon Chain 164
7.6. Induced Nucleation of Ice by Amphiphilic Alcohols Aggregated at the Oil/Water Interface 170
7.7. Induced Nucleation of Ice on Surfaces of Polar Crystals of α-Amino Acids 174
7.8. Conclusion and Outlook 177
Acknowledgments 179
References and Notes 180

Ion-Selective Monolayer Membranes Based on Self-Assembling Tetradentate Ligand Monolayers on Gold Electrodes: Nature of the Ionic Selectivity

Suzi Steinberg, Yitzhak Tor, Abraham Shanzer, and Israel Rubinstein

8.1. Introduction 183
8.2. Experiments 185
8.3. Results and Discussion 188
8.4. Conclusions 202
Acknowledgments 203
References and Notes 203

Specific Recognition at Functionalized Interfaces: Direct Force Measurements of Biomolecular Interactions

Deborah Leckband and Jacob Israelachvili

9.1. Introduction 207
9.2. Results 209
9.3. Discussion 220
9.4. Conclusion and Implications for Future Bioengineering Materials Technology 221
Acknowledgments 221
References 221

Formation of Recognition Patterns by Langmuir-Blodgett Techniques

Tilman Schwinn, Sven-Peter Heyn, Martin Egger, and Herman E. Gaub

10.1. Introduction 223
10.2. Materials and Methods 226

10.3. Conclusions 234
Acknowledgments. 235
References. 235

Electronic Properties

Photoinduced Electron Transfer in Monolayer Assemblies and Its Application to Artificial Photosynthesis and Molecular Devices

Masamichi Fujihira

11.1. Introduction 239
11.2. Construction of Charge Separation Units with LB Films 243
11.3. Photoelectric Conversion Efficiencies in Terms of the Electron Transfer Kinetics. 249
11.4. Simulation of the Primary Process by the Use of a Mixed Monolayer with Triad and Antenna Molecules 254
11.5. Construction of Nanodomains in Monolayers by the Use of Phase Separation 256
11.6. Artificial Photosynthesis With Multilayered LB FIlms 267
11.7. Conclusions 271
Acknowledgments. 272
References. 272

Photoelectric Behavior of Bacteriorhodopsin Thin Films at the Solid/Liquid Interface

Tsutomu Miyasaka

12.1. Introduction 279
12.2. Photoelectric Response Caused by bR Photocycle 280
12.3. Photoelectric Behavior of bR at Solid/Liquid Interfaces 283
12.4. Device Application for Optical Sensing and Image Processing . 287
12.5. Conclusions 290
Acknowledgments. 291
References. 291

Hole-Burning Spectroscopy of Dye-Doped Langmuir-Blodgett Films

Michel Orrit and Jacky Bernard

13.1. Introduction 293
13.2. Experimental Setup 296

13.3. The Hole Width and Its Temperature Dependence 299
13.4. The Hole as a Spectral Label 306
13.5. Conclusion and Prospects 312
Acknowledgments 313
References 313

Dynamic Processes

Evaluation of a Transfer Process for Langmuir-Blodgett Films by Means of a Quartz-Crystal Microbalance

Yoshio Okahata, Katsuhiko Ariga, and Kentaro Tanaka

14.1. Introduction 317
14.2. Quartz-Crystal Microbalances 318
14.3. Vertical Dipping Depositions of LB films on a QCM Plate 319
14.4. Horizontal Lifting Depositions of LB Films on a QCM 321
14.5. Conclusions 328
References 329

Translational Diffusion and Electron Hopping in Monolayers at the Air /Water Interface

Marcin Majda

15.1. Electrochemistry on the Water Surface 331
15.2. Translational Diffusion in Octadecyl Ferrocence Monolayers .. 334
15.3. Electron Hopping in Langmuir Monolayers 339
15.4. Conclusions 346
Acknowledgments 346
References 346

On-Line Structure Control of Langmuir-Blodgett Films

Hans Riegler and Karl Spratte

16.1. Introduction 349
16.2. Structural and Compositional Variety of Organic Multilayers .. 350
16.3. Schematic of the Preparation of Langmuir-Blodgett Films 351
16.4. Structure Control During the Transfer Due to Substrate/Monolayer Interactions 355

16.5. Conclusions 361
Acknowledgments 364
References 364

Structure

Phase Diagrams and Chain Order in Monolayers of Aliphatic Chains

Ian R. Peterson

17.1. Phases of Amphilic Monolayers 367
17.2. The Principle of Corresponding States 371
17.3. Unambiguous Tests of the Principle 375
17.4. Refinement of the Principle 378
17.5. Conclusions 382
References 383

Subject Index 385

Preface

In 1991 I published a book dedicated to the area of ultrathin organic films.[1] There, I covered the developments in the area since 1965, the year before Gaines published his classical book on Langmuir–Blodgett monolayers.[2] That Langmuir–Blodgett and self-assembled monolayers are important is clear from the number of articles published on these subjects. This number is so overwhelming—over one thousand each year—that writing a second edition to my book has become an impossible mission.

Reading through the literature, one finds it difficult to pinpoint the truly interesting and existing discoveries and developments in this area. Thus, some guidance may be useful, especially for those who find the area attractive and would like to start contributing. After discussions with a number of distinguished scholars—whose names can be found in the editorial board—I proposed a new series of books whose primary goal is to highlight important contributions. My subsequent discussions with the publisher, Academic Press, led to a conclusion that, instead of the creation of a new series, the volumes on organic thin films could be beneficially published within the existing, highly successful series *Physics of Thin Films*. Correspondingly, the scope of this series would be broadened and the title slightly modified to *Thin Films*. For this volume (Volume 20 in the *Thin Films* series), the editorial board suggested taking a general approach rather than focusing on a specific theme. The papers—all invited—are on many topics and represent some of the most interesting advancements in this area.

At the Fifth International Conference on Langmuir–Blodgett Films in Paris, Professor Hans Kuhn gave the closing remarks, summarizing the meeting and pointing to the most important contributions. These remarks expressed tremendous vision and enthusiasm and stimulated the correspondence between Professor Kuhn and myself that resulted in the opening article of this volume, "Supramolecular Assemblies—Vision and Strategy."

[1] Ulman, A. *An Introduction to Ultrathin Organic Films: From Langmuir–Blodgett to Self-Assembly*, Academic Press, Boston, 1991.

[2] Gaines, G. L., Jr. *Insoluble Monolayers at Liquid–Gas Interfaces*, Interscience, New York, 1966.

The papers on Langmuir–Blodgett monolayers deal with a number of important issues. Starting from the Langmuir layer at the air–water interface, Riegler and Spratte discuss the on-line structure control of Langmuir–Blodgett films, and Majda reports on the translational diffusion and electron hopping in monolayers at the air–water interface. Gaub *et al.* report on the formation of recognition pattern by Langmuir–Blodgett techniques, and Fendler reviews years of accomplishments in the area of oriented growth of nanocrystalline particulate films at monolayers. Moving to the transfer of monolayers from the air–water interface onto a solid substrate, Okahata *et al.* evaluate the transfer process of Langmuir–Blodgett films by using a quartz-crystal microbalance as a substrate. As we proceed to the structure of monolayers, Peterson discusses the phase diagrams and chain order in monolayers of aliphatic chains.

As we turn to materials and applications, Palacin *et al.* discuss building up two-dimensional polymers by the Langmuir–Blodgett technique, Miyasaka describes the photoelectric behavior of bacteriorhodopsin thin films at the solid–liquid interface, and Fujihira reviews many accomplishments in the area of artificial photosynthesis and molecular devices, describing photoinduced electron transfer in monolayer assemblies. Finally, Orrit and Bernard describe hole-burning spectroscopy of dye-doped Langmuir–Blodgett Films.

In the area of self-assembly, Sagiv *et al.* write on third-level self-assembly and beyond and describe the formation of polar hybrid superlattices via postassembly. Rubinstein *et al.* describe ion-selective monolayer membranes based upon self-assembling tetradentate ligand monolayers on gold electrodes. Finally, Calvert reviews many years of efforts in surface patterning, describing lithographically patterned self-assembled films.

There are three more papers, each dealing with an important subject. Lahav *et al.* review some years of work in the studies of Langmuir films of amphiphilic alcohols and surfaces of polar crystals as templates for ice nucleation. Leckband and Israelachvili report on the specific recognition at functionalized interfaces and describe direct force measurements of biomolecular interactions. Finally, McCarthy *et al.* review their work on polymer surface modification.

I would like to thank all the contributors who made this volume possible. Their work and contributions will help advance the area of organic thin films. Last, but not least, I thank the members of the editorial board who helped in the selection of the invited papers and assisted me throughout the process.

Abraham Ulman
Brooklyn, New York

Contributors

Numbers in parentheses indicate the pages on which the authors' contributions begin

JENS ALS-NIELSEN (145), Department of Physics, Ris o National Laboratory, DK4000 Roskilde, Denmark

KATSUHIKO ARIGA (317), Department of Biomolecular Engineering, Tokyo Institute of Technology, Yokohama 227, Japan

TIMOTHY G. BEE (85), Polymer Science and Engineering Department, University of Massachusetts, Amherst, Massachusetts 01003

GARRY BERKOVIC (41), Department of Materials and Interfaces, The Weizmann Institute of Science, Rohovot 76100, Israel

JACKY BERNARD (293), Centre de Physique Moleculaire Optique et Hertzienne, C.N.R.S. et Université Bordeaux 1, F-33405 Talence Cedex, France

JOAN V. BRENNAN (85), Polymer Science and Engineering Department, University of Massachusetts, Amherst, Massachusetts 01003

JEFFREY M. CALVERT (109), Center for Bio/Molecular Science & Engineering (Code 6900), Naval Research Laboratory, Washington, D.C. 20375

ELISA M. CROSS (85), Polymer Science and Engineering Department, University of Massachusetts, Amherst, Massachusetts 01003

ANTHONY J. DIAS (85), Polymer Science and Engineering Department, University of Massachusetts, Amherst, Massachusetts 01003

MARTIN EGGER (223), Boehringer Mannheim GmbH, 8182 Tutzing, Germany

JANOS H. FENDLER (11), Department of Chemistry, Syracuse University, Syracuse, New York 13244

NICOLE L. FRANCHINA (85), Polymer Science and Engineering Department, University of Massachusetts, Amherst, Massachusetts 01003

MASAMICHI FUJIHIRA (239), Department of Biomolecular Engineering, Tokyo Institute of Technology, Yokohama 227, Japan

HERMAN E. GAUB (223), Physikdepartment E22, Technische Universität München, D-8046 Garching, Germany

SVEN-PETER HEYN (223), VDI/BDE Technologiezentrum, 1000 Berlin 30, Germany

JACOB ISRAELACHVILI (207), Department of Chemical and Nuclear Engineering, and Materials Department, University of California at Santa Barbara, Santa Barbara, California 93106

KRISTIAN KJAER (145), Department of Physics, Ris o National Laboratory, DK4000 Roskilde, Denmark

HANS KUHN (1), MPI für Biophysikalische Chemie, D-3400 Göttingen-Nikolausberg, Germany

MEIR LAHAV (145), The Department of Materials and Interfaces, The Weizmann Institute of Science, Rehovot 76100, Israel

DEBORAH LECKBAND (207), Department of Chemical Engineering, State University of New York at Buffalo, Buffalo, New York 14260

KANG-WOOK LEE (85), Polymer Science and Engineering Department, University of Massachusetts, Amherst, Massachusetts 01003

LESLIE LEISEROWITZ (145), The Department of Materials and Interfaces, The Weizmann Institute of Science, Rehovot 76100, Israel

MARCIN MAJDA (331), Department of Chemistry, University of California at Berkeley, Berkeley, California 94720

JAROSLAW MAJEWSKI (145), The Department of Materials and Interfaces, The Weizmann Institute of Science, Rehovot 76100, Israel

RIVKA MAOZ (41), Department of Materials and Interfaces, The Weizmann Institute of Science, Rehovot 76100, Israel

THOMAS J. MCCARTHY (85), Polymer Science and Engineering Department, University of Massachusetts, Amherst, Massachusetts 01003

TSUTOMU MIYASAKA (279), Ashigara Research Laboratories, Fuji Photo Film Company, Ltd., Minamiashigara, Kanagawa 250-01, Japan

YOSHIO OKAHATA (317), Department of Biomolecular Engineering, Tokyo Institute of Technology, Yokohama 227, Japan

MICHEL ORRIT (293), Centre de Physique Moleculaire Optique et Hertzienne, C.N.R.S. et Université Bordeaux 1, F-33405 Talence, France

SERGE PALACIN (69), CEA/Service de Chimie Moléculaire, C.E. Saclay, 91191 Gif sur Yvette Cedex, France

IAN R. PETERSON (367), Institute für Physikalische Chemie, Johannes Gutenberg-Universität Jakob, DW-6500 Mainz, Germany

RONIT POPOVITZ-BIRO (145), The Department of Materials and Interfaces, The Weizmann Institute of Science, Rehovot 76100, Israel

FLORENCE PORTEAU (69), CEA/Service de Chimie Moléculaire, C.E. Saclay, 91191 Gif sur Yvette Cedex, France

HANS RIEGLER (349), Institute für Physikalische Chemie, Universität Mainz, D-6500 Mainz, Germany

ANNIE RUADEL-TEIXIER (69), CEA/Service de Chimie Moléculaire, C.E. Saclay 91191 Gif sur Yvette Cedex, France

ISRAEL RUBENSTEIN (183), The Department of Materials and Interfaces, The Weizmann Institute of Science, Rehovot 76100, Israel

JACOB SAGIV (41), Department of Materials and Interfaces, The Weizmann Institute of Science, Rehovot 76100, Israel

TILMAN SCHWINN (223), Physikdepartment E22, Technische Universität München, D-8046 Garching, Germany

ABRAHAM SHANZER (183), The Department of Organic Chemistry, The Weizmann Institute of Science, Rehovot 76100, Israel

MOLLY S. SHOICHET (85), Polymer Science and Engineering Department, University of Massachusetts, Amherst, Massachusetts 01003

KARL SPRATTE (349), Institut für Physikalische Chemie, Universität Mainz, D-6500 Mainz, Germany

SUZI STEINBERG (183), The Department of Materials and Interfaces, The Weizmann Institute of Science, Rehovot 76100, Israel

KENTARO TANAKA (317), Department of Biomolecular Engineering, Tokyo Institute of Technology, Yokohama 227, Japan

YITZHAK TOR (183), The Department of Organic Chemistry, The Weizmann Institute of Science, Rehovot 76100, Israel

ABRAHAM ULMAN (1), Imaging Research and Advanced Development, Eastman Kodak Company, Rochester, New York 14650

JINN-LUNG WANG (145), The Department of Materials and Interfaces, The Weizmann Institute of Science, Rehovot 76100, Israel

RUTH YAM (41), Department of Materials and Interfaces, The Weizmann Institute of Science, Rehovot 76100, Israel

Introduction

Supramolecular Assemblies: Vision and Strategy

HANS KUHN

MPI für Biophysikalische Chemie,
D-3400 Göttingen-Nikolausberg, Germany

ABRAHAM ULMAN

Polytechnic University, Department of Chemistry
Polymer Research Institute, Brooklyn, New York

1.1. Introduction

When deciding to climb a mountain we usually have a nebulous vision of how exciting it would be on top, of how important, as a unique experience, such an event would be. Then we develop a strategy to get us there, and before starting we equip ourselves well and undergo a conditional training so we are fit for the climb. Our work in supramolecular assemblies and nanoscopic films resembles the task of climbing a mountain; at this moment, we are nevertheless still at the bottom.[1]

Self-assembly is a process in which supermolecular hierarchical organization is established in complex systems of interlocking components. In fact, self-assembly is one of the most, if not *the* most, important phenomenon in nature, and the living cell is its best paradigm. It is important to emphasize, however, that self-assembly is not limited to the molecular level, and one can imagine self-assembly of molecular clusters, organic, and inorganic crystals, etc.

Taking ideas from nature (biological sciences), scientists prepare model systems and study their properties, thus establishing structure property relationships (physical sciences). Having established a working phenomenon and the understanding of both materials and mechanistic issues, scientists look for opportunities where these can be utilized for applications (engineering sciences). Figure 1.1 presents the relationship among the different sciences and emphasizes the fact that self-assembly is at their

ISBN 0-12-523485-6

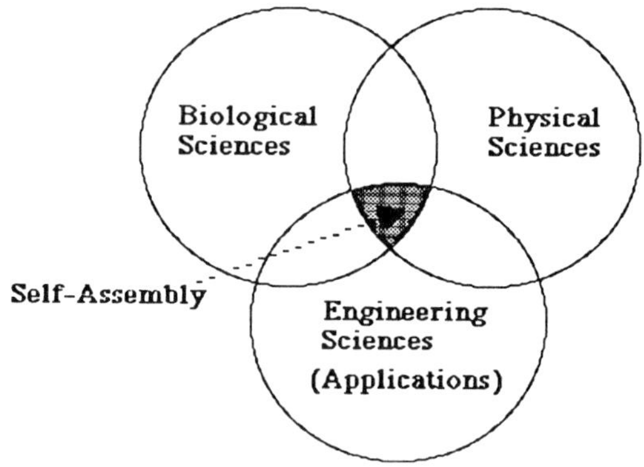

FIG. 1.1. Self-assembly is at the junction of biological, physical, and engineering sciences.

junction, which, we believe, makes it both an important and exciting area of research.

The *vision* that is unique to work in the area of supramolecular assemblies is molecular engineering. We envisage fabrication of supramolecular machines for information processing, memories, communicative systems, energy conversion, and water desalination. We desire microsensors, effectors, and molecular switches and wires to carry signals by photons, electrons, solitons, and ions. Though still at the very beginning of the realization of this vision, we have already witnessed chemistry in a fascinating transition from the synthesis and study of isolated molecules to supramolecular engineering. Taking individual molecules and assembling them into two- and three-dimensional systems provides a clear pathway from the molecular to the supramolecular, from chemistry to materials science and solid-state physics. Chemistry is thus changing focus, interfacing with physics on the one hand and with biology on the other.

The *strategy* we adopt is to assemble molecules that interlock in a planned manner, forming intended machineries. Molecules will self-assemble under programmed external influence to form modules. These modules then would self-assemble to form increasingly complex supramolecular systems. For these processes to happen, however, it is essential that each molecule be constructed to fit *perfectly* in a module and that the modules *uniquely* distinguish a distinct complementary partner.

Fabrication of the simplest supramolecular devices is accomplished by intelligently planning appropriately functionalized molecules to interlock in an intended manner, and by letting them self-organize at a water surface, where interlocking is assisted by pushing them together. The different kinds of monolayers thus obtained are superimposed in an intelligently planned sequence, thus forming increasingly more complex supramolecular devices by means of the Langmuir-Blodgett (LB) deposition method (spreading solution, solvent evaporation, monolayer compression, deposition). Another route is to let molecules self-organize at a solid/liquid interface. In the following we abbreviate the two routes of self-assembly (SA) by SA-l/g (liquid/gas) and SA-s/l (solid/liquid). (The distinction between one-component (LB) films and the SA-l/g films where several molecules are planned to interlock precisely should emphasize organized monolayer design by SA-l/g as a key feature of supramolecular architecture.) The question is one of what is the preferred strategy that can take us to the realization of our vision?

Simple molecules, such as dyes with long alkyl chains, were useful to start the game. They were confined to a water surface and pushed together (SA-l/g). Further developments in the design of complex amphiphiles include more and more sophisticated molecules self-organizing into purposely planned complex arrays and amphiphilic polymers. These amphiphilic polymers offer the most stable films so far, produced at the liquid/gas interface. SA-s/l and SA in solution, on the other hand, require an ever increasing sophistication for precise interlocking and will certainly be increasingly important.

Superimposition of differently functionalized monolayers in a programmed sequence is an important strategy today. One can construct a supramolecular assembly—a superlattice—where different groups of layers perform different physical functions. For example, part of this assembly can have nonlinear optical (NLO) properties, another can be photoconductive, and still another, at the very top, may serve as a biosensor. Together, and in concert, these parts can act like a machine, performing a series of complicated functions to achieve the desired objective. There are also simple manipulation techniques, seldom used today, that may be useful in the future. Assemblies can be manipulated in many ways, for instance, by cleaving between distinct monolayers and printing the entire assembly onto a given surface.

In thinking about future strategies for building supramolecular machines, a basic principle in molecular biology and in engineering must be considered. This is the assembly of modules of increasing hierarchic order and the testing of modules before each assembly step, rejecting incorrect samples (Fig. 1.2).

Having such a sequence is basic to avoiding more and more defects when increasing the complexity of the machine. Nature's trick to realize such self-repair in biosystems is the aggregation of modules. Hence, incorrect samples do not match; they are rejected and exchanged for the correct ones. Clearly, without adopting such an aggregation route, it is impossible to envisage the building of complex molecular machines.

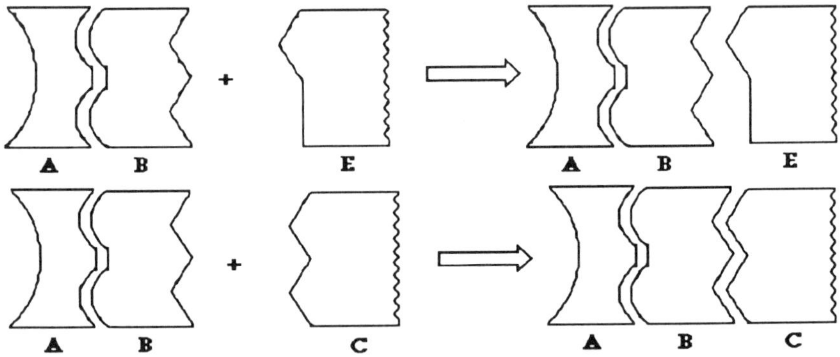

FIG. 1.2. Testing modules before each assembly step and rejecting incorrect samples: **E**, an erroneous copy is rejected (top), while **C**, a correct copy, is accepted (bottom).

What comes out of our arguments so far is that the advance of supramolecular assemblies depends on the capacity of scientists to develop thin films with optimized properties and a high degree of ordering. We have envisioned self-organization as a method by which one can provide solutions to this problem and develop the capacity to design organic films with predicted architectures and properties.

1.2. Efforts to Achieve Well-Defined Organized Molecular Assemblies

That molecules will spontaneously form a stable close-packed layer is vital to the success of any material design based on the molecular assembly approach. Of further importance is the ability of complex molecules to form two-dimensional (2-D) assemblies with long-range order, preferably in monodomain macroscopic films. For that to happen, the incommensurability of intra-assembly planes should be minimized. Clearly, 2-D assemblies with a perfect intra-assembly match of atomic planes are in a state of a thermodynamic equilibrium. Such molecular layers in space are closely packed like a normal crystal, exhibiting long-range order. Increasing the intra-assembly mismatch decreases the cohesive energy and should lead to enhanced defect density, ultimately decreasing the layered structure stability. Incommensurability of molecular parts does not mean, however, that a 2-D assembly cannot be formed. The stability of the layered assembly can be augmented by applying pressure, as is the case of SA-l/g in a Langmuir trough, or by chemisorption or physisorption of a solid surface (SA-s/l). Thus, though the unconstrained layered structure is not in a global minimum of the free energy, there are mechanisms of the types mentioned already that, together with conformational modification, may yield a layered assembly that is in a global minimum. Excellent examples for such systems are the suspended smectic liquid crystals recently demonstrated.

Fabricating and investigating organized molecular assemblies, organizates, is the goal common to SA-l/g, assisted by pushing aggregates together and SA-s/l, or in solution. Achieving this goal clearly requires a *fundamental* understanding of the relationships between the molecular structure of complex molecules and their organization in layered aggregates. Such an understanding would be achieved by joining theoretical and experimental efforts. Recent advances in computer modeling—still far from being satisfactory even for small organizates—offer a hope that it may be possible in the near future to carry out detailed analyses and examine their results experimentally. More efforts are needed in developing reliable force fields and fast algorithms and in testing increasingly more complex systems. However, what is really needed is the understanding of our scientific community that collaboration of theory and experiments is vital for achieving our goal.

Beyond self-organization of 2-D assemblies at interfaces, the next level of complexity requires controlling the third dimension. The different fabrication methods for organized molecular films offer a mechanism for building multilayer films, each with its advantages and disadvantages. Thus, while amphiphiles used in SA-l/g fabrication are reasonably stable, the resulting films are unstable thermally, maybe with the exception of those made of polymeric amphiphiles. Note, however, that the viscosity of these types of layers prevents fast deposition rates and may limit large-scale fabrication of useful devices. Understanding how the structure of a polymeric amphiphile, its molecular weight, and molecular weight distribution relate to the viscosity of its monolayer at the air/water interface is of crucial importance. Relating deposition rates to parameters such as surface viscosity and temperature is not straightforward and requires a large matrix of experiments. Analyzing the resulting films for defects and relating order parameters to deposition rates is also complicated. Nevertheless, without such efforts, it is hard to see a production line based on the SA-l/g technique.

Monolayers made by adsorption on solids (SA-s/l) do not require the use of a trough and, in principle, can be carried out in a beaker. However, in this case, since no external force is used to push molecular assemblies together, and the molecules are pinned to a solid surface, issues such as chemisorption energy, epitaxy, and the interlocking of individual molecules at the surface become important. Multilayer fabrication requires surface modification and functionalization. It also requires that layers assemble with high lateral order. The question of whether the overall time required for the fabrication of 0.5-μm film using SA-s/l would be significantly different from that required by the SA-l/g technique is of practical importance. Recently, there has been a report on the facile preparation of hydroxylated surfaces.[2] Clearly, applications based on multilayer films of alkyltrichlorosilane derivatives may be more realistic now. Another example is the intercalation of ions and molecules into noncentrosymmetric multilayer matrices of hydrogen-bonded silanes,[3] thus allowing manipulation of three-dimensional (3-D) structures after their formation. While these developments are promising, more work on surface modification and functionalization and on new SA-s/l systems is needed. The SA-s/l technique may be the technique of choice in every case where lateral structure is important, that is, when the film is comprised of ordered patterns of different molecules.

1.3. Perspectives in Supramolecular Engineering

Beyond today's self-assembly techniques that use 2-D assemblies to build a 3-D film, the next level of complexity requires controlling the folding and self-organization of macromolecules, much as is observed in proteins. This may be an exciting route for the fabrication of 3-D modules at the nanometer-size level. Such modules can be self-assembled at either solid/liquid or liquid/gas interfaces. The construction of composites of biosynthetic and synthetic molecules of increasing complexity is of great interest. It is crucial that preparative chemists envision the supramolecular machine—instead of the isolated molecular species—as the goal of synthetic work.

We are moving now in different, sometimes perceived as "parallel," directions. However, these directions would eventually converge to a comprehensive route, clearly pointing to the establishment of molecular materials as a viable part of modern technology. The search for novel SA techniques, using lithography to introduce assemblies of modules and developing advanced analytical tools for the study of molecular organizates, should be vigorously pursued. The development of surface treatments for making templates that allow positioning of individual macromolecules, inducing the formation of well-designed organizates, seems feasible and is a great challenge. The variety of experimental and theoretical investigations carried out by many groups all over the world are already providing encouraging landmarks. While the field is clearly full of excitement, staying patient, optimistic, and most importantly enthusiastic is the way to go. We cannot expect a fast technological revolution, but instead many important changes as a result of the possibility of manipulating single molecules. In the present state of the art, we must focus on shorter term projects. We must learn to walk before we run, since we are essentially in the conditional training stage, in preparation for the big journey. We must test assemblies and assembly techniques with all efforts to control and avoid undesired changes in the assembly architecture. It is important to make sure, in each case, that the intended architecture is really present.

Self-organization phenomena apparently are becoming a recognized part of materials science. The advantages of using self-organization for the fabrication of increasingly more sophisticated molecular organizates are clear. However, it may take a long time before we see production lines using organized molecular film techniques. The scientific community, and especially industry, should realize the enormous difficulties and the necessity of many small steps required to achieve the desired control on the structure of matter. Achieving this control will open up unlimited opportunities in materials science, electronics, electro-optics, etc. As we slowly climb the mountain toward the fascinating goal, chemistry is changing. Eventually, we shall master intermolecular forces (i.e., van der Waals and electrostatic interactions), hydrogen bonding, etc., as we have mastered chemical bonding, thus bringing self-organization to exciting new levels.

References

1. For leading references, see Swalen, J. D., Allara, D. L., Andrade, J. D., Chandross, E. A., Garoff, S., Israelachvili, J., McCarthy, T. J., Murray, R., Pease, R. F., Rabolt, J. F., Wynne, K. J., and Yu, H. (1987). *Langmuir* **3**, 932; Kuhn, H. (1989). *Thin Solid Films* **178**, 1; Ulman, A. (1991). *An Introduction to Ultrathin Organic Films From Langmuir-Blodgett to Self-Assembly*, Academic Press, Boston. Kuhn, H., and Möbius D. (1993). In: *Physical Methods of Chemistry* 2nd ed., (B. W. Rossiter and R. C. Baetzold, eds.). Wiley, New York, Vol. 9B, pp. 375–539.
2. Sukenik, H. Private Communication.
3. Maoz, R., Yam, R., Berkovic, G., and Sagiv, J., this volume.

Advanced Materials

Oriented Growth of Nanocrystalline Particulate Films at Monolayers: A Colloid Chemical Approach to Advanced Materials

JANOS H. FENDLER

Syracuse University, Department of Chemistry, Syracuse, New York 13244-4100

2.1. Introduction

Progress in material science and engineering has been spectacular in the past decade. Modern materials with vastly improved mechanical, thermal, electrical, optical, and magnetic properties have become available. Continued development of advanced materials is crucial to the economic well-being of our nation.[1]

Advanced electronic devices are made by varying the composition and doping of semiconductors within ultrasmall distances (ranging from submicrons to subnanometers) to desired band structures. *Bandgap engineering*, as this process is known in the semiconductor industry, has been approached empirically by various epitaxies in ultrahigh vacuum.[2,3] Development of a new generation of nanostructured devices requires innovative approaches based on a fundamental understanding of the chemistry and physics involved at the molecular level. Chemists have become increasingly involved in the various aspects of materials science. They have designed new synthetic methodologies for established materials and have created new ones with unique physical and chemical properties.

Construction of nanostructured advanced materials, based on modern "wet" colloid chemistry, has been the long-term research objective of our laboratories. Our approach has been based on membrane mimetic chemistry[4] and inspired by biomineralization.[5-10] Advantage has been taken of membrane mimetic systems to provide chemical and spacial control for the *in situ* generation and stabilization of ultrasmall catalytic, semiconducting,

ISBN 0-12-523485-6

and magnetic particles and particulate films. Reversed micelles,[11-17] surfactant vesicles,[18-21] bilayer lipid membranes,[22-24] Langmuir-Blodgett (LB) films,[25,26] and monolayers[27-37] have been used as membrane mimetic systems.

We have found monolayers to be particularly useful as templates for the *in situ* generation of nanocrystalline particulate films because they mimic two-dimensional (2-D) crystal growth at biological surfaces[9,10] and because precise crystallographic information has become available on the orientation and packing of surfactant headgroups from x-ray diffraction studies using synchrotron sources.[38-43] The use of monolayer matrices for nanoparticulate film generation has several advantages. First, stable, well-characterized, and long-lasting monolayers can be formed from a large variety of surfactants. Second, monolayer surface areas and charges are controllable two dimensionally, and the composition of the aqueous subphase is readily variable. Third, monolayers, along with the particulate films grown under them, can be conveniently transferred to solid supports (i.e., to substrates). Differences between inorganic particles generated between the headgroups of LB films[26] and particulate films formed under monolayers floating on an aqueous subphase[30-37] should be recognized. Available space between the headgroups of LB films limits the growth of the particles to 40 to 60 Å. Conversely, particulate films up to several thousand angstroms can be grown under monolayers; and, subsequent to their transfer to substrates, the surfactant monolayer can, if desired, be removed.

Formation of nanocrystalline particulate films under monolayers is surveyed in this chapter. As is customary, emphasis is placed on our own research. First, details are provided on the experimental methodologies established and the characterizations utilized. Second, a mechanism is proposed for the growth of the nanocrystalline particulate film. Third, the importance of size quantization and potential for bandgap engineering is delineated. Fourth and fifth, semiconductor and silver particulate films formed to date are critically evaluated. Sixth, the role of molecular organization and recognition is illustrated by the available data on epitaxial growth of nanocrystalline particulate films. Finally, conclusions are provided and future directions projected.

2.2. Experimental Methodologies and Characterizations

2.2.1. Monolayer Formation and Characterization

Monolayers (monomolecular layers) are formed by spreading naturally occurring lipids, synthetic surfactants, or certain polymers, dissolved in a volatile solvent, over water in a Langmuir trough. The polar headgroups of the amphiphiles are in contact with and hydrated by water, the subphase, while their hydrocarbon tails protrude in the air. Surface pressure π versus surface area A, surface potential versus A, and surface viscosity versus isotherms, determined at different temperatures, have been interpreted in terms of 2-D thermodynamics.[44,45] The spreading of a monolayer on a subphase in a film balance and a typical π-A isotherm are illustrated in Fig. 2.1.

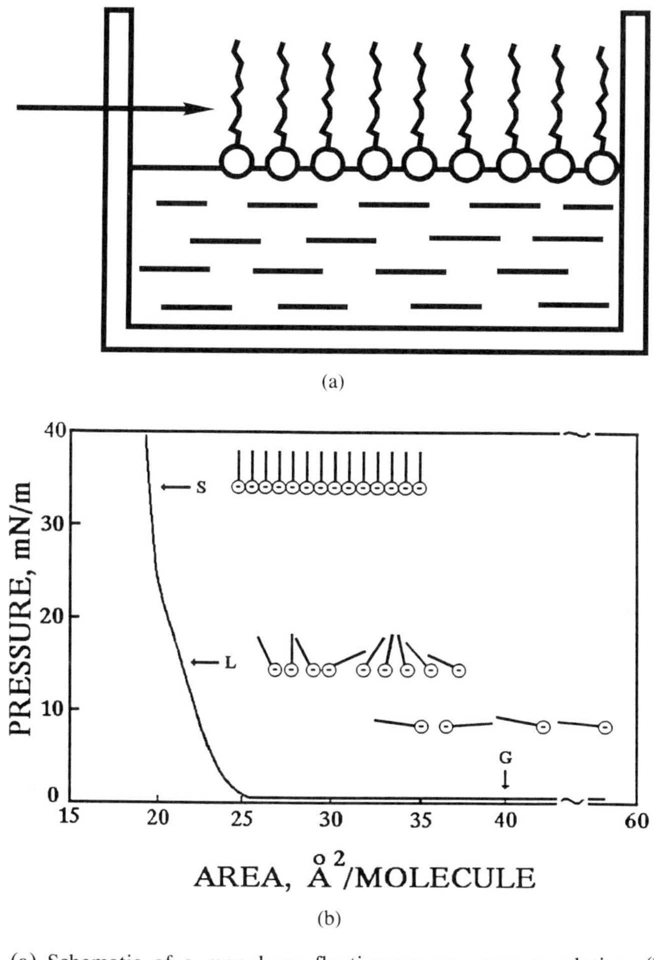

FIG. 2.1. (a) Schematic of a monolayer floating on an aqueous solution. (b) Electron diffraction pattern of the PbSe crystals shown in part (a). The area of the electron beam coverage is 2 μm in diameter.

Structural information on the molecular arrangements in monolayers have been obtained by *in situ*, in trough spectroscopic (uv[46,47] fluorescence,[47,48] Fourier transform infrared (FTIR),[49-52] and Raman[53,54]), microscopic (epifluorescence[55-58] and Brewster-angle[59,60]), and electro-optical[61,62] measurements. Particularly significant has been the utilization of intense and well-collimated variable-wavelength x-ray beams from synchrotron sources for obtaining direct structural information, at the atomic level, on surfactant packing and interaction in monolayers.[38,40,42]

2.2.2. Chemical Generation of Nanocrystalline Particulate Films

Experimental setups used for the generation and *in situ* monitoring of nanocrystalline particulate films are illustrated in Fig. 2.2. The arrangement shown in the top allows the injection of a precursor gas (H_2S, H_2Se,

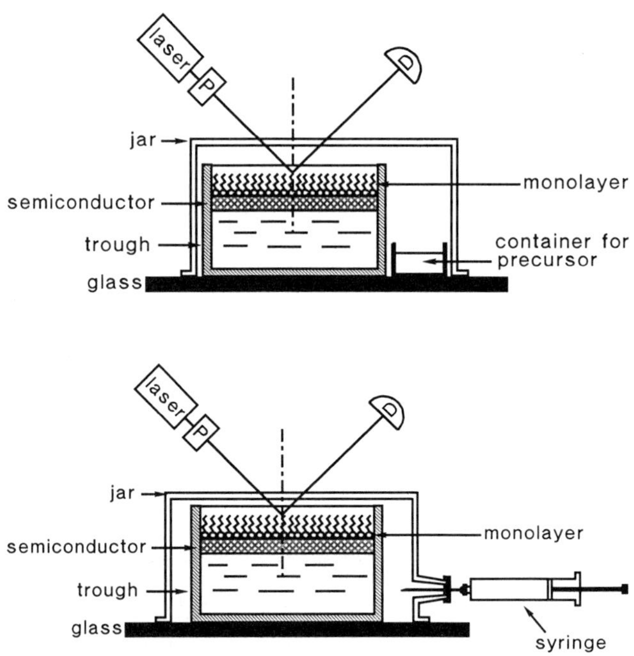

Fig. 2.2. Schematics of the experimental arrangements used for the generation of semiconductor particles at the negatively charged, surfactant headgroup/aqueous subphase interface and that used for the *in situ* monitoring of reflectivities. P = polarizer; D = detector.

NH_3, for example), while that in the bottom permits the generation of the desired gaseous precursor (Na_2S + dilute acid, for example, for generating H_2S). Facilities were available for determining surface pressure versus surface area and surface potential versus surface area isotherms in the film balance placed under the glass cover. Arrangements were also made for the continuous monitoring of reflectivities; angle-dependent reflectivities, Brewster-angle and fluorescence microscopies, and nonlinear optical parameters.

2.2.3. Electrochemical Generation of Nanocrystalline Particulate Films

Silver particulate films were generated in an experimental set-up (shown in Fig. 2.3). A 1.00-mm-diameter, 3-cm-long silver electrode was immersed into the subphase. Electrical connection was made through a 20-μm-diameter platinum electrode, which was floated, subsequent to the monolayer formation, on the aqueous silver ion solution surface in the middle of

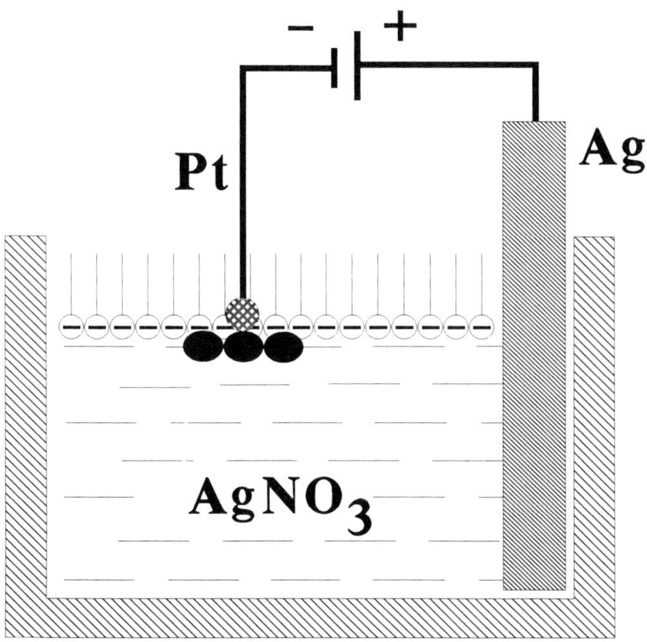

FIG. 2.3. Schematic of the circular trough (not drawn to scale) used for the electrochemical generation of silver particulate films at monolayer interfaces.

the trough. The application of a potential of 1.8 to 1.9 V resulted in the formation of a silver particulate film.

2.2.4. Optical Characterization of Nanocrystalline Particular Films *In Situ* at Monolayers and *Ex Situ* on Substrates

Evolution of semiconductor particulate films under the monolayer in the Langmuir trough was monitored by angle-dependent reflectivity measurements. The obtained data provided information on the thickness and porosity of the nanocrystalline particulate film being formed. Repetition of the measurements on substrates led to the assessment of transfer efficiencies.

Measurements of the absorption spectra and determinations of the absorption edges on quartz substrates allowed for the estimation of the diameters of the nanocrystals constituting the semiconductor particulate film. The relationship between absorption spectra of CdS particles and their sizes has been studied extensively.[63–67] In particular, Henglein and coworkers prepared CdS particles with different mean diameters and obtained a curvilinear relationship between their absorption thresholds λ_t and their independently determined sizes.[66] The thinnest (8-Å-thick) particulate CdS films, prepared from Cd-arachidate monolayers, had an absorption threshold λ_t of 450 nm. This value led to an assessment of 30 ± 5 Å for the mean diameter of CdS particles when using Henglein's λ_t versus particle size curve,[66] which is in good agreement with those obtained in transmission electron microscopic measurements.

We have launched Brewster-angle microscopic and nonlinear optical techniques for the more intimate characterization of the nanocrystalline semiconductor particulate films generated in our laboratories.[59,60]

2.2.5. Electrochemical Characterization of Nanocrystalline Particulate Films *In Situ* at Monolayers and *Ex Situ* on Substrates

Formation of nanocrystalline particulate films, under monolayers, was examined *in situ* in an electrochemical cell (shown in Fig. 2.4). The automatic "sample approach" system of the scanning tunneling microscope (STM) was used to bring the STM tip into contact with the arachidic acid (AA)-monolayer-supported CdS semiconductor particulate film. A relatively blunt, uninsulated Pt tip was rinsed in concentrated H_2SO_4 solution for 20 to 30 min, followed by rinsing with copious amounts of ultrapure water. The dried tip was placed close to the surface of the AA-monolayer-supported CdS particulate film and connected to the feedback loop circuit.

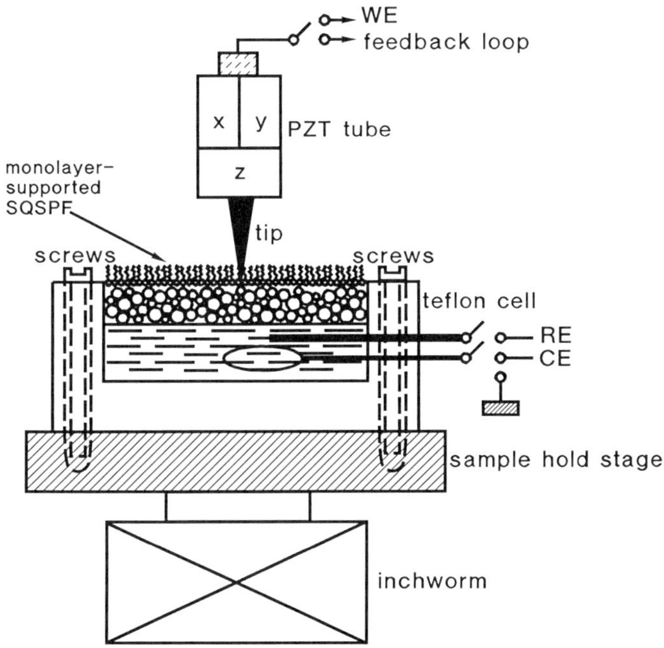

FIG. 2.4. Schematics of the cell used for the electrochemical characterization of monolayer-supported CdS particulate films. *WE* = working electrode; *RE* = reference electrode; *CE* = counter electrode.

The Pt ring was grounded to the sample hold stage (Fig. 2.4). The tip bias was set at +0.5 V with respect to the ground (i.e., the Pt ring) during the sample approach. The computer drove the continuous movement of the piezoelectric "inchworm," which brought the sample hold stage upward to the tip until it reached an acquired electrical current (by setting 1 to 2 nA in an "auto stop" mode). In this arrangement, a small ionic conduction current flows through the solution as a result of the contact between the tip and the AA-supported CdS particulate film. The electrochemical cell used for *in situ* measurements was set by switching the tip and Pt ring to the working electrode (WE) and counter electrode (CE), respectively, and connecting an Ag/AgCl wire to the reference electrode (RE) of the potentiostat. The entire system was covered by a plastic jar in order to diminish evaporation of water from the subphase. When needed, the system was purged by and kept under nitrogen.

The electrochemical properties of nanocrystalline particulate films were examined *ex situ* on substrates in an electrochemical cell (shown in Fig.

2.5). The arrangement allowed the measurements of current-voltage curves in the dark and under illumination.

2.2.6. Characterization of Nanocrystalline Particulate Films by Electron Microscopy, Scanning Tunneling Microscopy, and Atomic Force Microscopy

Electron microscopy and electron diffraction measurements have provided the most vital and direct information on the sizes and morphologies of nanocrystals constituting the particulate films grown under monolayers.[28-37]

Following the award of a Nobel prize to Binnig and Rohrer, the STM has evolved into a powerful and widely used instrument.[68-79] The tip of the STM is brought close to the surface of a sample by an *xyz* piezoelectric scanner. Development of a tunneling current, typically under constant-current mode, produces three-dimensional images (3-D) that represent the surface topology down to subnanometer scale. Scanning tunneling spectroscopy (STS) has been shown to provide a powerful method for the nanometer-scale resolution of electronic structures at metallic and semi-

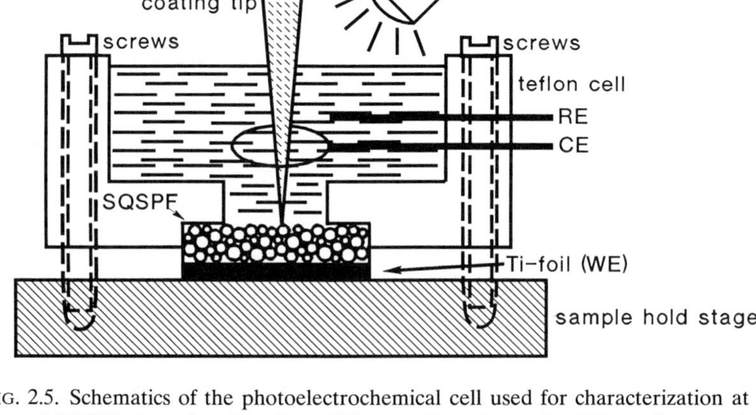

Fig. 2.5. Schematics of the photoelectrochemical cell used for characterization at Ti-foil-supported SQSPF-electrolyte junctions. *WE* = working electrode; *RE* = reference electrode; *CE* = counter electrode.

conductor surfaces.[74,76] The STS measurements essentially involve keeping the metal tip of the microscope at a constant distance from the sample (0.5 to 15 Å), applying a voltage between the tip and the sample, and measuring the tunneling current. In the STS measurements, the tip of the microscope is kept at a constant distance from the sample (0.5 to 15 Å), a voltage is applied between the tip and the sample, and then the tunneling current is measured. Both current I versus voltage V and differential conductance dI/dV versus voltage V are determined. Scanning electrochemical microscopy (SECM) provides information on the nanoscale topography of a sample and on its electrical and chemical properties in contact with electroactive species. Atomic resolution electrochemistry can also be investigated by atomic force microscopy (AFM).[77] The tip of an electrode with a radius of 10 μm or smaller is moved into close proximity to the substrate bathed in an electrolyte solution. The tip current is controlled by electrochemical reactions and by the substrate, and is related to the tip-to-sample distance and to the electrochemical nature of the substrate. Thus, all of the classical electrochemical experiments can be performed by SECM. Time-resolved luminescence measurements provide important structural information.

2.3. Growth of Nanocrystalline Particulate Films

Evolution of a nanocrystalline particulate film, illustrated by the formation of sulfide semiconductor particulate films (Fig. 2.6), has been discussed in terms of the following steps[30,31]:

1. formation of metal-sulfide bonds at a large number of sites at the monolayer/aqueous interface
2. downward growth of well-separated nanocrystalline metal sulfide particles
3. coalescence of clusters into interconnected arrays of semiconductor particles
4. formation of a "first layer" of a porous sulfide semiconductor particulate film composed of 20- to 40-Å-thick, 30- to 80-Å-diameter particles
5. diffusion of fresh metal ions to the monolayer headgroup area
6. formation of a "second layer" of a porous sulfide semiconductor particulate film (by steps 1, 2, and 3)
7. build up of "subsequent layers" of the sulfide semiconductor particulate film (by steps 1, 2, and 3) up to a plateau thickness (about 300 Å for CdS and about 3500 Å for ZnS) beyond which the film cannot grow.

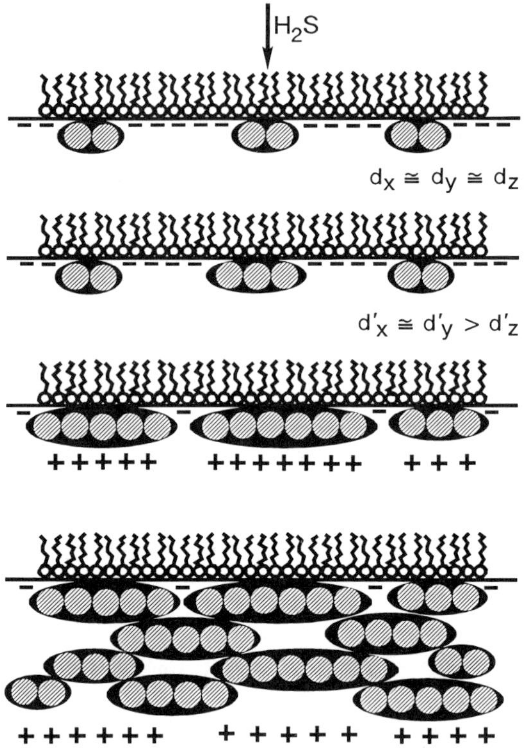

FIG. 2.6. Proposed schematics for the initial and subsequent growth of a monolayer-supported, porous SQSPF. The d_x and d_y dimension are in the plane and the d_z dimension is normal to the plane. They refer to the earliest observable particles; whereas d'_x, d'_y, and d'_z are dimensions in the plane and are normal to the plane. They refer to particles observed at later stages of their growth.

The presence of a monolayer with an appropriate surface charge is an essential requirement for sulfide semiconductor particulate film formation. Infusion of H_2S over an aqueous metal ion solution, in the absence of a monolayer, led only to the formation of large, irregular, and polydisperse metal-sulfide particles that precipitated and settled at the bottom of the trough. Furthermore, no sulfide semiconductor particulate film formation could be observed on the infusion of H_2S to a positively charged monolayer (dioctadecyldimethylammonium bromide, for example) floating on an aqueous metal ion ($CdCl_2$, for example) subphase.[30] Finally, we need to reemphasize that sulfide semiconductor particulate films can be transferred to solid substrates at any stage of their growth by horizontal lifting.[31]

2.4. Size Quantization and Bandgap Engineering

Demonstration of size quantization in semiconductor particulate films is the most significant result of the present work. Confinement of the electron and the hole in a particle that is smaller than the exciton diameter (i.e., the de Broglie wavelength) of the bulk semiconductor results in the quantization of the energy levels. This is in contrast to a bulk semiconductor in which the conduction bands constitute virtual continua. The length of the exciton diameter depends on the extent of electron delocalization and on the effective mass of the charge carrier. In the crystal lattice of semiconductors (CdS, ZnS, and GaAs, for example), the effective mass of the charge carrier is substantially smaller than that in free space and, hence, the exciton diameter can be quite long (60 to 80 Å for CdS, for example).[65] Consequently, size quantization becomes observable in CdS particles whose diameters are equal or smaller than 60 to 80 Å. It is manifested in the transformation of the continuous bands of the bulk semiconductor into a series of discrete energy levels and in the shifting of the lowest allowed absorptions to higher energy.

Several size regimes have been recognized in the progressive transition of n number of individual molecules to bulk semiconductors.[2] In the first regime, at relatively small n (< 100, typically) properties of the clusters are molecular. Although band structures begin to develop in the second regime (typically $10^2 < n < 5 \times 10^3$), referred to as the "Q state,"[80] electrons remain in discrete molecular orbitals. The most pronounced effects of size quantization are observed here. In the small quantum confinement regime (typically 20- to 30-Å-diameter spherical particles), absorption spectra can be blue shifted by 0.5 eV or more and electrons remain in discrete excited states. In the intermediate quantum confinement regime (typically 40- to 50-Å-diameter spherical particles), bulk electronic properties are not fully developed and absorption edges begin to shift to higher energies. Finally, in the third regime, spherical particles in the 100- to 150-Å-diameter range behave like bulk semiconductors.

Not only size, but also dimensionality reductions have been recognized in epitaxially grown semiconductors.[81] Electron-hole confinement in nanosized spherical particles, discussed earlier, results in 3-D quantum-size effects or, in the language of physicists, in the formation of "quantum dots," "quantum crystallites," or "zero-dimensional excitons." In one-dimensional (1-D) size quantization, the exciton movement is confined in the growth direction, but it is unrestricted in the other two directions[81]; in the language of solid-state physics, "quantum wells" or "2-D excitons" are formed. In quantum wells, size quantization manifests in the growth direction, while bulk properties prevail in the other two dimensions.

Finally, 2-D confinement of the charge carrier (i.e., providing the exciton only with 1-D mobility) results in quantum well wires.

The present "wet" colloid chemical approach is readily amenable to bandgap engineering. Several means are available to control the sizes, size distributions, compositions, and morphologies of the nanocrystalline particulate films formed under monolayers. The amount and rate of infusion of the gaseous precursor through the monolayer profoundly influence the size of particles in the incipient semiconductor particulate film. As expected, drastic limitation of the reagent results only in the formation of molecular clusters (the first regime). A somewhat longer exposure of the monolayer to the reagent brings about the formation of size-quantized semiconductor particulate films (the second regime).[81,82] Dimensionality reduction can be brought about by generating domains in the monolayers. Formation of string-like CdS particles has been demonstrated under monolayers prepared from mixtures of positively and negatively charged surfactants.[82] In a different experiment, rod-like CdSe particles were epitaxially grown under arachidic acid monolayers.[37] In some cases, particularly at high monolayer surface pressures, formation of rod-like PbSe particles was observed (Fig. 2.7). Typically, the thickness of the rods was 10 nm or less and the length was of the order of 100 nm. In addition to the rods, there were many additional dot-like particles in the 10-nm range (Fig. 2.7). Electron diffraction of the area shown in Fig. 2.7 resulted in a pseudo sixfold symmetrical pattern caused by the alignment of crystals with their [110] direction parallel to the electron beam, in addition to some of them being rotated 60 and 120 deg around the [110] axis. The single crystal diffraction pattern, originating in the alignment of the crystals with their [110] direction parallel to the electron beam, would only give a twofold symmetry. Exciton confinement, within the length of a PbSe rod, could lead to 2-D size quantization (i.e., to the formation of quantum wires).

Concurrent and/or consecutive introduction of different cations can lead to the formation of mixed or sandwich-type semiconductors. Mixed CdS-ZnS semiconductor particulate films have been prepared, for example, by the infusion of hydrogen sulfide over monolayers floating on aqueous $CdCl_2$ and $ZnCl_2$ mixtures in different compositions.[82] Similarly, infusion of different gases (H_2S and H_2Se, for example) concurrently, in different compositions, or consecutively should lead to mixed or sandwich-type semiconductor particle formation.

Opportunities also exist for doping. Once again, dopants can be introduced into the subphase or can be carried by the infusing gases. Introduction of copper ions into the aqueous subphase, under an already grown CdS particulate film, resulted,[30] for example, in the formation of copper sulfide centers in CdS.

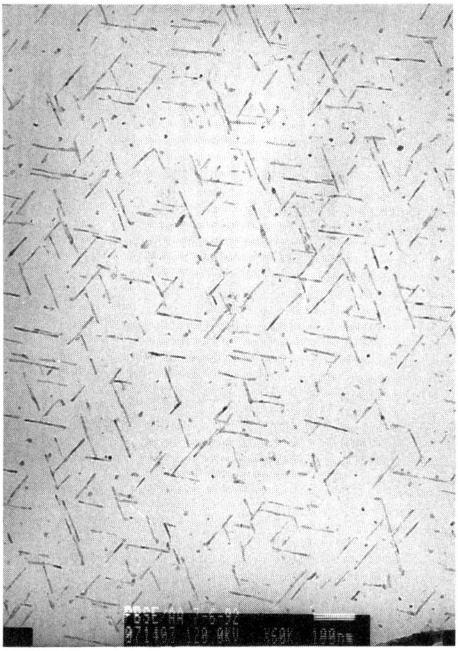

FIG. 2.7. Transmission electron micrograph of a PbSe particulate film. The film was formed by the 25-min infusion of H_2Se (50 μL) over an AA monolayer, kept at 40 mN/m, which was floating on an aqueous 5.0×10^{-4} *M* $Pb(NO_3)_2$ solution in the Lauda trough. The PbSe particulate film was transferred to an amorphous-carbon-coated 200-mesh copper grid. The bar = 100 nm.

In situ nanofabrication, of monolayer-supported semiconductor particulate films, by STM or AFM, also provides an exciting possibility for bandgap engineering.

2.5. Semiconductor Particulate Films

Cadmium sulfide, zinc sulfide, lead sulfide, cadmium selenide, and lead selenide semiconductor particulate films have to date been grown *in situ* under monolayers.[30–37] Absorbances A increased linearly with increasing thicknesses of the CdS and ZnS particulate films. Absorption coefficient σ values calculated from

$$\sigma = A/d'_s \tag{2.1}$$

Absorption coefficients of 2.4×10^5 cm^{-1} at 239 nm and 5.8×10^4 cm^{-1} at 475 nm were obtained for the CdS particulate film. This absorption coefficient agreed well with that determined for electrodeposited CdS films [σ (435 nm) = 2.06×10^4 cm^{-1}].[83] Similarly, an absorption coefficient of 5.8×10^4 cm^{-1} at 315 nm was determined for the ZnS particulate film. Knowledge of absorption coefficients allowed the assessment of direct bandgap energies E_g from

$$(\sigma h \omega)^2 = (h \omega - E_g)C \tag{2.2}$$

where $h\omega$ is the photon energy. Typical plots of the data, according to Eq. 2.2, are shown in Fig. 2.8. Values for E_g for CdS particulate films of d'_s = 63, 125, 163, 204, 263, and 298 Å were assessed to be 2.54, 2.48, 2.46,

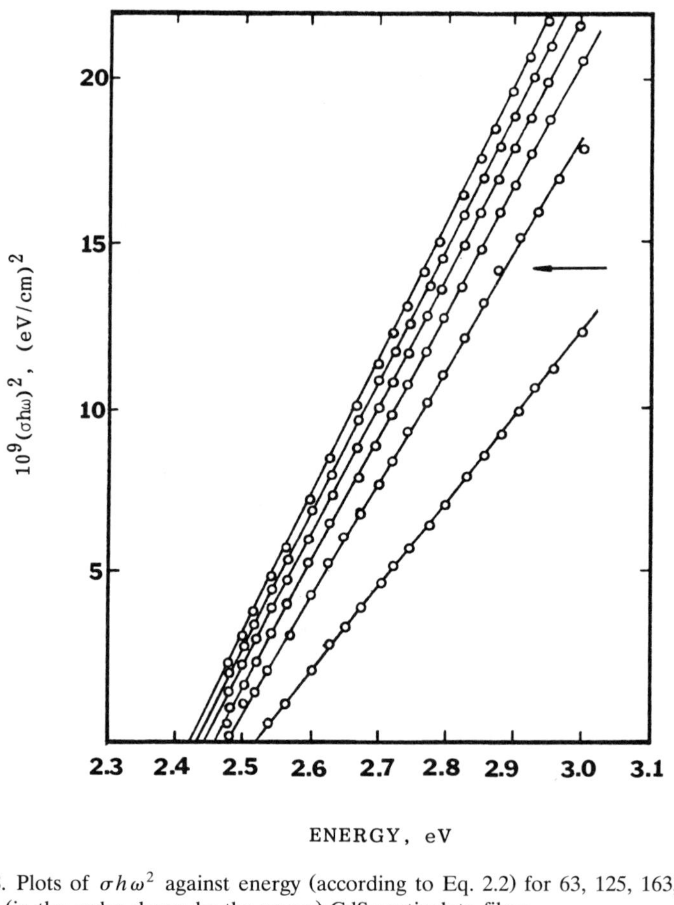

FIG. 2.8. Plots of $\sigma h \omega^2$ against energy (according to Eq. 2.2) for 63, 125, 163, 204, 263, and 298 Å (in the order shown by the arrow) CdS particulate films.

2.44, 2.43, and 2.42 eV, respectively (Fig. 2.8). Taking advantage of Henglein's published E_g versus particle-size curve,[63] we estimate the average diameter of the 63-Å-thick CdS particles to be about 50 Å. Increasing the thickness of the CdS particulate film resulted in progressively decreased direct bandgaps and, hence, in progressively larger CdS particles. At the thickest CdS particulate film, the measured CdS direct bandgap corresponded to that reported[63] for bulk CdS semiconductors (2.4 eV). A direct bandgap of 3.75 eV was assessed for the 359-Å-thick ZnS particulate film.

Prolonged heating of the semiconductor particulate films at high temperatures resulted in pronounced changes in their absorption spectra. The absorbance of a 192-Å-thick CdS particulate film (vacuum dried at 10^{-3} Torr for three days) decreased on heating at 490°C for 5, 15, and 25 min. Five minutes of heating shifted the direct bandgap from 2.47 to 2.40 eV (or to an absorption edge of 515 nm), which was attributable to bulk CdS[2]. A similar behavior was noted for 359-Å-thick ZnS particulate films; heating at 300°C for 15 min shifted the direct bandgap from 3.75 to 3.64 eV (or to an absorption edge of 340 nm). Ignoring the weak absorption tails results[65] in an uncertainty of about 5%. Prolonged heating of semiconductor particulate films has, therefore, two important consequences. First, their properties become similar to those found for bulk semiconductors. Second, they are annealed. Prolonged heating annealed the semiconductor particulate film to the substrate. Annealed semiconductor particulate films could not be washed or wiped away from their substrates. In contrast, vertical dipping of untreated semiconductor particulate films into water resulted in a partial loss of material from the subphase.

Transmission electron micrographs of 30- to 50-Å-thick CdS particulate films indicated the presence of 20- to 80-Å-diameter particles in a relatively narrow size distribution with an average diameter of 47 Å. STM established highly oriented pyrolytic graphite (HOPG) to provide an atomically flat surface with periodic roughnesses on the order of 1 Å. In 2-D STM images of HOPG-supported ZnS and CdS particulate films, the presence of 10- to 20-Å-thick, 30- to 40-Å-diameter ZnS and 20- to 30-Å-thick, 40- to 50-Å-diameter CdS particles is clearly discernable. The widths of the semiconductor particles observed by STM agree well with the corresponding diameters determined by transmission electron microscopy.

Electrical and photoelectrical measurements were carried out on CdS particulate films deposited on glass substrates or Teflon sheets. The resistivity ρ of a semiconductor particulate film, measured between two parallel copper electrodes, is given by

$$\rho = R\frac{Ld'_s}{a} \tag{2.3}$$

where R is the measured resistivity in ohms, L is the length of the copper electrodes, a is the distance between them, and d'_s is the thickness of the semiconductor particulate film. Resistivities of 200- to 300-Å-thick CdS particulate films were determined to be (3 to 6) $\times 10^7$ Ω cm. The range represents measurements of 10 samples of different thicknesses and is due in part to the presence of different amounts of water in the films. The ρ values determined for CdS particulate films are some six orders of magnitude higher than those observed for materials having intrinsic conductivity.

The dark resistance of CdS particulate films was found to decrease with increasing temperature exponentially (Fig. 2.9). Illumination decreased the resistivity (i.e., increased the conductivity) of CdS particulate films by some two orders of magnitude (Fig. 2.10) and matched the absorption spectrum of the corresponding CdS particulate film nicely (Fig. 2.10). Photoconductivity originates, therefore, in the production of conduction band electrons, e^-_{CB}, and valence band holes, h^+_{VB}, in band-gap irradiation of CdS:

$$\mathrm{CdS} \xrightarrow{h\nu} e^-_{CB} + h^+_{VB} \tag{2.4}$$

Steady-state irradiation of CdS particulate films also resulted in the development of photovoltage. Irradiation by a 10-ns, 343-nm laser pulse gave rise to a transient photovoltage. The magnitude of the photovoltage (1 to 8 mV) was found to increase linearly with the energy of the laser pulse (0.1 to 1.0 mJ). The rise time of the transient signal, corresponding to Eq. 2.4, was faster than the response time of the instrument used (10 ns). The decay time of the signal was on the order of 3×10^{-4} s. This decay corresponds to charge recombination.

2.6. Silver Particulate Films

Two alternative methods have been developed for silver particulate film formation under monolayers. In the first method, silver ions, electrostatically attached to the headgroups of negatively charged monolayers, were reduced by formaldehyde in the experimental setup of Fig. 2.2 (bottom). In the second method, silver ions were electrochemically reduced at the monolayer interface in the setup illustrated in Fig. 2.3.[84]

In the chemical generation, the time of exposure to formaldehyde had a dramatic effect on the structure of the silver particulate films formed under monolayers, as illustrated in Fig. 2.11.[85] At relatively short exposure times, only small clusters are formed. Longer exposure to formaldehyde results in the formation of larger aggregates whose absorption maxima broaden and appear increasingly at longer wavelengths. After very long

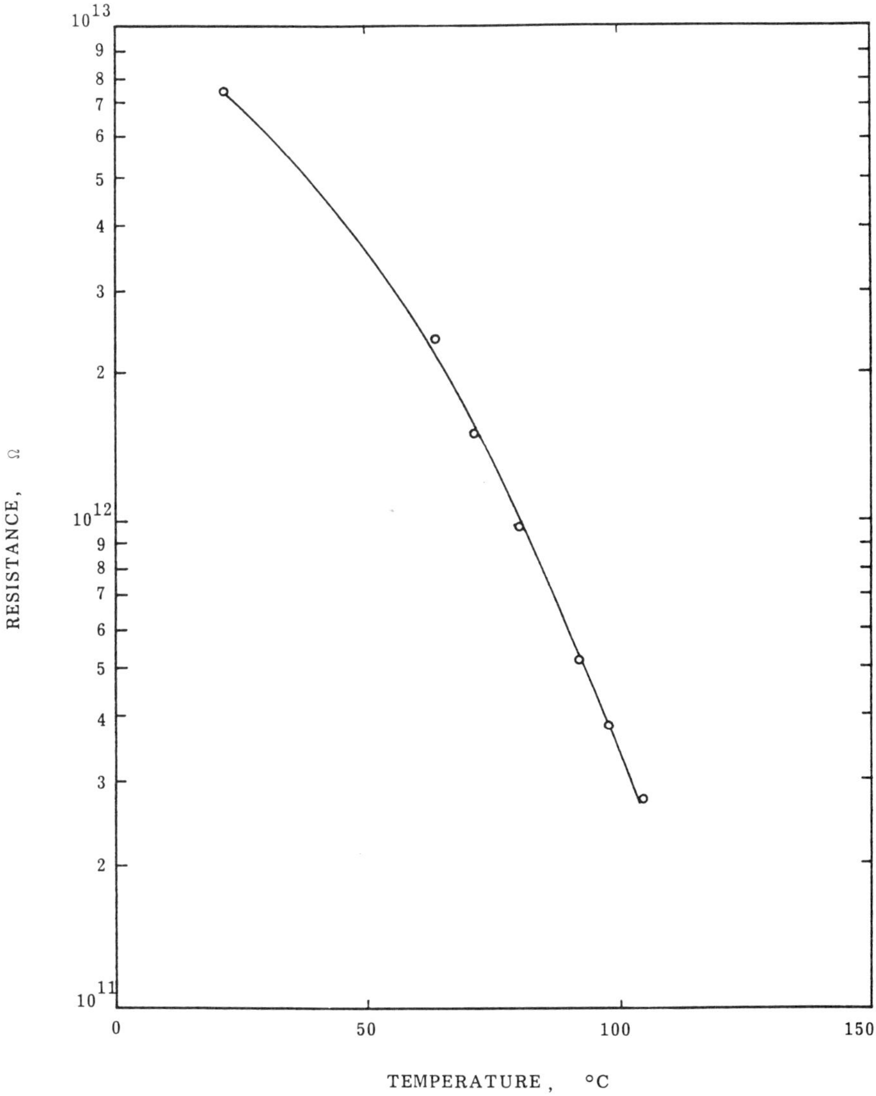

FIG. 2.9. A plot of dark resistance of a 298-Å-thick CdS particulate film against temperature.

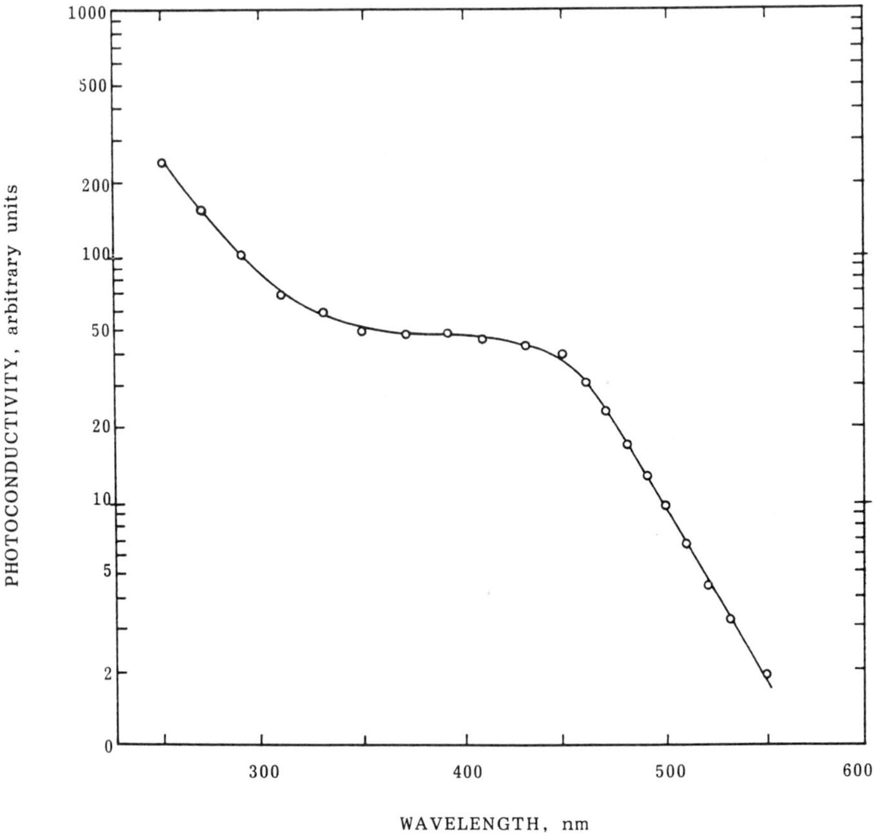

FIG. 2.10. Photoconductivity action spectra of a 300-Å-thick CdS particulate film.

absorption (typically days to weeks), an apparent negative absorption appears with a minimum at 320 nm (Fig. 2.11). This interband transition is characteristic for rough silver particles having large surface area-to-volume ratios.[85]

Interband formation was also observed in the electrochemical generation of silver particles under monolayers.[86] Importantly, no silver particles could be observed on applying the same potential to the water surface in the absence of surfactants or to monolayers prepared from positively charged surfactants. The electron diffraction pattern of the silver particulate film on a grid has been analyzed for a face-centered cubic polycrystalline structure with lattice constants of 4.08 Å and *hkl* indices of (111), (200), (220), (311), (222), (400), (311), and (402).

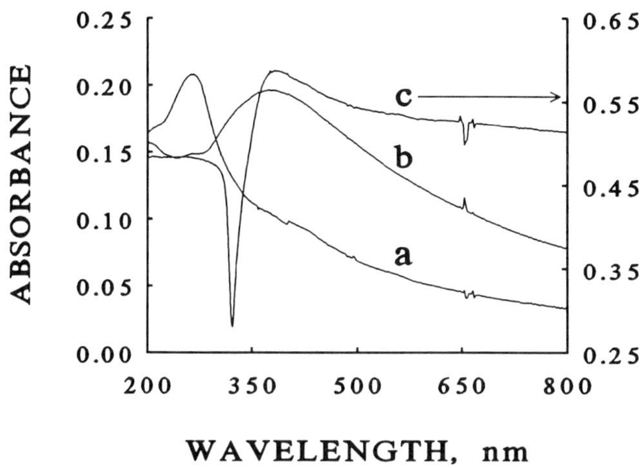

FIG. 2.11. Absorption spectra of silver particulate films exposed to formaldehyde for (a) 2 h, (b) 12 h, and (c) one week.

Resistivity ρ of the silver particulate films on quartz was measured by a multimeter connected to two silver electrodes pressed on to the film at a distance of 0.2 cm (L) to be 40 Ω (R). Taking the cross section of the silver particulate film to be 2.25×10^{-6} cm^2 gave 4.5×10^{-4} Ω cm for ρ, which is some two orders of magnitude larger resistivity than that of bulk silver (1.586×10^{-6} Ω cm).[87]

Preferential 2-D silver ion electroreduction is likely to be mediated by the lateral conductivity at the monolayer headgroups. The initially formed silver particles extend the cathode and continue to reduce silver counter ions at the monolayer surface.

The ultrathin silver films, prepared either chemically or electrochemically, contain interconnected roughened metal particles that can be exploited for catalysis, electron transfer, and nonlinear and surface-enhanced optics and related applications.

2.7. Organizational Imperative: Molecular Recognition and Epitaxial Growth

Molecular recognition between the monolayer headgroups and incipient semiconductor nanocrystallites mimics biomineralization[5-10] and represents an important milestone in the realization of the potential of a colloid chemical approach to bandgap engineering. Lead sulfide particulate films composed of highly oriented, equilateral triangular crystals have been generated *in situ* by the exposure of AA-monolayer-coated aqueous lead nitrate [$Pb(NO_3)_2$] solutions to H_2S (Fig. 2.12). AA monolayers in their solid states consist of $CH_3(CH_2)_{18}COOH$ molecules, two dimensionally arrayed at the air/water interface. Spread over the aqueous subphase, the carboxyl or the carboxylate groups of AA are aligned perpendicularly to the water surface. The alkyl chains of AA, fully extended in the air in a planar zigzag conformation, are oriented[38,88] approximately normal to the surface in a triangular lattice of hexagonal close packing with a lattice constant of $a = 4.85$ Å and $d_{(100)}$ spacing of 4.20 Å. Combined synchrotron

FIG. 2.12. Transmission electron micrograph at limiting aperture coverage of PbS crystals formed by the slow (30-min) infusion of H_2S to an AA monolayer in the Lauda film balance (kept at $\pi = 26$ mN/m surface pressure) floating on an aqueous 5.0×10^{-4} *M* $Pb(NO_3)_2$ solution. The PbS particulate film was deposited on a formvar-coated, 200-mesh copper grid.

x-ray reflection and diffraction data established a structural model for AA monolayers at air/water interfaces.[38] The model required the hydrocarbon chains to be well packed in a pseudohexagonal lattice and tilted toward their nearest neighbor.[38]

Rationalization of the packing of the AA headgroups at the aqueous subphase/air interface is, unfortunately, less than straightforward.[40] The absence of information on the extent of headgroup ionization (at a bulk pH of 5.5), counter ion binding, and water hydration hinders the interpretation of experimental results and the development of a reliable theoretical approach for predicting headgroup organization at the monolayer/subphase interface. Use of experimentally determined value for the surface area of one AA molecule (20.0 $Å^2$/molecule) permitted assessment of the lattice constant to be 4.81 Å (a) and the $d_{(100)}$ spacing to be 4.16 Å [$d_{(100)} = a \sin 60\,\mathrm{deg}$]. These values are in good agreement with those determined for AA monolayers by synchrotron x-ray scattering (a = 4.85 Å and d = 4.13 Å).[38] They are also similar to those determined for cadmium stearate (a = 4.89 Å and $d_{(100)} = 4.20 \pm 0.10$ Å)[38,88] and other fatty acid[89,90] monolayers.

Reliable assessment of the arrangement and crystallinity of Pb^{2+} at the AA/monolayer interface is equally elusive. Our data are best accommodated in terms of an AA : Pb^{2+} = 3 : 4 ratio (Fig. 2.13). Grazing incidence x-ray diffraction measurements of lead arachidate monolayers demonstrated[89] the existence of long-range ordering (250 Å) of Pb^{2+}. Evidence has also been obtained for the presence of an ordered Cd^{2+} layer, with a 1000-Å coherence length, under uncompressed cadmium arachidate monolayers.[91] Data obtained in the present work are in accord with such orderings of arachidate monolayers (*vide infra*).

PbS is known[92] to crystallize in a cubic crystalline lattice with a lattice constant of a = 5.9458 Å. Atomic coordinates are $(0, 0, 0)$ and $(\frac{1}{2}, \frac{1}{2}, \frac{1}{2})$ for Pb and they are $(\frac{1}{2}, \frac{1}{2}, \frac{1}{2})$ and $(1, 1, \frac{1}{2})$ for S. Distances between the closest Pb–Pb and S–S atoms, 4.20 Å, fit the $d_{(100)}$ network spacing of the AA monolayer. This fit implies the alignment of PbS along its (111) plane to the (100) plane of the AA monolayer (Fig. 2.13). A comparison of the interatomic Pb to Pb distance of the (111) plane of the PbS crystal (4.20 Å) with that of the $d_{(100)}$ spacing of the AA monolayers (4.16 Å) revealed a mismatch of only 1% between these two crystals.

The epitaxial growth of PbS under well-compressed AA monolayers is explicable in terms of the geometrical complementarity between PbS and the AA headgroups. Strong electrostatic interaction results in a very high Pb^{2+} concentration at the monolayer interface. The extremely low solubility of PbS in water ($K_{SP} = 8.81 \times 10^{-29}$ at 25°C) will drive its rapid and random nucleation at the monolayer interface unless the rate of reaction is

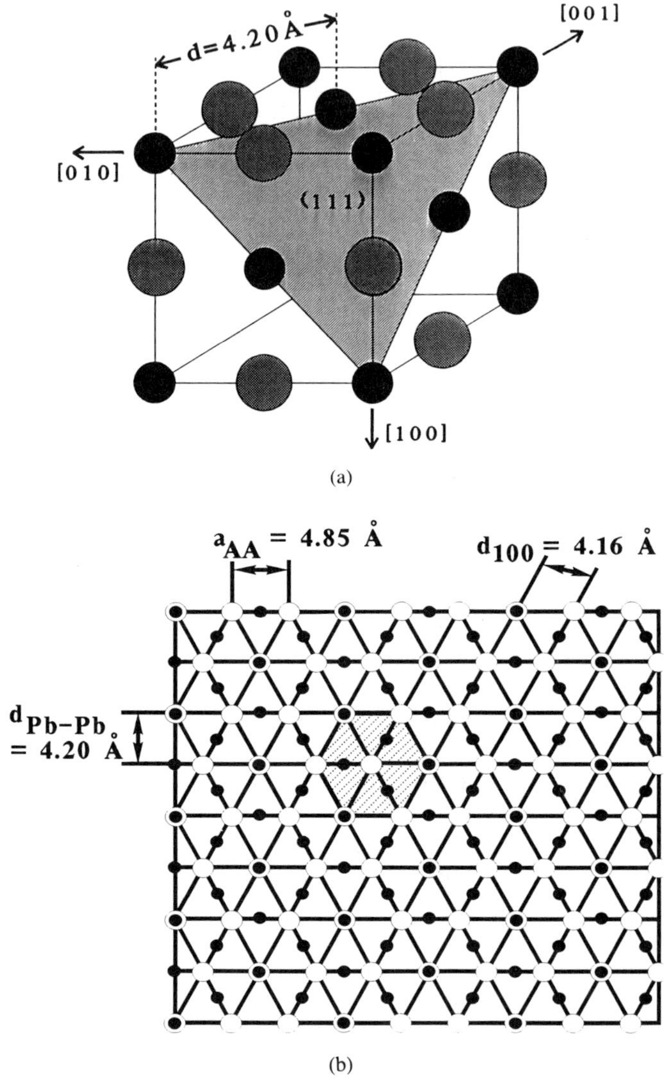

FIG. 2.13. (a) A 3-D representation of the PbS crystal lattice and (111) plane. (b) Schematic 2-D representation of the proposed overlap between Pb^{2+} ions and AA headgroups; ○ = AA headgroup, ● = Pb^{2+}, and ◉ = Pb^{2+} and AA headgroups. A unit cell is highlighted by the dotted area, which is enclosed by heavy lines.

drastically diminished. Encounters of the PbS precursors have been controlled by interposing the monolayer between them and by limiting the rate and amount of H_2S exposure. These measures have ensured the formation of a critically sized nucleus and the subsequent ion-by-ion heteroepitaxial growth of the PbS crystals.

Significantly, equilateral triangular PbS crystals have been grown under compressed monolayers in the same orientation (Fig. 2.12).

Epitaxial growth of nanocrystalline lead selenide particles under AA monolayers has also been demonstrated.[37] Exposure of an AA-monolayer-coated $Pb(NO_3)_2$ solution to H_2Se resulted in the formation of reasonably uniform crystals, which constituted the particulate film. A typical transmission electron micrograph of PbSe particulate films is shown in Fig. 2.14. This film was generated by the infusion of H_2Se for 10 min, over an AA monolayer, kept at 35 mN/m surface pressure, which was floating on an aqueous 5.0×10^{-4} M $Pb(NO_3)_2$ solution. The PbSe particulate film formed under the monolayer was transferred via an amorphous-carbon-coated, 200-mesh copper grid for electron microscopy. The presence of equilateral triangles, with 50.1-nm mean lengths ($\sigma = 12.3$ nm), is clearly discernible in Fig. 2.14. Interestingly, almost all crystals are in the same orientation and are aligned in their edge directions. The edges of each triangle are perpendicular to $[\bar{1}\bar{1}2]$, $[\bar{1}2\bar{1}]$, $[2\bar{1}\bar{1}]$. The transmission electron diffraction (TED), taken in the area shown in Fig. 2.14 by confining the beam size to 2 μm, showed diffraction arches that were indexed as (220), (422), and (440) faces. The faces (111), (200), (311), (222), (400), and (420) were conspicuously absent. The hexagonal symmetry and systematical cancellation of diffraction indicate that all crystals align upward and have their (111) planes perpendicular to the electron beam.

Additional information on the morphology of PbSe crystals was obtained by AFM. A typical 3-D 470-nm $\times$ 470-nm image of PbSe crystallites is shown in Fig. 2.15(a). The thickness of these crystals was determined to be around 65 Å by sectioning [Fig. 2.15(b)]. Interestingly, the equilateral triangular PbSe crystals were not uniform; along their corners they were somewhat thicker than in their middles [Fig. 2.15(a)].

In some cases, particularly at high monolayer surface pressures, formation of rod-like PbSe particles was observed (Fig. 2.7). Typically, the thickness of the rods was 10 nm or less and the length was of the order of 100 nm. In addition to the rods, there were many additional dot-like particles in the 10-nm range (Fig. 2.7). Electron diffraction of the area shown in Fig. 2.7 resulted in a pseudo sixfold symmetrical pattern caused by the alignment of crystals with their [110] direction parallel to the electron beam, in addition to some of them being rotated 60 and 120 deg around the [110] axis. The single crystal diffraction pattern, originating in

FIG. 2.14. Transmission electron micrograph of a PbSe particulate film. The film was formed by the 10 min infusion of H_2Se (50 μL) over an AA monolayer, kept at 35 mN/m, which was floating on an aqueous 5.0×10^{-4} *M* $Pb(NO_3)_2$ solution in the Lauda trough. The PbSe particulate film was transferred to an amorphous, carbon coated, 200-mesh copper grid. The bar = 100 nm.

the alignment of the crystals with their [110] direction parallel to the electron beam, would only give a twofold symmetry.

Appearance of rod-like PbSe particles (Fig. 2.7) is rationalized by the good fit between the Pb^{2+} ions in the (110) plane of the hexagonally close-packed AA. Thus, the preferred orientation of the [110] axis in PbSe is perpendicular to the monolayer, while the $[\bar{1}10]$ and $[00\bar{1}]$ axes are parallel to the monolayer and are related to the short and long axes of the PbSe rod, respectively. The overlap of Pb^{2+} ions with the unstrained crystal lattice of arachidic acid is schematically shown in Fig. 2.16(a). Using values of a_{PbSe} = 6.125 Å (Ref. 92) and a_{AA} = 4.81 Å and $d_{(100)}$(AA) = 4.16 Å (Ref. 36) for the intrinsic PbSe and AA crystal lattices

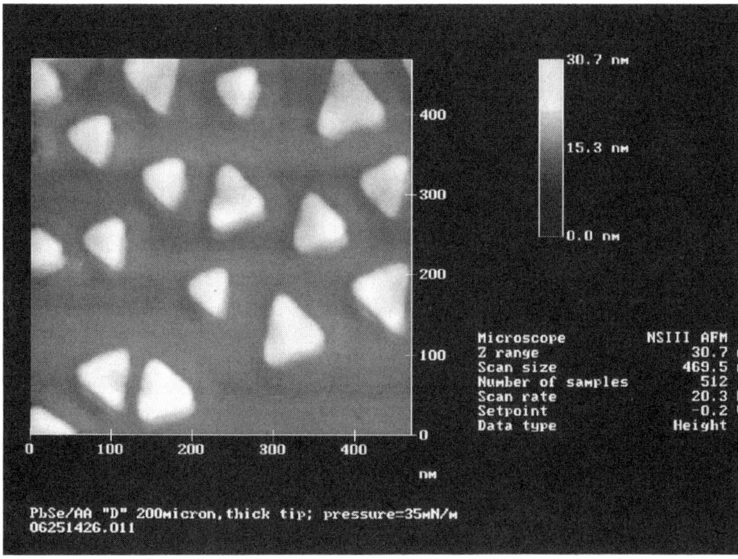

(a)

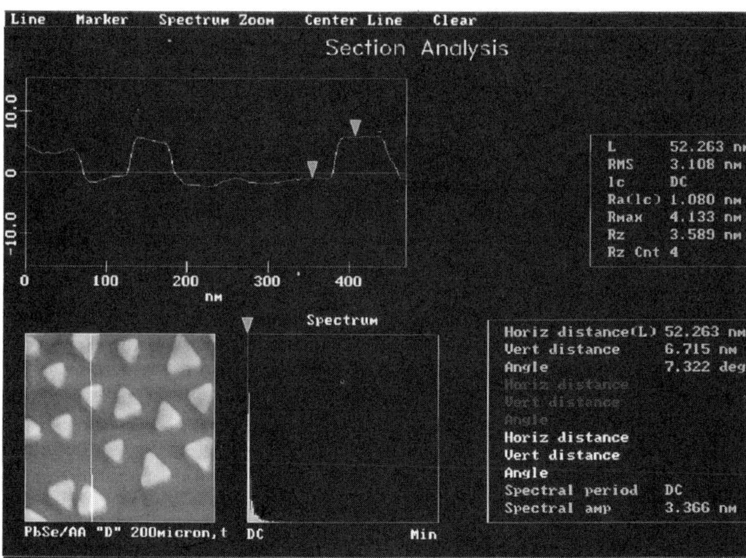

(b)

FIG. 2.15. (a) A 3-D side view of the PbSe particulate film shown in Fig. 2.14, imaged on a 470-nm × 470-nm section by AFM. (b) Profile of the PbSe crystals by AFM sectioning. The vertical line in the image shows the position of the sectioning.

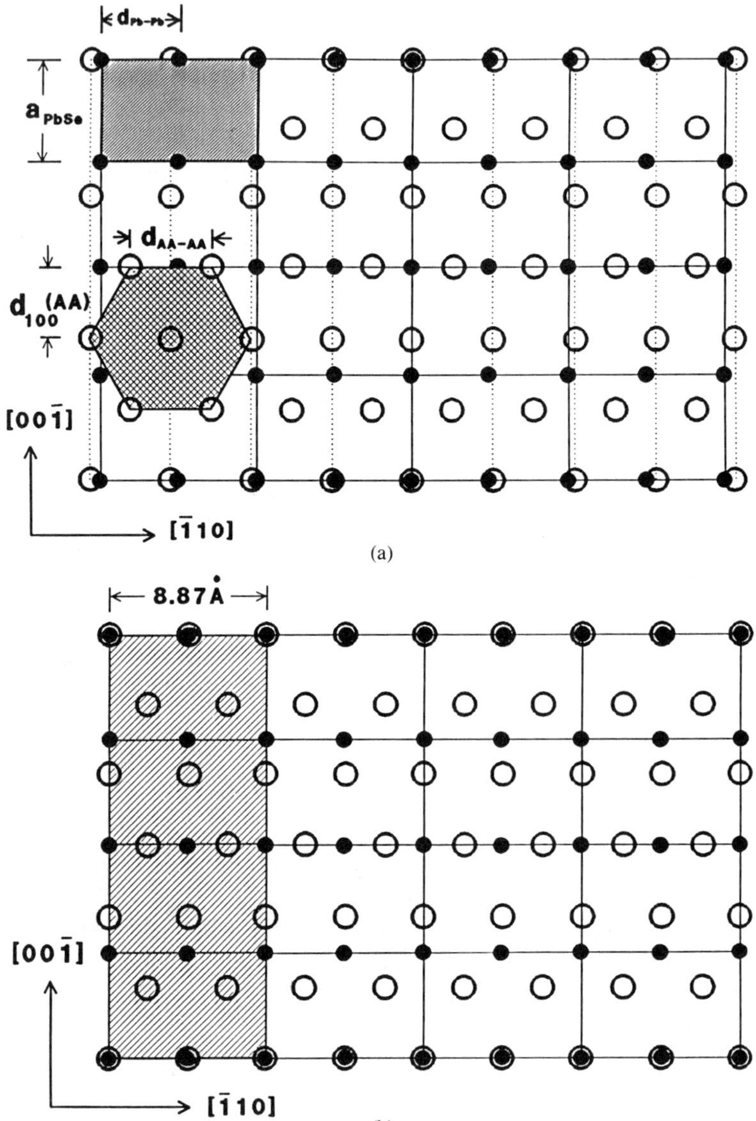

FIG. 2.16. (a) Schematic 2-D representation of Pb^{2+} (●) in the (110) plane of unstrained lattice (a = 6.125 Å) and hexagonally close-packed AA (○) headgroups. Two directions are indicated as [00$\bar{1}$] and [$\bar{1}$10] in the PbSe lattice. Both unit cells of AA headgroups and Pb^{2+} ions are highlighted by hexagonal and rectangular shaded areas, respectively. The mismatch between PbSe and AA crystal lattices is evident. (b) Schematic two-dimensional representation of the improved matching achieved by stretching PbSe lattice along [$\bar{1}$10] direction. The strained PbSe crystal lattice was experimentally determined as a = 6.27 Å.

led to 2% and 10% mismatch along the vertical $[00\bar{1}]$ and the horizontal $[\bar{1}10]$ axes between PbSe and AA. Stretching the PbSe crystal lattice to $a_{PbSe} = 6.27$ Å (experimentally determined in the present system) led to an improved overlap between PbSe and AA along the vertical (0.5% mismatch) and the horizontal (8% mismatch) axes [Fig. 2.16(b)]. The better overlap in the $[00\bar{1}]$ than in the $[\bar{1}10]$ direction favors, of course, the growth of PbSe crystallites along the vertical direction and leads to the formation of the observed rod-like structures. The overlap between the AA template and PbSe crystals was analyzed in terms of the observed $d_{(100)}(AA) = 2a_{PbSe}/3$ relationship.

2.8. Conclusions and Future Directions

Oriented growth of nanocrystalline particulate films under monolayers represents the initial phases in our research directed at the development of colloid chemical methodologies for bandgap engineering. The method of preparation permits convenient thickness, morphology, and dimensionality controls. Manipulation of the surfactant composition and organization, as well as the chemistries at the monolayer interphase and subphase, permits the generation of periodically and spatially modulated superlattices. The relatively simple experimental procedure is versatile and allows for the convenient monitoring of the physical and chemical changes that accompany the *in situ* generation of nanocrystalline particulate films. Importantly, monolayer-supported nanocrystalline particulate films can be readily transferred onto solid substrates. The method developed here enables the characterization of the same system at an aqueous/solid interface (i.e., at an electrolyte–solid-state junction) and in the solid state. It provides, therefore, a molecular-level insight into currently utilized epitaxies. Most significantly, the colloid chemical approach to bandgap engineering will open the door to construction of highly innovative nanofabricated electronic devices. The future will be exciting and scientifically rewarding. Only lack of imagination will hinder our progress.

Acknowledgments

Support of this research by grants from the National Science Foundation and the Department of Energy is gratefully acknowledged. Thanks is also due to the Alexander von Humboldt Foundation for an award during the tenure of which this account was written. The gracious hospitality of Professor Arnim Henglein at the Bereich Photochemische

Energieumwandlung, Hahn-Meitner Institute, Berlin, has been especially appreciated. The true credit is due, of course, to my co-workers (whose names appear in the cited joint publications) for their creative, skillful, and dedicated work.

References

1. National Research Council (1989). "Materials science and engineering for the 1990s," Report of the National Academy Press, Washington, DC.
2. Andres, R. P., Averback, R. S., Brown, W. L., Brus, L. E., Goddard, W. A., Kaldor, A., Louie, S. G., Moskovits, M., Percy, P. S., Riley, S. J., Siegel, R. W., Spaepen, F., and Wang, Y. (1989). *J. Mater. Res.* **4,** 704.
3. Capasso, F. (1991). *MRS Bull.* **23.**
4. Fendler, J. H. (1982). *Membrane Mimetic Chemistry.* John Wiley, New York.
5. Westbroek, P., and de Jong, E. W. (1983). *Biomineralization and Biological Metal Accumulation.* Reidel, Dordrecht, Holland.
6. Simkiss, K., and Wilbur, K. M. (1989). *Biomineralization.* Academic Press, San Diego.
7. Lowenstam, H. A., and Weiner, S. (1989). *On Biomineralization.* Oxford University Press, New York.
8. Addadi, L., and Wiener, S. (1992). *Angew. Chem. Int. Ed. Eng.* **31,** 153.
9. Mann, S., Archibald, D. D., Didymus, J. M., Heywood, B. R., Meldrum, F. C., and Wade, V. J. (1992). *MRS Bull.* **17**(10), 32.
10. Mann, S., Webb, J., and Williams, R. J. P. (1989). *Biomineralization, Chemical and Biological Perspectives.* VCH, Weinheim, Germany.
11. Meyer, M., Wallberg, C., Kurihara, K., and Fendler, J. H. (1984). *J. Chem. Soc. Chem. Commun.* 90.
12. Lianos, P., and Thomas, J. K. (1986). *Chem. Phys. Lett.* **125,** 299; Lianos, P., and Thomas, J. K. (1987). *Colloid Interface Sci.* **117,** 505.
13. Towey, T. F., Khan-Gadhi, A., and Robinson, B. H. (1990). *J. Chem. Soc. Faraday Trans.* **96,** 3757.
14. Motte, L., Petit, C., Boulanger, L., Lixon, P., and Pileni, M. P. (1992). *Langmuir* **8,** 1049.
15. Dannhauser, T., O'Neil, M., Johansson, K., Whitten, D., and McLendon, G. (1986). *J. Phys. Chem.* **90,** 6074.
16. Steigerwald, M. L., Alivisatos, A. P., Gibson, J. M., Harris, T. D., Kortan, R., Muller, A. J., Thayer, A. M., Duncan, T. M., Douglas, D. C., and Brus, L. E. (1988). *J. Am. Chem. Soc.* **110,** 3046.
17. Marcus, M. A., Flood, W., and Steigerwald, M. L. (1991). *J. Phys. Chem.* **95,** 1572.
18. Tricot, Y. M., and Fendler, J. H. (1986). *J. Phys. Chem.* **90,** 3369.
19. Youn, H.-C., Tricot, Y.-M., and Fendler, J. H. (1987). *J. Phys. Chem.* **91,** 581.
20. Watzke, H. J., and Fendler, J. H. (1987). *J. Phys. Chem.* **91,** 854.
21. Youn, H.-C., Baral, S., and Fendler, J. H. (1988). *J. Phys. Chem.* **92,** 6320.
22. Baral, S., Zhao, X. K., Rolandi, R., and Fendler, J. H. (1987). *J. Phys. Chem.* **91,** 2701.
23. Zhao, X. K., Baral, S., Rolandi, R., and Fendler, J. H. (1988). *J. Am. Chem. Soc.* **110,** 1012.
24. Zhao, X. K., Herve, P. J., and Fendler, J. H. (1989). *J. Phys. Chem.* **93,** 908.
25. Zhao, X. K., Xu, S., and Fendler, J. H. (1990). *J. Phys. Chem.* **94,** 2573.
26. Xu, S., Zhao, X. K., and Fendler, J. H. (1990). *Adv. Mater.* **2,** 183.

27. Yuan, Y., Cabasso, I., and Fendler, J. H. (1990). *Macromolecules* **23,** 3198.
28. Yi, K. C., and Fendler, J. H. (1990). *Langmuir* **6,** 1519.
29. Zhao, X. K., Yuan, Y., and Fendler, J. H. (1990). *J. Chem. Soc. Chem. Commun.* 1248.
30. Zhao, X. K., Xu, S., and Fendler, J. H. (1991). *Langmuir* **7,** 520.
31. Zhao, X. K., and Fendler, J. H. (1991). *Chem. Mat.* **3,** 168.
32. Zhao, X. K., and Fendler, J. H. (1991). *J. Phys. Chem.* **95,** 3716.
33. Zhao, X. K., McCormick, L. D., and Fendler, J. H. (1991). *Langmuir* **7,** 1255.
34. Zhao, X. K., McCormick, L. D., and Fendler, J. H. (1991). *Chem. Mater.* **3,** 922.
35. Zhao, X. K., McCormick, L. D., and Fendler, J. H. (1992). *Adv. Mater.* **4,** 93.
36. Zhao, X. K., Yang, J., McCormick, L. D., and Fendler, J. H. (1992). *J. Phys. Chem.* **96,** in press.
37. Yang, J., Fendler, J. H., Jao, T.-C., and Laurion, T. (1992). *J. Electron. Microsc. Tech.* in press.
38. Kjaer, K., Als-Nielsen, J., Helm, C. A., Tippmann-Krayer, P., and Möhwald, H. (1989). *J. Phys. Chem.* **93,** 3200.
39. Als-Nielsen, J., and Möhwald, H. (19xx). In: *Handbook on Synchrotron Radiation* (S. Ebashi, M. Koch, and E. Rubinstein, eds.). Elsevier Science Publishers, The Netherlands, Vol. 4, pp. 1–53.
40. Jacquemain, D., Wolf, S. G., Leveiller, F., Deutsch, M., Kjaer, K., Als-Nielsen, J., Lahav, M., and Leiserowitz, L. (1992). *Angew. Chem. Int. Ed. Eng.* **31,** 130.
41. Möbius, D., and Möhwald, H. (1991). *Adv. Mater.* **3,** 19.
42. Jacquemain, D., Leveiller, F., Weinbach, S. P., Leiserowitz, L., Kjaer, K., and Als-Nielsen, J. (1991). *J. Am. Chem. Soc.* **113,** 7684.
43. Tippmann-Krayer, P., and Möhwald, H. (1991). *Langmuir* **7,** 2303.
44. Gaines, G. L., Jr. (1966). *Insoluble Monolayers at Liquid-Gas Interfaces.* Interscience, New York.
45. Ulman, A. (1991). *An Introduction to Ultrathin Organic Films. From Langmuir Blodgett to Self-Assembly.* Academic Press, New York.
46. Kuhn, H., Möbius, D., and Bücher, H. (1972). In: *Physical Methods in Chemistry.* (A. Weissberger and B. W. Rossiter, eds.). John Wiley, New York, Vol. 1, Part 3b.
47. Johnson, E., Aroca, R., and Nagao, Y. (1991). *J. Phys. Chem.* **95,** 8840.
48. Ahuga, R. C., and Möbius, D. (1992). *Langmuir* **8,** 1136.
49. Bain, C. D., Davis, P. B., Ong, T. H., and Ward, R. N. (1991). *Langmuir* **113,** 1563.
50. Ikeura, Y., Kurihara, K., and Kunitake, T. (1991). *J. Am. Chem. Soc.* **113,** 7342.
51. Frey, S., and Tamm, L. K. (1991). *Biophys. J.* **60,** 922.
52. Su, W.-F. A., Kurata, T., and Koezuka, H. (1992). *Langmuir* **8,** 915.
53. Bryant, M. A., and Pemberton, J. E. (1991). *J. Am. Chem. Soc.* **113,** 8284.
54. Battisti, D., and Aroca, R. (1992). *J. Am. Chem. Soc.* **114,** 1201.
55. Löche, M., and Möhwald, H. (1984). *Rev. Sci. Instr.* **55,** 1968.
56. Moy, V. T., Keller, D. J., and McConnell, H. M. (1988). *J. Phys. Chem.* **92,** 5233.
57. Rice, P. A., and McConnell, H. M. (1989). *Proc. Natl. Acad. Sci. USA* **86,** 6445.
58. McConnell, H. M., Rice, P. A., and Benvegnu, D. J. (1990). *J. Phys. Chem.* **94,** 8965.
59. Henon, S., and Meunier, J. (1991). *Rev. Sci. Instr.* **62,** 936.
60. Hönig, D., and Möbius, D. (1991). *J. Phys. Chem.* **95,** 4590.
61. Vogel, V., Mullin, C. S., and Shen, Y. R. (1991). *Langmuir* **7,** 1222.
62. Xiao, X.-K., Vogel, V., Shen, Y. R., and Marowski, G. (1991). *J. Chem. Phys.* **94,** 2315.
63. Henglein, A. (1988). *Topics Curr. Chem.* **143,** 113.
64. Rosetti, R., Hull, R., Gibson, J. M., and Brus, L. E. (1985). *J. Chem. Phys.* **82,** 552.
65. Wang, Y., Suna, A., Mahler, W., and Kasowski, R. (1987). *J. Chem. Phys.* **87,** 7315.

66. Spanhel, L., Haase, M., Weller, H., and Henglein, A. (1987). *J. Am. Chem. Soc.* **109,** 5647.
67. Zunger, A., and Freeman, A. (1978). *J. Phys. Rev.* **B17,** 4850.
68. Binnig, G., and Rohrer, H. (1982). *Helv. Chim. Acta* **55,** 726.
69. Hansma, P. K., and Tersoff, J. (1987). *J. Appl. Phys.* **61,** R1.
70. Hansma, P. K., Drake, B., Marti, O., Gould, S. N. C., and Prater, C. B. (1989). *Science* **243,** 641.
71. Garcia, R., Sancez, J. J., Soler, J. M., and Garcia, N. (1987). *Surf. Sci.* **181,** 69.
72. Feenstra, R. M., and Stroscio, J. A. (1987). *J. Vac. Sci. Technol. B* **5,** 923.
73. Feenstra, R. M., Stroscio, J. A., and Fein, A. P. (1987). *Surface Sci.* **181,** 295.
74. Fan, F.-R. F., and Bard, A. J. (1990). *J. Phys. Chem.* **94,** 3761.
75. Hodes, G., Albu-Yaron, A., Decker, F., and Motiuke, P. (1987). *Phys. Rev. B* **36,** 4215.
76. Kuk, Y., and Silverman, P. (1989). *Rev. Sci. Inst.* **60,** 165.
77. Manne, S., Hansma, P. K., Massie, J., Elings, V. B., and Gewirth, A. A. (1991). *Science* **151,** 183.
78. Prater, C. B., Hansma, P. K., Tan, I.-H., Lishan, D. G., and Hu, E. L. (1992). *J. Vac. Sci. Technol. B* **10,** 1211.
79. Weisenhorn, A. L., Maivald, P., Butt, H.-J., and Hansma, P. K. (1992). *Phys. Rev. B* **45,** 11226.
80. Henglein, A. (1989). *Chem. Rev.* **89,** 1861.
81. Ploog, K. (1988). *Angew. Chem. Int. Ed. Engl.* **27,** 593.
82. Yuan, Y., and Fendler, J. H. (1992). Unpublished results.
83. Preusser, S., and Cocivera, M. (1990). *Solar Energy Mater.* **20,** 1.
84. Yi, K. C., and Fendler, J. H. (1992). Unpublished results.
85. Raetner, H. (1988). *Surface Plasmons on Smooth and Rough Surfaces and on Gratings.* Springer-Verlag, Berlin.
86. Zhao, X. K., and Fendler, J. H. (1990). *J. Phys. Chem.* **94,** 3384.
87. *Handbook of Chemistry and Physics*, 67th ed., 1986–1987, CRC Press, Boca Raton, FL, p. F120.
88. Garoff, S., Deckman, H. W., Dunsmuir, J. H., Alvarez, M. S., and Bloch, J. M. (1986). *J. Phys.* (*Paris*) **47,** 701.
89. Dutta, P., Peng, J. B., Lin, B., Ketterson, J. B., Prakash, M., Georgopoulos, P., and Ehrlich, S. (1987). *Phys. Rev. Lett.* **58,** 2228.
90. Bunnerat, A., Chollet, P. A., Frishby, H., and Hoclet, A. (1985). *Chem. Phys.* **97,** 365.
91. Leveiller, F., Jacquemain, D. L., Lahav, M., Leiserowitz, L., Deutsch, M., Kjaer, K., and Als-Nielsen, J. (1991). *Science* **252,** 1532.
92. *Landolt-Börnstein Handbook.* 1986, New Series III/14b2, Springer Verlag, Berlin/Heidelberg, p. 301.

Third-Level Self-Assembly and Beyond: Polar Hybrid Superlattices via Postassembly Intercalation into Noncentrosymmetric Multilayer Matrices of Hydrogen-Bonded Silanes

RIVKA MAOZ, RUTH YAM, GARRY BERKOVIC, AND JACOB SAGIV

Weizmann Institute of Science
Department of Materials and Interfaces
Rehovot, Israel

3.1. Introduction

Pointing to the future prospects of Langmuir-Blodgett (LB) film research, Hans Kuhn concluded the introduction to his opening lecture at the 4th International Conference on LB Films (Tsukuba, Japan, April 1989) with the following statement[1]:

> This possibility should be of interest in building planned structures at several hierarchical levels. Assembly modules on a two-dimensional (2-D) base to form three-dimensional (3-D) complex structures should be of interest in future supramolecular engineering strategies. According to these considerations the challenging new goal in chemistry may be given by the paradigm: synthesize molecules that are tailored to interlock and interact to form a planned solid complex superstructure; use the superstructure as the matrix for chemical reactions—highly specific syntheses or reactions linking discrete parts of superstructure—thus increasing its complexity and intricacy in a planned manner.

In this chapter we report on a series of recent experimental results demonstrating significant progress toward the actual implementation of

ISBN 0-12-523485-6

such a "hierarchical modular" strategy in the construction of some new types of artificial supramolecular structures. Our approach, however, is exclusively based on processes of chemically controlled self-assembly, the significance of which has also been recognized by Kuhn,[1] rather than on the initially proposed LB methodology.

Within the present context, the term *third-level self-assembly* is thus used to signify a third hierarchical level of supramolecular architectural complexity, built up through a sequence of successive steps of controlled molecular self-assembly: spontaneous monolayer self-assembly at the solid/liquid interface (first level) is utilized as the elementary process in the stepwise construction of an ordered multilayer structure (second level), in a manner such that each last deposited monolayer acts as a 2-D template for the self-assembly of the next layer, the entire multilayer then being utilized as a preorganized 3-D host matrix template for the postassembly intercalation of various selected guest species. The resulting intercalated superlattice (third level) may, finally, be subjected to additional physicochemical modification steps, affecting either or both the intercalants and the matrix, eventually leading to even *higher levels* (beyond the third level) of compositional/structural complexity.

The possible formation of high-quality organic monolayers by the spontaneous self-assembly of various molecular species from a fluid phase onto a suitable solid substrate is currently accepted as a well-established fact.[2] Although apparently less convincing, at least as far as considerations of general applicability and film quality are invoked,[3] the feasibility of controlled multilayer formation by the sequential self-assembly of successive discrete monolayers, one on top of the other, has also been demonstrated.[2,4,5] The main purpose of this chapter is to report briefly on a number of recent advances in mono/multilayer self-assembly that now permit (1) the preparation of high-quality multilayer structures of laterally interconnected silanes with interlayer couplings of the hydrogen bond type (Fig. 3.1), and (2) the utilization of such H-bonded multilayers as preorganized host matrices with expandable interlayer polar spaces into which suitable mobile guest species may be incorporated by lateral permeation from a contacting fluid phase. A large number of different kinds of ions, polar molecules, and combinations of both such species, organic as well as inorganic, were found to behave as efficient intercalants, which thus paves the way for the construction of a wide variety of new types of artificial (self-assembling) superlattice structures with rather unusual composition, molecular architecture, and structure-derived properties.

As demonstrated by the examples described later, the H-bonded host matrix plays the role of a 3-D layered template that controls the entrance, lateral mobility, and final mode of spatial organization of the intercalated

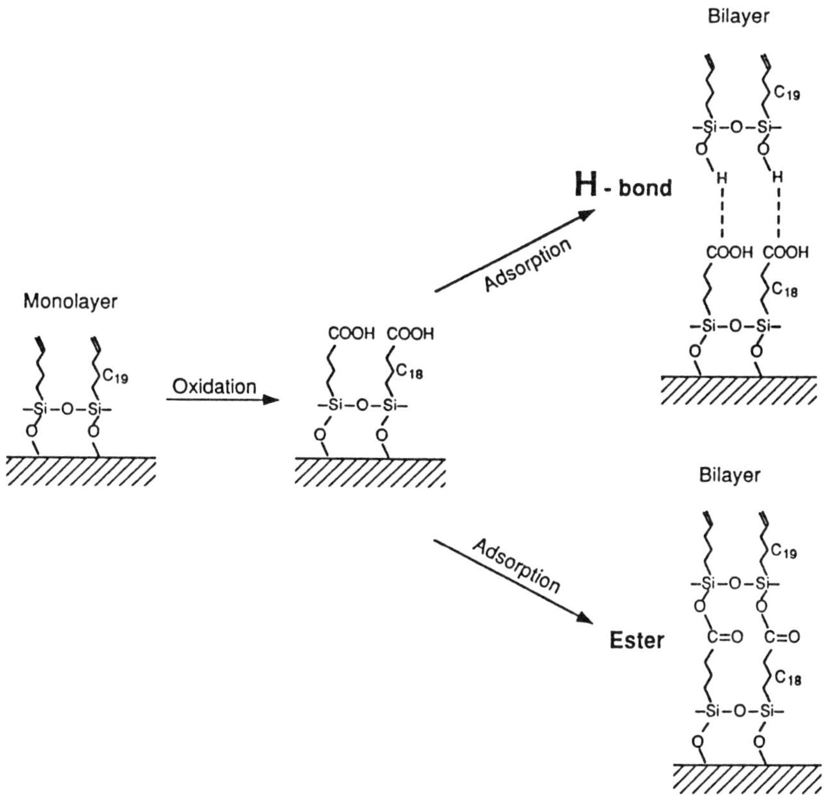

FIG. 3.1. Scheme of the controlled self-assembly of bilayer films of laterally interconnected silanes with interlayer coupling via either covalent (ester) or multiple hydrogen bonds. For reasons of clarity, only 3 out of the 17 chain methylenes of the C = C-terminated C_{19} silane (NTS) and –COOH-terminated C_{18} silane (NTS_{ox}) are shown (see text for further details).

guests. This is achieved through direct guest/matrix interactions, matrix-mediated guest/guest interactions, as well as more intricate forms of guest/guest/matrix interactive synergism. Among others, the intrinsic noncentrosymmetry of the multilayer host matrices (the –COOH and –SiOH functions are separately located in alternating adjacent planes—see Fig. 3.1) described here imposes a unique polar axis on the spatial alignment of intercalated dipolar guests. We focus here on this particular aspect, which is obviously of considerable interest both within the context of the general understanding of molecular organization in artificial systems and for a range of possible applications requiring thin-film materials with such structural characteristics.

Figure 3.2 depicts the formation of a noncentrosymmetric organic-inorganic hybrid superlattice via the simultaneous co-intercalation of *p*-aminobenzoic acid (PABA) molecules and Cu^{2+} ions into a H-bonded host multilayer matrix assembled according to the method depicted in Fig. 3.1. The dipolar PABA chromophores are shown to assume a perpendicular orientation on the layer planes, with unidirectional alignment of their dipoles along a unique preferred direction. Under the given conditions, alignment of the PABA dipoles along the polar axis indicated in Fig. 3.2 is expected to occur spontaneously in the intercalation process. Indeed, as

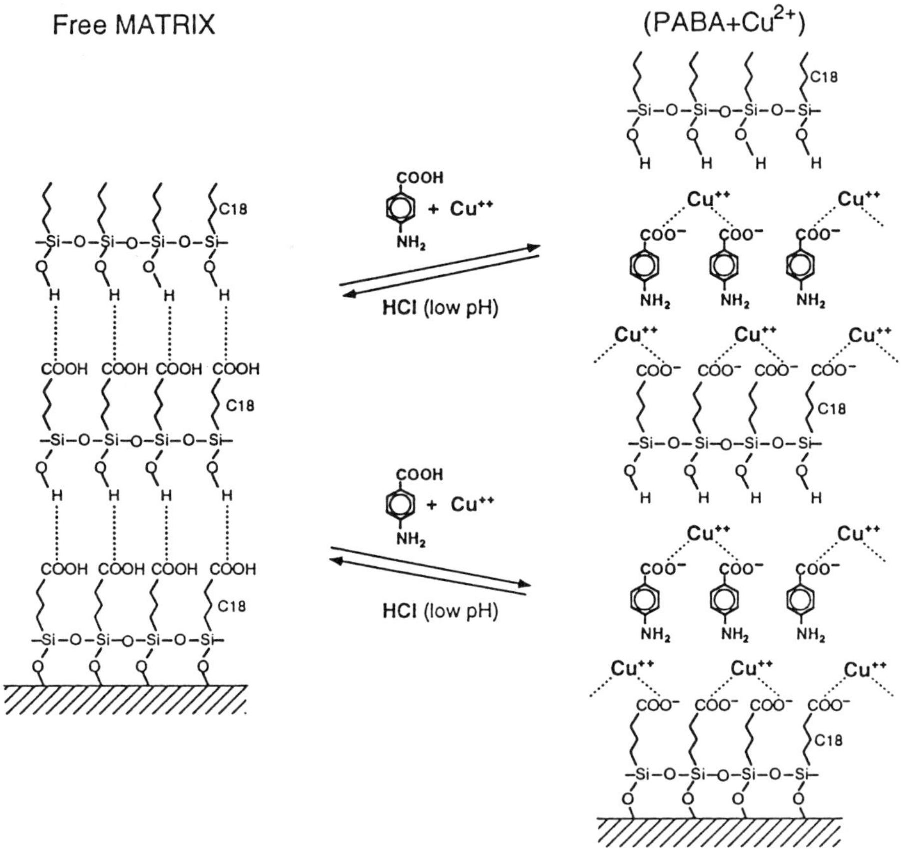

FIG. 3.2. A schematic view of the reversible postassembly intercalation of PABA molecules and Cu^{2+} ions into the interlayer polar spaces of an expandable H-bonded multilayer matrix structure with top OTS monolayer (C_{18} silane, saturated) assembled according to the synthetic route indicated in Fig. 3.1. Each Cu^{2+} ion is shown to be formally bound to a pair of either aromatic or aliphatic $-COO^-$ functions.

demonstrated later, the quantitative formation of the Cu^{2+} salt of both the matrix and guest carboxylates ($-COO^-$), coupled with the strong specific interactions of the guest amino ($-NH_2$) groups with the matrix bound Cu^{2+} ions (leading to the energetically favored coordination of each of these Cu^{2+} ions by both matrix carboxylate and guest amino groups), appears to provide in this case the necessary driving force for the spatial alignment of the PABA dipoles suggested in Fig. 3.2.

Because of size limitations, this chapter focuses mainly on one representative example, (PABA + Cu^{2+}), which was selected as a convenient model system for the study of some of the key features of the formation, composition, and structure of the new class of complex superlattices described. To complete the discussion, as well as to emphasize the broader scope of the present synthetic approach, we also refer to some additional experimental data, the details of which are provided elsewhere.

3.2. The Multilayer Matrix

Hydrogen-bonded multilayer films were prepared on synthetic quartz slides, double-side polished Si wafers and attenuated total reflection (ATR) Si prisms by repeating the stepwise chemical process outlined in Fig. 3.1. Effective control of the preferred mode of interlayer coupling—covalent (ester formation) versus hydrogen bond—was found to be achieved through the careful adjustment of a number of experimental parameters that appear to affect mainly the wettability of the outer exposed carboxylic acid surfaces generated prior to the adsorption of each new monolayer. In general, very hydrophilic surfaces (H_2O contact angle close to 0-deg) tend to promote H-bond formation.[6] The oxidation of the terminal double bonds to terminal carboxyls may be carried out via two different routes leading to virtually identical results—a slow process using the $KMnO_4$-crown ether complex in benzene,[6-9] and a rapid photo-oxidative one, with O_2 and $FeCl_3$ as catalyst.[6] All film specimens discussed in this chapter were assembled with the latter method, which enables[6] quantitative conversion of the double bonds to $-COOH$ functions within less than 3 min. Immersion in the silane solution (in pure bicyclohexyl) for about 30 s, immediately after the generation of the surface $-COOH$ groups, is sufficient to produce a complete silane monolayer, anchored by hydrogen bonds to the one underneath it.[6,8] By means of this procedure, it was possible to assemble manually high-quality H-bonded multilayers at the rate of approximately 5 min/monolayer.[10]

The quality of the H-bonded multilayer films assembled by this process, with NTS (18-nonadecenyltrichlorosilane, $CH_2{=}CH{-}(CH_2)_{17}{-}SiCl_3$, gift sample kindly supplied by Dr. K. Ogawa, Matsushita Electric Industry Company Ltd., Osaka) as the basic monolayer component, and OTS (*n*-octadecyltrichlorosilane, CH_3–$(CH_2)_{17}$–$SiCl_3$, Merck-Schuchardt, for synthesis) as the stabilizing top layer component,[8] was assessed by quantitative FTIR spectroscopy,[6–8] wettability,[6] and x-ray[8] measurements. For example, the plots in Fig. 3.3 show the perfect linear dependence of the infrared (IR) peak absorbance at about 2916 to 2917 cm^{-1} and 2849 to 2850 cm^{-1} (the hydrocarbon chain $-CH_2-$ stretch vibrations) on the number of stacked monolayers in a series of multilayer film specimens on synthetic quartz slides containing between 1 and 24 monolayers per stack. Particularly remarkable in these plots is the fact that the displayed data points represent different slide specimens, some of the points represent more than one specimen (10 for the monolayer, 3 for the bilayer, 2 for the trilayer, 2 for the 11-layer film), and the spectra of part of the specimens were recorded prior as well as after tens of intercalation and subsequent extraction cycles involving a variety of different guest species and corrosive solvents (*vide infra*), over periods of time exceeding a year after their preparation.

The peak positions of all measured film samples in Fig. 3.3 fall within 1 cm^{-1}, at 2849 to 2850 cm^{-1} for the symmetric H–C–H stretch and 2916 to

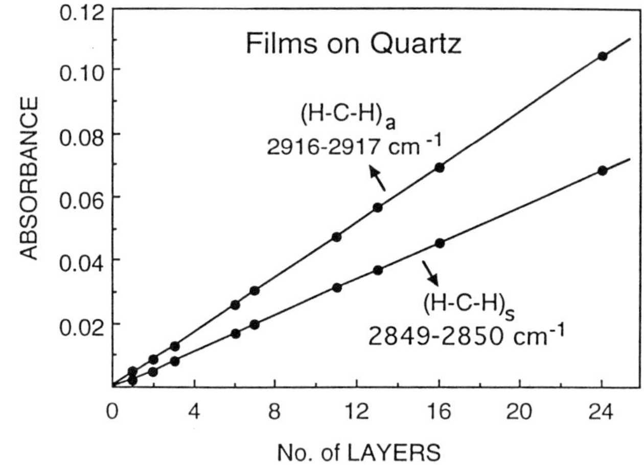

FIG. 3.3. Plot of the 2917- and 2850-cm^{-1} IR peak absorbance (measured in transmission at normal incidence) versus the number of stacked layers in a series of H-bonded silane multilayer films on synthetic quartz (0.1-cm-thick slides, double-side coating). The one-layer points represent both NTS and OTS monolayers, while each of the thicker films is made of oxidized NTS terminated with a top OTS monolayer. Note that the actual spread of the experimental data in the plot is considerably smaller than what might be inferred from the size of the displayed data points.

2917 cm^{-1} for the antisymmetric one, which is indicative of a densely packed, solid-like structure, with an ordered all-*trans* general arrangement of the paraffinic chains.[11] Considering the fact that each of the molecular components in these films—OTS, NTS, and oxidized NTS (NTS_{ox})—contains the same number of $-CH_2-$ groups (17) per chain, the perfect linearity of the plots in Fig. 3.3, for both the antisymmetric and symmetric vibrations, together with the invariance of the peak positions and peak widths with the number of superimposed monolayers, demonstrate their virtually identical structure in terms of molecular packing density and orientation, as well as perfection of surface coverage, regardless of their location within the film and total film thickness.

A comparison with identically measured IR spectra of LB monolayers of Cd^{2+}-arachidate (C_{20}; 18 $-CH_2-$ groups per chain) transferred on glass slides in their highest compression state reveals a higher peak absorbance per $-CH_2-$ group in the silane monolayers by about 2% to 4%. Thus, the monolayer surface density achieved in the currently described self-assembled silane films is similar to or even somewhat higher than that of highly compressed LB monolayers of saturated long-chain fatty acids.[6,12] This finding is consistent with the x-ray data obtained for OTS and NTS monolayers and a five-layer silane film produced by a similar, though not identical, experimental procedure on smooth Si wafers. These data indicate[8] average perpendicular orientation of the paraffinic chains and a surface density of about 21 Å^2 chain. Finally, the remarkably high equilibrium (static) contact angles (114 to 115 deg for H_2O, 54 to 55 deg for bicyclohexyl, and 51 to 52 deg for *n*-hexadecane, no detectable hysteresis) measured on each of the OTS-terminated films in Fig. 3.3, independent of the number of layers in the film or the number and nature of intercalation/extraction cycles to which some of these films were subjected (see the section on wetting behavior later in this chapter), together with the complete wetting of the outer $-COOH$ surfaces generated during the assembly of the films by each of the mentioned liquids, including water ($\sim$ 0-deg advancing contact angle on freshly prepared $-COOH$ surfaces), furnish additional supporting evidence for the dense packing and essentially perpendicular orientation of the paraffinic chains,[13–15] the quantitative conversion of the terminal double bonds to carboxyls,[6–9,13i,13j,14,15] and the high structural regularity and robustness of these multilayer structures.[6,9,12,15]

To complete the analysis of the data summarized in Fig. 3.3, it would be instructive to draw attention to the fact that accumulation of structural defects during film construction of the order of less than 0.5% of a monolayer per deposited monolayer should be easily detectable by means of current IR measurements. The statistical reproducibility of each of the data points in Fig. 3.3, including both deviations in the preparation of the films and in the IR measurement, is evaluated to about $\pm 1.2\%$ and $\pm 1.8\%$ of a monolayer, considering the antisymmetric and symmetric peaks, respectively. Accumulation of defects—missing material—at the rate of about 0.5% of a monolayer per each deposited monolayer should result in a systematic deviation in the measured surface coverage of the order of about 5% of a monolayer per 10 monolayers of deposited film, or about 12% of a monolayer for the 24-layer film. Obviously, no systematic deviations may be detected at all in the plots in Fig. 3.3. This is because statistical deviations per monolayer of deposited material, calculated from the statistical spread of the absorbance values of all displayed data points, are within $\pm 0.4\%$ and $\pm 1.4\%$ of a monolayer for the antisymmetric and symmetric peaks, respectively. The IR data thus indicate remarkable structural and surface coverage regularity and reproducibility, independent of the number of stacked layers, which implies a high degree of intralayer lateral bridging and self-healing capability.[16] As also evidenced by the x-ray results, the individual monolayers in such silane multilayer films are smoothly deposited, one on top of the other, with no interlayer epitaxial relationship.[8]

3.3. Intercalated Superlattices

The intercalation of ionic or ionizable polar guests in the interlayer polar spaces of the current H-bonded multilayer matrices occurs via a process of lateral transport coupled with reversible exchange of mobile ionic species belonging to the matrix backbone—the exchangeable acid protons of the immobile (matrix bound) $-COOH$ functions (see Figs. 3.2 and 3.4). Full preparative details and a critical analysis of the incorporation/exchange/extraction of a variety of guest species, such as metal ions, ionizable organic molecules, and combinations of more than one such species, carried out in one or several consecutive steps, will be furnished elsewhere.[6,17] Here we wish to draw the attention to two particularly remarkable aspects of these complex self-assembly processes: (1) their two-dimensionality, i.e., effective confinement of the transport to the interlayer polar planes of the host matrix, and (2) the striking disproportion between the degree of structural complexity achievable through the judicious manipulation of such self-assembly processes and their basic simplicity at the experimental level.

For example, the present (PABA + Cu^{2+}) hybrid superlattices (Fig. 3.2) are produced by simply dipping multilayer matrices with the desired

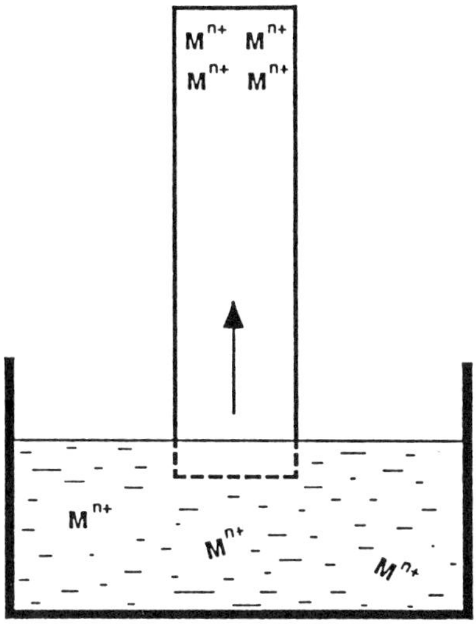

FIG. 3.4. Schematic illustration of the intercalation of metal ions (M^{n+}) into a matrix film supported on a rectangular solid slide, utilizing solution-film contacts established along a slightly immersed edge of the slide. To fill the entire available film area, the metal ions must migrate laterally from the points of contact with the solution along the interlayer polar planes of the film.

number of layers, preassembled on the solid support of interest (quartz slides, Si wafers, etc.), in a millimolar aqueous solution of PABA and copper(II) acetate at the ambient temperature, for 10 to 15 min. Extraction of the guests and regeneration of the initial H-bonded matrix is performed in a similar manner, by dipping the intercalated structure in aqueous HCl at the ambient temperature. By controlling the pH of the HCl solution it is possible to extract selectively each of the two intercalated species, in two consecutive steps. Thus, dipping for 5 to 10 min in a 5.0×10^{-6} M HCl solution will remove only the Cu^{2+} ions from the film, whereas quantitative extraction of the PABA molecules requires dipping for about 5 min in 5 M HCl.

All incorporation and extraction operations mentioned must be carried out as indicated in Fig. 3.4. Attempts to perform analogous intercalation experiments by bringing the liquid in contact with the outer surface of the multilayer matrix while avoiding direct contacts with edges of the substrate were total failures. These observations confirm the expected high impenetrability of the matrix-forming silane monolayers[9] and demonstrate the effective two dimensionality of the process of lateral transport responsible for the intercalation of guest species in such structures.

The incorporation and extraction of the various intercalants were *ex situ* followed by FTIR and UV-visible spectroscopic techniques.[6,7,17] Such techniques were also employed to provide quantitative evidence for the composition and structure of the intercalated superlattices, in terms of guest/host and guest/guest stoichiometries, packing densities, and orientation relative to the matrix, and the chemical transformations occurring upon intercalation and extraction of the guests.[6,7,17] Second harmonic generation (SHG) measurements, were used to characterize the intercalated structures with respect to the unidirectional alignment of dipolar guest chromophores imposed by the intrinsic noncentrosymmetry of the matrix. Finally, some rather unusual wettability effects, static as well as dynamic, were found to accompany the intercalation and extraction of guest species from the films (*vide infra*). If correctly interpreted, these effects might prove highly informative as a tool for the qualitative evaluation of electrical charge distribution and mobility in such composite structures.

3.3.1. FTIR Spectroscopy

The Brewster's angle IR spectra[8] obtained with the 10-layer superlattice structure (11 matrix layers including the top OTS monolayer; see Fig. 3.2) on Si wafer given as example in Fig. 3.5 furnish evidence for the presence of the (PABA + Cu^{2+}) guest species, the chemical transformations affecting the matrix and the guest, and the stability of the matrix upon intercalation.

A comparison in Fig. 3.5 of curves a, b, and c in the low-frequency spectral region reveals the quantitative conversion of the H-bonded −COOH functions of the matrix (C=O stretch peak[6–8,18] at 1713 cm^{-1}) into the respective ionized copper(II) carboxylates ($-COO^-$ antisymmetric and symmetric stretch modes around 1560 to 1590 cm^{-1} and 1440 cm^{-1}, respectively[6,7] and the appearance of the characteristic aromatic PABA band around 1602 to 1606 cm^{-1} (associated with the skeletal stretch vibration of an unsymmetrically *p*-substituted benzene ring[19a]), following immersion of the matrix into the (PABA + Cu^{2+}) solution. Additional vibration bands of interest, partially overlapping with those of the matrix carboxylates and the 1605-cm^{-1} aromatic band, are those due to the stretch modes of the aromatic (PABA) $-COO^-$ functions around 1544 and 1426 cm^{-1}, respectively,[17] and the PABA $-NH_2$ deformation mode[17,19b] around 1600 cm^{-1}. Together, all these vibrations give rise to the broad, complex features observed in curve c around the three distinct peaks at 1440, 1554, and 1602 cm^{-1}. The quantitative conversion of both the aliphatic (matrix) and aromatic (PABA) −COOH functions into their Cu^{2+}-carboxylates is evident from the complete absence of either aliphatic

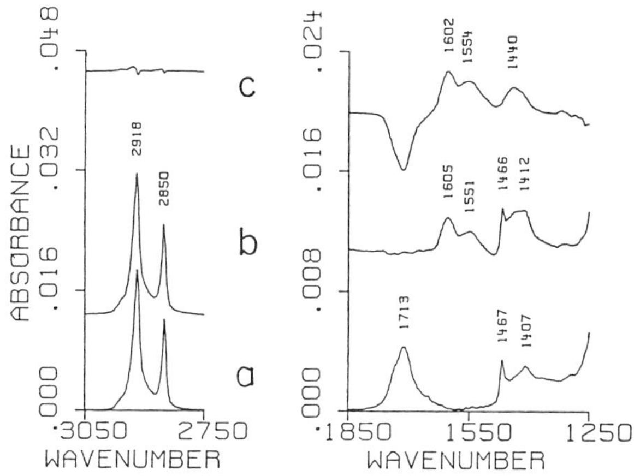

FIG. 3.5. FTIR spectra (recorded with a Nicolet 730 system at a resolution of 4 cm^{-1}) of a superlattice film structure consisting of 10 H-bonded NTS_{ox} layers and a top OTS monolayer, coated on both sides of a 0.3-mm-thick double-side polished Si wafer. Curve a: the free matrix film (before intercalation); curve b: after the simultaneous co-intercalation of (PABA + Cu^{2+}), as in Fig. 3.2; and curve c: difference spectrum (b − a). The spectra were taken in transmission with a *p*-polarized beam (electric field parallel to the plane of incidence) at the Si Brewster's angle of incidence (θ_B = 73.7 deg).[8] The Brewster-angle configuration minimizes the strong interference fringes caused by reflection of the IR beam at the front and back Si/air interfaces, which tend to obscure the spectral features due to the film coating.[8] All curves are corrected for the spectral contributions of the bare silicon substrate.

or aromatic free acid bands following the incorporation of the PABA and Cu^{2+} guest species (curve b).

The C–H stretch bands in the high-frequency region of the spectrum are affected only very slightly by intercalation. While their absorbance intensities remain practically unchanged (curves a and b), there is a small blue shift of the 2917- and 2850-cm^{-1} peaks, of the order of less than 0.5 cm^{-1}, mainly evident from the characteristic first-derivative-like shape of the difference curve c. This points to a small tendency toward a somewhat less rigid packing of the paraffinic chains,[11] no major modification of the initial structure of the multilayer matrix being, however, caused by intercalation. The structural stability of the matrix layers upon intercalation is further evident from the perfect mathematical cancellation of the 1467-cm^{-1} $-CH_2-$ bending bands in the difference curve c in the low-frequency region of the spectrum (compare with corresponding curves a and b).

We should, finally, stress the perfect reversibility of the intercalation of PABA and Cu^{2+}, as evidenced by the repetitive quantitative conversion of curves b into curves a (in both the low- and high-frequency regions of the spectrum) and vice versa, upon repetitive extraction and reincorporation of these intercalants.

The orientation of the intercalated PABA chromophores relative to the matrix layer planes was determined by both IR and UV-visible spectroscopies.[17] For example, the polarized ATR spectra of the five-layer (PABA + Cu^{2+}) intercalated superlattice on silicon (six matrix layers) displayed in Fig. 3.6 provide clear evidence for the perpendicular orientation of the long axes of the PABA chromophores on the matrix layer planes. Thus, the 1602-cm^{-1} aromatic band (compare with Fig. 3.5)[20] arising from a benzene ring mode with the transition dipole moment parallel to the line connecting the two *para*-positioned substituents[19a] appears only in the *p*-polarized curve (compare curves *p*, *s*, and $p - s$), i.e., the curve obtained with incident radiation having electrical field components both parallel and perpendicular to the matrix layer planes.[13a,17] All other major vibration modes contributing to the spectra in Fig. 3.6 have transition moments with large components parallel to the matrix layer planes (the surface of the substrate), which, as expected, results in weak dichroic effects, with the *s*-polarization (parallel to the matrix layer planes) producing somewhat stronger absorbance signals.[13a] The lack of a detectable contribution from the 1602-cm^{-1} mode to the *s*-polarized curve (similar results were also

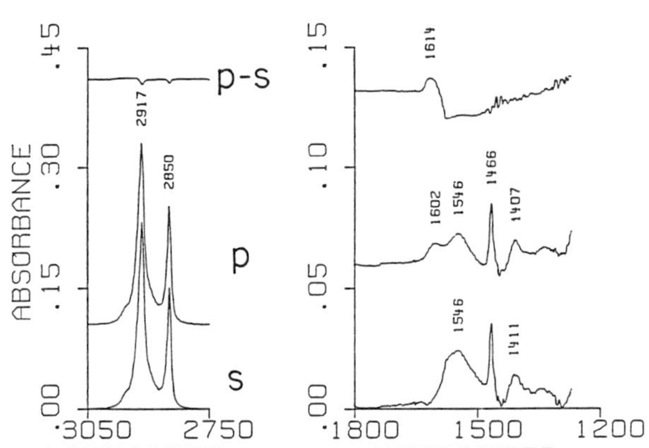

FIG. 3.6. Polarized FTIR-ATR (45-deg incidence) spectra of a five-layer (PABA + Cu^{2+}) intercalated superlattice film (six matrix layers, including the top OTS monolayer; see Fig. 3.2) on Si recorded with same instrument and at the same resolution as in Fig. 3.5. Curve *s*: incident beam with electric field perpendicular to the plane of incidence (parallel to the film layer planes)[13a]; curve *p*: incident beam with electric field parallel to the plane of incidence (components both parallel and perpendicular to the film layer planes)[13a]; curve $p - s$: difference spectrum. All curves are corrected for the spectral contributions of the uncoated Si substrate.

obtained with analogous Cu^{2+}-free PABA structures[17]) is, therefore, strong evidence for the highly perfect perpendicular orientation of the intercalated PABA molecules. This conclusion is further corroborated by the UV-visible and SHG data (*vide infra*).

3.3.2. UV-Visible Spectroscopy

UV-visible spectroscopy was applied, in addition to FTIR, to the structural characterization of intercalated superlattices containing intercalants with characteristic electronic transitions in this spectral region.[17] In the case of such superlattices on quartz substrates, which are not transparent in the infrared below about 2500 cm^{-1}, UV-visible was used as the main analytical tool. Due to the reversibility of the intercalation process it was also possible to evaluate directly the quantity of intercalated material by its extraction and spectral determination in solution. UV-visible spectroscopy in solution has been extensively used for this purpose, providing a simple and unambiguous means for the straightforward assessment of the surface concentrations of intercalated guest species as well as for the determination of the guest/matrix and guest/guest binding stoichiometries. A number of examples, illustrating each of these applications, are given in the following.

Selective removal of the Cu^{2+} ions from a hybrid (PABA + Cu^{2+}) superlattice is seen (Fig. 3.7) to leave a featureless absorption curve representing the analogous copper-free PABA structure[21] [(PABA + Cu^{2+}) − Cu^{2+}]. The peak at about 251 nm, also observed in intercalated structures containing only Cu^{2+} ions,[6,7] appears thus to be associated mainly with the Cu^{2+} component. In clear contradistinction to intercalated PABA, these molecules when isotropically dispersed in solution (PABA/H_2O) give rise to a strong absorption peak at about 275 nm, due to the characteristic $\pi - \pi^*$ charge transfer transition of a *para*-substituted benzene chromophore.[22] Because this transition is polarized along the line connecting the two *para*-positioned substituents,[23] if PABA intercalates with its long molecular axis oriented perpendicularly on the matrix layer planes, the transition moment vector associated with the 275-nm band should be perpendicular to the surface of the quartz substrate, or parallel to the direction of propagation of the incoming light beam (in measurements at normal incidence, as in Fig. 3.7). This implies orthogonality of the molecular transition moment and electrical field of the light beam, so that excitation of this benzenoid charge transfer transition cannot occur. The missing charge transfer band of PABA in the intercalated state is thus strong evidence for its highly perfect perpendicular orientation, in agreement with the polarized IR-ATR data obtained for the

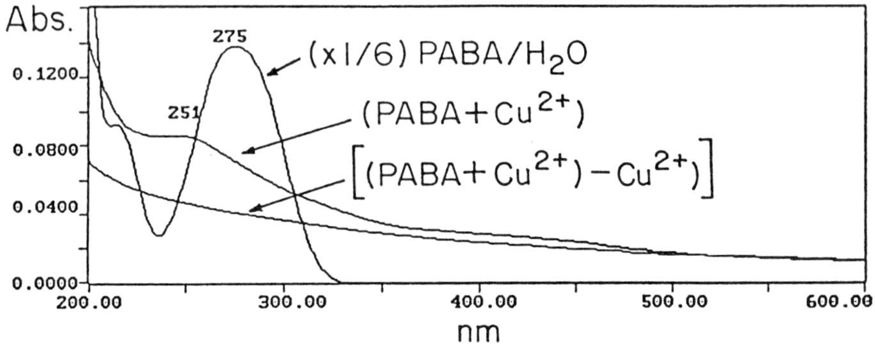

FIG. 3.7. UV-visible spectra (measured in transmission at normal incidence) of a 10-layer intercalated superlattice (11 matrix layers; included in the plot of Fig. 3.3) on synthetic quartz (3.75-× 1.2-× 0.1-cm slide, double-side coating) containing either (PABA + Cu^{2+}) as intercalated species or PABA, left upon the selective extraction[21] of the copper ions from the initial hybrid structure [(PABA + Cu^{2+}) − Cu^{2+}]. PABA/H_2O is the spectrum of a solution of PABA in water measured in a UV cell with 1-cm optical path, at a concentration equal to that resulting from the quantitative extraction of PABA from the 10-layer intercalated film into 1.0 cm^3 of water. All spectra were obtained with a Cary 3 double-beam spectrophotometer, with the displayed curves corrected for the residual absorption of the uncoated quartz slide or the water-filled UV cell, respectively.

(PABA + Cu^{2+}) system discussed earlier as well as for an analogous Cu^{2+}-free film on silicon.[17]

The discrete incorporation of the intercalants by the matrix, according to the number of intercalable matrix layers, resulting in virtually equivalent composition of each of the intercalated layers, is demonstrated by the spectra of the (PABA + Cu^{2+}) hybrid structures displayed in Fig. 3.8 and the linearity of the corresponding plot of the 251-nm peak absorbance versus number of intercalated layers (Fig. 3.8, insert). This is confirmed by the linear plot of the 267-nm peak absorbance measured for the (PABA + Cu^{2+}) mixture extracted from a series of hybrid superlattices in 5 M aqueous HCl (Fig. 3.9). A similar linear plot obtained for material extracted from a series of analogous hybrid superlattice films containing *p*-nitroaniline (PNA) and Cu^{2+} as guest species is also shown in Fig. 3.9.

The amount of intercalated Cu^{2+} and PABA material was determined by selectively extracting first the Cu^{2+} and then the PABA components, and measuring their respective concentrations in solution. This was done spectroscopically using calibration curves constructed with Cu^{2+} or PABA solutions of known concentrations. Knowing the geometrical area of the substrate and the number of extracted layers, it was then possible to evaluate the surface densities of both Cu^{2+} and PABA in the intercalated

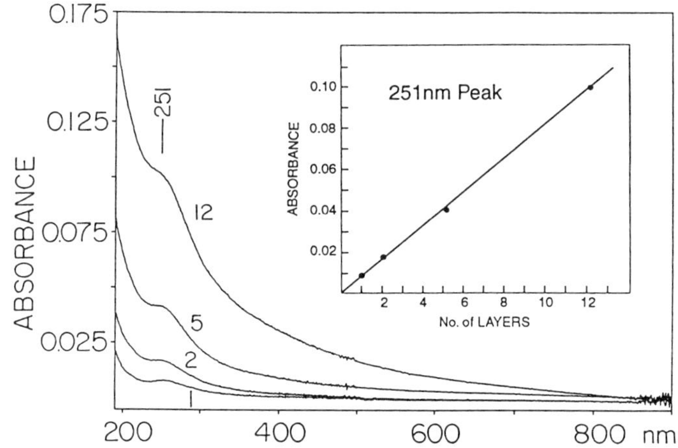

FIG. 3.8. UV-visible spectra, as in Fig. 3.7, of a series of (PABA + Cu^{2+}) hybrid superlattices on synthetic quartz containing between 1 to 12 intercalated layers (2 to 13 matrix layers; included in the plot of Fig. 3.3). Insert: Plot of the 251-nm peak absorbance versus the number of intercalated layers.

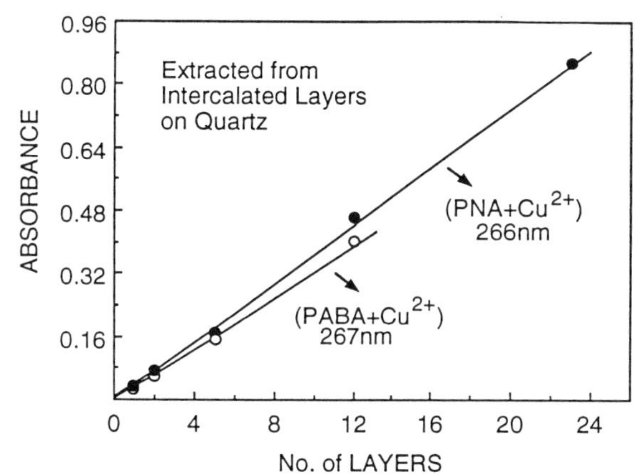

FIG. 3.9. Plots of the 267- and 266-nm peak absorbance of (PABA + Cu^{2+}) and (PNA + Cu^{2+}) solutions, respectively, extracted from a series of hybrid superlattices on synthetic quartz slides (as those of Fig. 3.8) versus the number of intercalated layers. All extractions were done in volumes of 1 cm^3 aqueous HCl (5 M) and the spectra measured, as in Fig. 3.7, with a UV cell with 1-cm optical path, versus a matched cell filled with the pure HCl solvent. Same set of matrix specimens (also included in the plot of Fig. 3.3) were employed for both the PABA and PNA structures.

state. The surface densities of PABA in pure PABA superlattices (assembled without Cu^{2+} ions) and of PNA and Cu^{2+} in co-intercalated (PNA + Cu^{2+}) structures were also determined by this procedure. All these surface coverage measurements were performed using a matrix film with 10 intercalable layers (11 matrix layers) constructed on a double-side polished 0.3-mm-thick Si wafer of about 14 cm^2 total geometrical area.[24,25] The results, summarized in Table 3.1, show the following:

- There is considerably more Cu^{2+} in the (PABA + Cu^{2+}) than in the (PNA + Cu^{2+}) structure.
- The presence of Cu^{2+} ions has a strong effect on the surface density of intercalated PABA molecules.

Assuming a binding stoichiometry of the copper ions such as that suggested in Fig. 3.2, i.e., one Cu^{2+} ion per each pair of carboxylate functions, aliphatic as well as aromatic, the surface density of Cu^{2+} in (PABA + Cu^{2+}) should be determined by that of the matrix carboxylates (1 Cu^{2+}/2 matrix $-COO^-$ groups $\cong$ 40 $Å^2/Cu^{2+}$) plus that of the intercalated PABA molecules (1 Cu^{2+}/2 PABA $-COO^-$ groups $\cong$ 52 $Å^2/Cu^{2+}$). Together, these make about a 22.6 $Å^2/Cu^{2+}$ ion. PNA, on the other hand, has no ionizable carboxylic acid functions that might bind Cu^{2+} ions, which should thus limit the copper content of (PNA + Cu^{2+}) to the binding capacity of the matrix only, i.e., about 40 $Å^2/Cu^{2+}$ ion. The two experimentally obtained figures in Table 3.1, 23 $Å^2/Cu^{2+}$ and 37 $Å^2/Cu^{2+}$, respectively, for the (PABA + Cu^{2+}) and (PNA + Cu^{2+}) systems, compare well with those evaluated on the basis of this preceding model. A related observation emerging from the data in Table 3.1 concerns the packing density of PABA in (PABA + Cu^{2+}) structures, which is considerably higher than that reached by PABA when intercalated in the absence of Cu^{2+} ions (26 $Å^2$/molecule versus 37 $Å^2$/molecule). The lateral coupling of PABA carboxylates through the mediation of the divalent copper ions (Fig. 3.2) appears, therefore, to counterbalance efficiently the

TABLE 3.1

SURFACE DENSITIES OF GUEST SPECIES IN A 10-LAYER INTERCALATED SUPERLATTICE STRUCTURE (11 MATRIX LAYERS) ON A DOUBLE-SIDE POLISHED Si WAFER

Guest (Superlattice Composition)	PABA (PABA + Cu^{2+})	Cu^{2+} (PABA + Cu^{2+})	PABA (Pure PABA)	PNA (PNA + Cu^{2+})	Cu^{2+} (PNA + Cu^{2+})
Area/Molecule, Ion	26 $Å^2$	23 $\pm$ 1.5 $Å^2$	37 $Å^2$	29 $Å^2$	37 $Å^2$

dipolar repulsion between the parallel PABA dipoles, thus facilitating their dense packing in the hybrid (PABA + Cu^{2+}) structure. Once formed, the dense PABA structure does, however, survive the selective removal of the Cu^{2+} ions without relaxation to a less ordered state (also see the SHG results given later).

The interesting conclusion emerging from the preceding analysis is that specific guest/guest/matrix interactions in such complex organized systems may give rise to significant cooperativity effects, strongly affecting their composition, structure, and properties. This is nicely demonstrated by the lateral binding effect exerted by the Cu^{2+} ions in the formation of dense PABA structures, and the final stability of such structures on subsequent removal of the metal ions that played a key role in their initial formation. The selective extraction of Cu^{2+} from (PABA + Cu^{2+}), resulting in the formation of a dense copper-free PABA structure that is not obtainable by the direct self-assembly of its constituent molecules, may be regarded as a fourth level of controlled self-assembly.

3.3.3. Second Harmonic Generation

Second harmonic signals were generated from double-side coated synthetic quartz 0.1-cm-thick slides using the 1064-nm fundamental beam of a Nd^{3+} : YAG laser in a transmission geometry at variable angles of incidence around 45 deg. Several maxima and minima of the interference fringe pattern produced under these conditions[26] were recorded and the square root of the second harmonic maxima closest to the nominal angle of incidence (45 deg) plotted versus the number of intercalated layers in the respective film samples. Figure 3.10 shows two such plots obtained with one set of matrix samples containing either (PABA + Cu^{2+}) or (PNA + Cu^{2+}) as intercalated guest species.

The quadratic dependence of the second harmonic intensity on the number of intercalated layers, evident from the plots in Fig. 3.10, confirms the compositional and structural equivalence of each of the layers in the stack, as determined by the IR and UV-visible measurements described earlier. Furthermore, the *p*-polarization of the fundamental (electric field components both parallel and perpendicular to the matrix layer planes) generates signals ($p_{in} - p_{out}$) about one order of magnitude higher than the *s*-polarization ($s_{in} - p_{out}$), which has the electric field parallel to the matrix layers. This points, in agreement with the IR and UV-visible findings, to the perpendicular orientation of the long axes of the aromatic chromophores on the matrix layer planes.

SHG data were also collected from the free matrix structures themselves as well as from the respective pure Cu^{2+} intercalated films (without

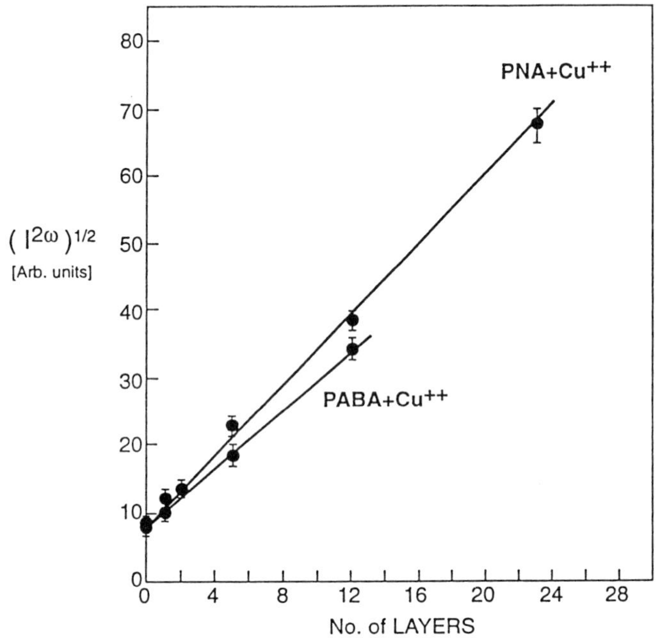

FIG. 3.10. Plot of square root of second harmonic maxima ($p_{in} - p_{out}$) versus number of intercalated (PABA + Cu^{2+}) and (PNA + Cu^{2+}) layers obtained with a series of hybrid superlattices on synthetic quartz slides. Both PABA and PNA films were produced using the same set of matrix specimens, also employed to obtain the data of Fig. 3.9.

chromophore molecules). In both these cases, the second harmonic signals were much lower than those generated in the presence of the PABA or PNA chromophores, which demonstrates that most of the observed effect is indeed associated with the chromophores and originates in their noncentrosymmetrical spatial alignment. The SHG results thus provide direct evidence for the preferred unidirectional alignment of the intercalated chromophores, as expected on the basis of the intrinsic noncentrosymmetry of the matrix multilayer.

A further important observation is that selective removal of the Cu^{2+} ions from the hybrid (PABA + Cu^{2+}) structures does not result in a weakening of the SHG effect. We actually observed a small but definite increase in the intensity of the corresponding second harmonic signals, thus indicating not only that the surface density and orientation of the chromophores are not affected, but also that the noncentrosymmetry of their alignment is preserved on removal of the Cu^{2+} ions. The SHG results confirm, therefore, the conclusion derived from the UV-visible and polarized IR-ATR data that no major structural relaxation occurs on removal of the Cu^{2+} component (the fourth-level self-assembly), although pure PABA structures of similar surface density cannot self-assemble in the absence of the binding Cu^{2+} ions.

3.3.4. Wettability Effects

The wetting behavior of the currently described film systems is complex, pointing to the possible interplay of a number of different types of interfacial forces, both short and long ranged, as the source of the observed effects. Here, we merely wish to draw attention to the existence of certain rather unusual patterns of behavior, a more complete description of these phenomena being left for a future publication.

Table 3.2 summarizes a series of static contact angle data obtained by the sessile drop method[13a] for two film structures with a top OTS mono-

TABLE 3.2

Static Advancing and Receding Contact Angles Measured on Some OTS-Terminated Superlattice Structures on Synthetic Quartz Slides

2-Matrix Layers (1-Intercalated)			16-Matrix Layers (15-Intercalated)		
H_2O adv.—rec.	BCH[a] adv.—rec.	HD[a] adv.—rec.	H_2O adv.—rec.	BCH[a] adv.—rec.	HD[a] adv.—rec.
Free Matrix (No Intercalant)[b]			Free Matrix (No Intercalant)[b]		
115°—115°	55°—55°	52°—52°	115°—115°	55°—55°	51°—51°
Cu^{2+} Intercalated			Cu^{2+} Intercalated		
111°—106°	52°—50°	48°—44°	108°—86°	48°—46°	43°—40°
(PABA + Cu^{2+}) Intercalated			(PABA + Cu^{2+}) Intercalated		
87°—66°	50°—47°	48°—43°	82°—67°	48°—45°	44°—42°
[(PABA + Cu^{2+}) − Cu^{2+}][c] Intercalated			[(PABA + Cu^{2+}) − Cu^{2+}][c] Intercalated		
115°—106°	53°—52°	49°—47°	115°—95°	50°—48°	47°—44°
(PNA + Cu^{2+}) Intercalated			(PNA + Cu^{2+}) Intercalated		
115°—97°	50°—48°	48°—45°	105°—71°	49°—47°	44°—40°
[(PNA + Cu^{2+}) − Cu^{2+}][c] Intercalated			[(PNA + Cu^{2+}) − Cu^{2+}][c] Intercalated		
118°—110°	54°—53°	49°—48°	113°—70°	50°—49°	48°—46°

[a] *BCH*, bicyclohexyl; *HD*, *n*-hexadecane.

[b] Same contact angles were measured before and after the intercalation and subsequent extraction of many different intercalants, including Cu^{2+}, (PABA + Cu^{2+}), (PNA + Cu^{2+}), etc.

[c] After selective extraction of the Cu^{2+} ions from the respective hybrid structures.[21]

layer, containing 2-matrix (1-intercalated) and 16-matrix (15-intercalated) layers, respectively. We should emphasize that although the absolute values of the contact angles in Table 3.2 may reflect the somewhat subjective nature of data obtained by this method, the relative trends are real, because all measurements were performed by the same person with a reproducibility of the order of ± 1 deg. With this consideration in mind, the following comments may be made regarding the main qualitative trends evident in the data of Table 3.2:

- The wetting behavior of the free matrix structures—very high contact angles, no hysteresis—is virtually identical to that of complete OTS monolayers on atomically smooth mica substrates.[12] This points to a very smooth, stable, and homogeneous outer surface of densely packed $-CH_3$ groups.[12,13,15]
- Significant contact angle hysteresis develops in all measured samples upon intercalation, and is also accompanied by a significant drop in the advancing angles in most samples containing Cu^{2+} ions. Differences between advancing and receding angles vary between 2 to 4 deg for bicyclohexyl (BCH) and *n*-hexadecane (HD) to values as large as 5 to 40 deg for H_2O. We also note significant differences between the bilayer and the 16-layer matrix structures, the latter displaying lower contact angles and larger hysteresis.

  A possible simple interpretation of such behavior is to ascribe it to disordering of the hydrocarbon tails, with a larger fraction of chain methylenes reaching the outer film surface, and to an overall deterioration of the ordered multilayer structure upon intercalation, so that penetration of the liquid, particularly water, into the inner core of the superlattice film becomes possible.[9] This simplistic interpretation is, however, incompatible with a number of other important trends evident in the data collected in Table 3.2, as well as with some additional interesting observations.
- As mentioned, the present intercalation/extraction processes are reversible and highly reproducible. Each intercalation/extraction cycle may be repeated many times, with no detectable change in any of the measured properties of the respective systems, including their wetting characteristics. Thus, once the intercalants are removed from a particular film structure, the free matrix film will always display the same high equilibrium contact angles (no hysteresis) as before intercalation, while the incorporation of a different intercalant into a particular matrix will always result in the characteristic wetting pattern associated with the respective intercalant, regardless of the intercalation history of the matrix.

- Selective removal of the Cu^{2+} ions from the (PABA + Cu^{2+}) systems raises the advancing H_2O contact angle to the value initially displayed by the free matrix (115 deg), while in the case of the corresponding [(PNA + Cu^{2+}) − Cu^{2+}] system an unusually high value of 118 deg, exceeding that of the free matrix, is observed with the bilayer. Yet, with the multilayer, the corresponding advancing angle is only 113 deg, and large hysteresis for the H_2O angles is observed in all these systems, with [(PNA + Cu^{2+}) − Cu^{2+}] showing the largest effect (113 to 70 deg).
- Finally, it was possible to monitor real-time contact angle variations (dynamic contact angle experiments) induced by the *in situ* extraction of the intercalant. For example, one can watch the spectacular growth of a water droplet (contact angle increase by 20 to 40 deg) placed on the outer surface of a (PABA + Cu^{2+}) or (PNA + Cu^{2+}) superlattice film as the Cu^{2+} ions are being extracted into a large drop of aqueous HCl brought in contact with an edge of the film coated substrate. It is difficult to imagine a presumably damaged, water containing, multilayer structure spontaneously recovering its initial high order, with the concomitant expulsion of the water incorporated into the film, under such conditions.

Taken together, these intricate contact angle variations are difficult to rationalize by simply invoking structure deterioration arguments. Penetration of liquid from the sessile drop into the inner core of the film is also rather improbable considering the high impenetrability of these films—no intercalation or extraction of guest species was observed under the present conditions in "across the film" configurations, i.e., by placing drops of liquid on the outer film surface (see earlier description of the intercalation/extraction experiments and also Fig. 3.4)—and the fact that no measurable incorporation of water into such films could be detected upon either intercalation or extraction of any of the investigated guest species.[6,7,17] A more plausible possibility to consider would, therefore, be that the peculiar wetting effects observed with the present systems originate in electrostatic interactions involving the various kinds of electric charges, net or induced, present in these intercalated films. The films contain fixed and mobile ions such as $-COO^-$ and Cu^{2+}, respectively, and dipolar entities such as PABA and PNA that may be expected to exhibit variable degrees of translational and rotational mobility, depending on their packing density, presence of other charges, etc. Liquid water is a hydrogen-bonded network of dipolar molecules that, most probably, assume a preferred spatial orientation in contact with the highly hydrophobic outer surface of the film.[27] Electrostatic interactions between surface-oriented water dipoles and the various electrical charges present in the intercalated films, leading to certain characteristic minimum-energy patterns of charge redistribution, with consequent modifications of the solid/vapor and solid/liquid interfacial free energies in the contact region of the film with the water drop, may explain some of the observed effects.[28] Consistent with this view is also the fact that the effects observed with the two organic liquids, BCH and HD, both devoid of permanent electric dipoles, are significantly smaller than those produced by water. Eventual differences in the deformability of the films under the vertical surface tension forces operating at the periphery of the sessile drop (the perpendicular component of the liquid/vapor surface tension at the liquid/solid/gas three-phase contact line),[28c,29] dependent on the content of the film and number of stacked layers, may also contribute to the observed effects.

It is obvious that considerable experimental and theoretical efforts will be needed to reach a reasonable understanding of the complex wetting phenomena observed with the current film structures.[30] However, even at this stage, and on a purely empirical basis, such wettability observations may be informative and useful. For example, as different metal ions give rise to different characteristic patterns of wetting behavior,[31] one may correlate many of the observed trends with the presence or absence of a particular ionic intercalant and its lateral mobility. It is also apparent that incorporation of dipolar intercalants producing an excess of negative charge near the outer film surface, such as the unidirectionally aligned PNA and PABA dipoles, may result in high advancing water contact angles, possibly reflecting a repulsive interaction with the oriented water dipoles at the water/film interface. If confirmed on a broader experimental basis, such effects may provide a direct and simple means for the assessment of the absolute sense of orientation of intercalated dipolar species.

3.4. Conclusion

The schematic model describing the composition and structure of the (PABA + Cu^{2+}) intercalated superlattice (Fig. 3.2) emphasizes a number of elements that follow consistently from the entire body of experimental evidence reported in this paper. Features of particular significance in the model are the spatial segregation of the matrix and guest components into well-defined alternating strata, the noncentrosymmetric perpendicular alignment of the PABA dipoles, and the lateral binding role of the divalent Cu^{2+} ions. The detailed packing motif of the aromatic molecules within each layer is an important structural aspect that remains to be elucidated in future studies. At the present stage, both geometrical considerations and the experimentally determined surface densities of the various components (Table 3.1) point to the largely uncorrelated lateral organization of the aliphatic (matrix) and aromatic (guest) portions of the film. Another interesting finding, not obvious from Fig. 3.2, is that once formed, a dense array of intercalated PABA molecules preserves its initial ordered structure even in the absence of the binding copper ions, without which its formation is not possible. A fourth level of controlled molecular self-assembly is thus achieved through the judicious exploitation of specific guest/guest/matrix cooperativity effects arising at various stages during the film self-assembly process.

Remarkable features of the intercalation processes responsible for the formation of present described composite films are (1) their reversibility and possible selective incorporation/removal of specific components, as in the [(PABA + Cu^{2+}) − Cu^{2+}] system, and (2) the effective anisotropic confinement of the guest transport to the interlayer polar planes of the matrix. Thus, regardless of the size of intercalated guest species, an ordered matrix film may be repeatedly utilized with no detectable deterioration of its initial composition and structure—tens of intercalation/extraction cycles of PABA, PNA, or even larger molecules, did not affect the chemical and structural integrity of the silane matrix layers in multilayer films ranging from 2 to 30 superimposed layers. The stability of such silane multilayers is particularly remarkable considering that the extraction of intercalated guests such as PABA and PNA requires exposure to a highly corrosive environment—5 M HCl.

An analysis of the formation and properties of the intercalated (PABA + Cu^{2+}) superlattice and related film structures allows us to propose the present synthetic strategy as a viable approach to the modular self-assembly of planned layered organizates at several hierarchical levels of structural and compositional complexity. In this respect, systems based on polymerizable silanes represent a particular case; various other chemistries offer desirable and useful synthetic alternatives. For example, *n*-octadecylamine (C_{18}-amine) self-assembles on the surface of an oxidized NTS monolayer with outer −COOH functions to form an ordered monolayer attached to the underlying one as the respective amine-carboxylic acid salt. Preparation of stable complex structures by the spontaneous intercalation of various ions and molecules, such as Cu^{2+}, PABA, and PNA, into ionic bilayers of this kind was found to proceed smoothly, with preservation of the structural integrity of the top amine monolayer. Thus, the incorporation of Cu^{2+} ions appears to be driven by the stoichiometric formation of the Cu^{2+}-carboxylate salt, accompanied by the concomitant complexation of the copper by the free amino groups of the top amine monolayer, whereas the incorporation and unidirectional alignment of PABA molecules is energetically favored by the mutually reinforcing acid-amine interactions operating at both the PABA-bottom monolayer and PABA-top monolayer interfaces. Similarly, the incorporation and noncentrosymmetric alignment of PNA into the bilayer is driven by the strong acid-amine and nitro-amine interactions operating at the boundaries with the bottom and top monolayers, respectively.[32] Taken together, these findings demonstrate the practical feasibility of modular self-assembly as a useful new approach to the construction of ordered thin films with a planned noncentrosymmetric arrangement of their molecular components. The inherent spatial segregation, in the intercalated systems, of the matrix and guest constituents, with the consequent decoupling of their lateral packing constraints, offers a natural solution to the structural incompatibility problems that tend to plague both the LB and self-assembly approaches whenever construction of high-quality films made of molecular components differing in size or geometry, such as an alkyl chain connected to an aromatic chromophore, is attempted.[3]

A conceptually related modular strategy was successfully employed in the self-assembly of a centrosymmetric H-bonded acid-acid bilayer, by first converting the outer exposed −COOH functions of an oxidized NTS monolayer into the respective Cu^{2+}-carboxylates (with aqueous copper acetate), then forming a second highly ordered acid-salt monolayer by the self-assembly of arachidic acid (C_{20}-acid) onto the Cu^{2+}-rich surface of the first layer and, finally, removing the Cu^{2+} ions from this bilayer structure by extraction into aqueous HCl—all without causing any damage to the integrity and order of the top acid monolayer.[32] As shown previously,[9,13a] adsorption of fatty acids from hydrocarbon solutions on solid surfaces through the direct formation of acid-surface hydrogen bonds is not a thermodynamically favored process, presumably because acid molecules in such nonpolar solutions already exist as H-bonded dimers.[33] Adsorbing the acid in the form of a metal carboxylate and then converting the resulting carboxylate bilayer into its free acid form, as described earlier, leads to the desired H-bonded bilayer structure, which, although thermodynamically stable, may not be directly produced from its component species. This is another obvious example of the application of modular self-assembly as an advanced synthetic strategy for the preparation of novel types of artificial layered organizates that may be difficult or impossible to obtain by other means.

Combining the high order and structural robustness of the host matrix with the mobility of exchangeable guest species, intercalated films produced by the present approach offer attractive new opportunities as artificial systems endowed with both static and dynamic properties that may be tailored for special purposes. Among others, the dynamic effects associated with the guests transport—large observed differences in the rates of guest incorporation (from seconds to hours), depending on the nature of the guest and other experimentally controllable parameters[6,7,17]—suggest interesting possibilities for the development of competitive processes relevant in microseparation and sensing, while the complex wetting behavior displayed by such composite films should make them attractive as advanced model systems in basic studies of wetting phenomena. In this context, our observations suggest that the depth sensitivity of contact angles to the inner content of a composite multilayer structure may, in general, be much more intricate than that usually observed with single monolayer systems.[34] The facile application of the present modular approach to the assembly of complex film structures on various solid materials, differing in size or shape, is another attractive feature, particularly important for practical applications.

Finally, we should emphasize again that lateral covalent coupling (as in the silane systems) may be desirable for reasons of structural stability, however, it is not an essential requirement for the successful postassembly chemical manipulation of complex film structures. As the latter examples[32] suggest, other chemistries, notably those involving ionic interactions,[4g,5,32] metal ion coordination,[4c,4f,5,17,32,35a] and multiple hydrogen bonds,[6–8,35] are likely to provide the diversity and flexibility often required, besides structural tenacity, for many interesting new applications.[36]

Acknowledgments

We wish to express our thanks to Mrs. Herut Sharan for her helpful technical assistance in the preparation of the manuscript.

References and Notes

1. Kuhn, H. (1989). *Thin Solid Films* **178,** 1.
2. Ulman, A. (1991). *An Introduction to Ultrathin Organic Films From Langmuir-Blodgett to Self-Assembly.* Academic Press, Boston, and references therein.
3. See, for example, Ref. 2, p. 279 and pp. 418–419.
4. (a) Maoz, R., Netzer, L., Gun, J., and Sagiv, J. (1988). *J. Chim. Phys.* **85,** 1059, and references cited therein. (b) Tillman, N., Ulman, A., and Penner, T. L. (1989). *Langmuir* **5,** 101. (c) Lee, H., Kepley, L. J., Hong, H.-G., Akhter, S., and Mallouk, T. E. (1988). *J. Phys. Chem.* **92,** 2597. (d) Ogawa, K., Mino, N., Tamura, H., and Hatada, M. (1990).

Langmuir **6,** 851. (e) Li, D.-Q., Ratner, M. A., Marks, T. J., Zhang, C. H., Yang, J., and Wong, G. K., (1990). *J. Am. Chem. Soc.* **112,** 7389. (f) Katz, H. E., Scheller, G., Putvinski, T. M., Schilling, M. L., Wilson, W. L., and Chidsey, C. E. D. (1991). *Science* **254,** 1485. (g) Decher, G., and Hong, J.-D. (1991). *Makromol. Chem., Macromol. Symp.* **46,** 321.

5. The possible construction of ordered multilayer structures of long chain amphiphiles and metal ions via various combinations of ionic, coordination, and covalent modes of interlayer coupling (involving chemical species such Ca^{++} or Al^{+++} carboxylates, phosphates, and silanes) has been reported by us (R. M. and J. S.) at a number of international meetings starting with the *Gordon Res. Conf. on Org. Thin Films*, Oxnard, California, February 15–19, 1988. We (R. M. and J. S.) have also repeatedly reported, starting with the *199th National Meeting of the Am. Chem. Soc.,* Boston, Massachusetts, April 22–27, 1990, on the possible construction of ordered multilayers of aromatic chromophores devoid of aliphatic long chain substituents and metal ions (such as Ag^{+} or Cu^{++} *p*-methoxybenzoates).
6. Maoz, R., and Sagiv, J. Paper in preparation giving full experimental details of the methodology employed in the assembly of the present described films as well as a detailed discussion of the modes of interlayer coupling achievable in the self-assembly of silane multilayers of the kind depicted in Fig. 3.1.
7. Maoz, R., and Sagiv, J., (1992). *Proc. Int. Symp. Functional Langmuir-Blodgett Films—In Memory of the Late Dr. Y. Kawabata*, Tokyo, Japan, July 1992, . 23.
8. Maoz, R., Sagiv, J., Degenhardt, D., Möhwald, H., and Quint, P. submitted for publication.
9. (a) Maoz, R., and Sagiv, J. (1985). *Thin Solid Films* **132,** 135. (b) Maoz, R., and Sagiv, J. (1987). *Langmuir* **3,** 1034. (c) *Ibid.*, 1045.
10. The preparation of such multilayer films should be easily adaptable to automation because, except for strict requirements regarding materials purity and environmental cleanliness, both the adsorption and oxidation steps involve rather simple chemicals and straightforward preparative conditions.[6]
11. (a) Casal, H. L., Cameron, D. G., and Mantsch, H. H. (1983). *Can. J. Chem.* **61,** 1736. (b) Casal, H. L., Mantsch, H. H., Cameron, D. G., and Snyder, R. G. (1982). *J. Chem. Phys.* **77,** 2825. (c) Doucet, J., Denicolo, I., and Craievich, A. (1981). *J. Chem. Phys.* **75,** 1523.
12. A comparative study of a number of different self-assembling monolayer systems on atomically smooth mica sheets and other solid substrates, with variable degrees of surface roughness (such as unpolished glass, polished quartz, ATR Si prisms, polished Si wafers), indicates considerable lateral bridging over surface irregularities in silane monolayers produced by the present method (R. Maoz and J. Sagiv, to be published).
13. (a) Maoz, R., and Sagiv, J. (1984). *J. Colloid Interface Sci.* **100,** 465. (b) Gun, J., Iscovici, R., and Sagiv, J. (1984). *J. Colloid Interface Sci.* **101,** 201. (c) Allara, D. L., and Nuzzo, R. G. (1985). *Langmuir* **1,** 45. (d) Gun, J., and Sagiv, J. (1986). *J. Colloid Interface Sci.* **112,** 457. (e) Cohen, S. R., Naaman, R., and Sagiv, J. (1986). *J. Phys. Chem.* **90,** 3054. (f) Finklea, H. O., Robinson, L. R., Blackburn, A., Richter, B., Allara, D., and Bright, T. (1986). *Langmuir* **2,** 239. (g) Sabatani, E., Rubinstein, I., Maoz, R., and Sagiv, J. (1987). *J. Electroanal. Chem.* **219,** 365. (h) Tilman, N., Ulman, A., Schildkraut, J. S., and Penner, T. (1988). *J. Am. Chem. Soc.* **110,** 6136. (i) Wasserman, S. R., Tao, Y.-T., and Whitesides, G. M. (1989). *Langmuir* **5,** 1074. (j) Bain, C. D., Troughton, E. B., Tao, Y.-T., Evall, J., Whitesides, G. M., and Nuzzo, R. G. (1989). *J. Am. Chem. Soc.* **111,** 321. (k) Chen, S. H., and Frank, C. W. (1989). *Langmuir* **5,** 978. (l) Laibinis, P. E., Whitesides, G. M., Allara, D. L., Tao, Y.-T., Parrikh, A. N., and Nuzzo, R. G. (1991). *J. Am. Chem. Soc.* **113,** 7152.
14. (a) Bain, C. D., Evall, J., and Whitesides, G. M. (1989). *J. Am. Chem. Soc.* **111,** 7155. (b) Bain, C. D., and Whitesides, G. M. (1989). *Langmuir* **5,** 1370. (c) Laibinis, P. E., and Whitesides, G. M. (1992). *J. Am. Chem. Soc.* **114,** 1990.

15. The wetting behavior of the present mono/multilayer films is very similar to that of analogous stable films adsorbed on atomically smooth mica,[12] thus suggesting the formation of highly smooth and stable outer surfaces of densely packed $-CH_3$ groups. These film systems display the highest equilibrium (no measurable hysteresis) contact angles so far reported for artificial surfaces with outer exposed $-CH_3$ groups.[2,4a,4b,9,13] Although comparably high *advancing* contact angles have been reported for a number of CH_3-terminated thiol monolayers on gold, silver, and copper, they were invariably accompanied by significant contact angle hysteresis (lower receding angles).[13j,13l,14c] As shown in a number of publications by Whitesides *et al.*,[13i,13j,14] high-quality thiol monolayers with outer −COOH functions adsorbed on gold or silver are wetted by water—advancing contact angles below 15 to 10 deg. On the other hand, phosphate and silane monolayers with outer exposed −COOH functions generated via the *in situ* oxidation of terminal ethylenic double bonds were found to display significantly higher water contact angles (60 to 30 deg).[4a,5,9,13i] This was attributed to a less ordered alignment of the hydrocarbon tails in the silane systems, incomplete oxidation of the double bonds, or both.[13i] Accurate infrared data obtained by us in a number of more recent studies indeed point to the lower order and compactness of the silane monolayers reported from this laboratory in the past[4a,9,13a,13b,13d,13e] as compared with those produced by the present method,[6–8,12] as well as to the possible incomplete double-bond reaction in some of the previously reported systems (about 90%)[9] versus the quantitative oxidation of the double bonds in the present films.[6–8]
16. Direct evidence for lateral bridging in silane monolayers was obtained from an investigation of the formation of silane bilayers in which the first monolayer is produced with variable density of outer binding functions (R. Maoz and J. Sagiv, unpublished results).
17. Maoz, R., and Sagiv, J. Paper in preparation providing details on the intercalation of chromophoric species and on some of the spectroscopic characteristics of the resulting superlattice intercalates.
18. Nuzzo, R. G., Dubois, L. H., and Allara, D. L. (1990). *J. Am. Chem. Soc.* **112,** 558.
19. (a) Bellamy, L. J. (1975). *The Infrared Spectra of Complex Molecules*, 3rd ed. John Wiley & Sons, New York, pp. 78–83. (b) *Ibid.*, p. 286.
20. The observed differences in relative band intensities between the spectra in Figs. 3.5 and 3.6 arise from the different modes of measurement by which the respective spectra were obtained—transmission at the Brewster's angle of Si (73.7 deg) in Fig. 3.5 (Ref. 8), versus total internal reflection at 45 deg incidence in Fig. 3.6 (Ref. 13a).
21. The notation [(PABA + Cu^{2+}) − Cu^{2+}] is used for PABA structures generated by the quantitative extraction of the Cu^{2+} ions from hybrid structures produced via the initial joint intercalation of both PABA and Cu^{2+}. Such structures are to be distinguished from pure PABA ones produced via initial intercalation of PABA molecules only.[17]
22. (a) Doub, L., and Vandenbelt, J. M. (1947). *J. Am. Chem. Soc.* **69,** 2714. (b) Scott, A. I. (1964). *Interpretation of the Ultraviolet Spectra of Natural Products*. Pergamon Press, Oxford, pp. 100–120.
23. Sagiv, J. (1977). *Tetrahedron* **33,** 2303, and references therein.
24. The inherent assumption that the film spans the geometrical area of the underlying solid substrate is consistent with the mentioned observations concerning the effective lateral bridging of H-bonded silane monolayers.[12,16]
25. The quantitative extraction of the various film components from intercalated films on silicon was confirmed by IR spectroscopy.
26. See, for example, Ledoux, I., Josse, D., Fremaux, P., Piel, J.-P., Post, G., Zyss, J., McLean, T., Hann, R. A., Gordon, P. F., and Allen, S. (1988). *Thin Solid Films* **160,** 217.
27. (a) Goh, M. C., Hicks, J. M., Kemnitz, K., Pinto, G. R., Bhattacharyya, K., Eisenthal, K. B., and Heinz, T. F. (1988). *J. Phys. Chem.* **92,** 5074. (b) Superfine, R., Huang, J. Y., Du,

Q., and Shen, Y. R. (1992). *Laser Spectrosc. 10th Int. Conf., 1991* World Science, Singapore, p. 117. (c) Du, Q., Freysz, E., and Shen, Y. R. (1994). *Phys. Rev. Lett.* **72**, 238.
28. For electrical charge effects in wetting, see (a) Holly, F. J. (1977). *J. Colloid Interface Sci.* **61,** 435. (b) Morcos, I. (1971). *J. Colloid Interface Sci.* **37,** 410. (c) Finch, J. A., and Smith, G. W. (1979). *Minerals Sci. Eng.* **11,** 36. (d) Ponter, A. B., and Yekta-Ford, M. (1984). *J. Colloid Interface Sci.* **101,** 282.
29. (a) Bikerman, J. J. (1970). *Physical Surfaces*. Academic Press, New York, p. 242. (b) Lester, G. R. (1961). *J. Colloid Sci.* **16,** 315. (c) Michaels, A. S., Sheldon, W. D., Jr. (1962). *J. Phys. Chem.* **66,** 1790. (d) Adamson, A. W. (1990). *Physical Chemistry of Surfaces*, 5th ed. John Wiley & Sons, New York, p. 403. (e) Hautman, J., and Klein, M. L. (1991). *Phys. Rev. Lett.* **67,** 1763.
30. Additional experimental observations that could not be included here appear to support the proposed electrostatic origin of the present reported contact angle variations (R. Maoz and J. Sagiv, to be published).
31. The effects produced by different intercalated metal ions are not identical, characteristic patterns of wetting behavior being ascribable to each particular ion.[30]
32. The construction and properties of complex films with interlayer coupling via acid-amine and acid-acid interactions are topics under current investigation (R. Maoz and J. Sagiv, to be published).
33. The possible deposition of carboxylic acids from the vapor phase on a carboxylic acid surface, with formation of hydrogen bonds between the adsorbate and surface functions, has been recently reported: Sun, L., Kepley, L. J., and Crooks, R. M. (1992). *Langmuir* **8,** 2101.
34. Whitesides, G. M, and Laibinis, P. E. (1990). *Langmuir* **6,** 87.
35. (a) Lehn, J.-M. (1990). *Angew. Chem. Int. Ed. Engl.* **29,** 1304. (b) Folkers, J. P., Zerkowski, J. A., Laibinis, P. E., Seto, C. T., and Whitesides, G. M. (1992). *Am. Chem. Soc. Symp. Ser.* **499,** 10. (c) For a comprehensive review of the role of hydrogen bonding in molecular assemblies, see Bernstein, J., Etter, M. C., and Leiserowitz, L., (1994), *Structure Correlation* (H. B. Bürgi and J. D. Dunitz, eds.), VCH, Weinheim, pp. 431–507.
36. Studies of the self-assembly of a series of complex hybrid organic-inorganic superlattices with alternating layers of various organic molecules and metal or semiconductor species, generated through the *in situ* reduction of intercalated metal ions or their further combination with other intercalable chemical species, are under way in our laboratory.

Building Two-Dimensional Polymers by the Langmuir-Blodgett Technique

SERGE PALACIN, FLORENCE PORTEU, and ANNIE RUAUDEL-TEIXIER

CEA / Service de Chimie Moléculaire, C.E. Saclay, 91191 Gif sur Yvette Cedex, France

4.1. Introduction

We know from the early works by Schmidt[1] that highly ordered polymers can be obtained via solid-state polymerization. The topochemical control of the polymerization has also been widely used in the peculiar example of Langmuir-Blodgett (LB) films: amphiphilic polyethylenes,[2] polydiacetylenes,[3] polyimides,[4] or polycondensates[5,6] have already been formed via solid-state polymerizations or polycondensations of the corresponding monomeric LB films. Following the topochemical rule, the final structure of the amphiphilic polymer is fully controlled by the geometry of the monomer within the precursor film. However, in all cases the pristine two-dimensional (2-D) lattice is converted into one-dimensional (1-D) polymers since the functionality of the monomers does not exceed 2.

In this present work, 2-D networks will be built through lateral linking between multifunctional amphiphilic monomers preorganized as a regular paving of the plane by the LB technique (Fig. 4.1). Thus, the peculiar geometry provided by the LB technique will be fully taken into account.

4.2. General Properties of Amphiphilic Porphyrins in LB Films

The building of 2-D networks requires three essential elements: (1) an engineering method able to preorganize the monomers in the suitable geometry, according to the topochemical rule; (2) designed monomers, fully compatible with the engineering method and able to offer more than

ISBN 0-12-523485-6

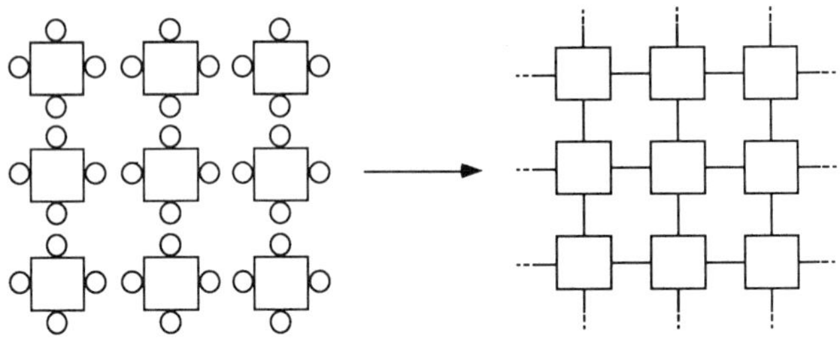

FIG. 4.1. Strategy for building 2-D networks.

two reacting sites per molecule within the same single plane; and (3) an efficient coupling reaction able to link adjacent reacting sites within the common plane. As already mentioned, the engineering method used is the LB technique, which is obviously the better way to organize designed molecules into planar assemblies. The monomers will be amphiphilic or semiamphiphilic porphyrins; indeed, numerous results already published have shown that, owing to suitable chemical modifications on the periphery of the aromatic nucleus, porphyrins can be successfully organized as a regular paving of the plane by the LB technique. Figure 4.2 gives selected molecular formulas for the mentioned porphyrins, the synthesis of which has already been published.[7-10] The main results concerning these molecules follow.

Amphiphilic porphyrins, such as P_1 or P_2, and semiamphiphilic ones, such as P_3, form stable monolayers at the air/water interface. The molecular area is around 160 Å^2 for the amphiphilic ones (Table 4.1). The

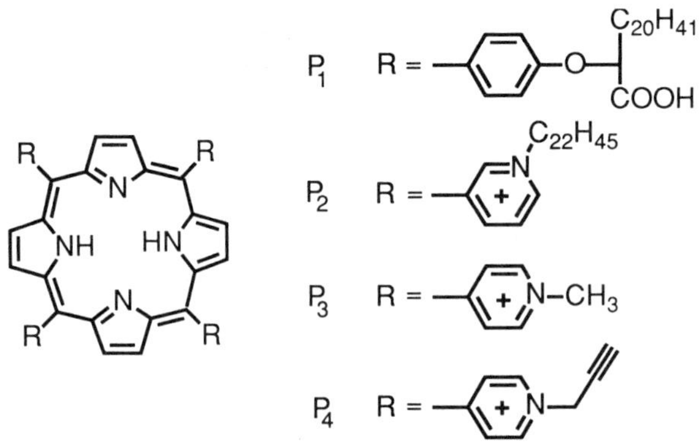

FIG. 4.2. Amphiphilic and water-soluble porphyrins for LB films.

TABLE 4.1

THERMODYNAMIC AND ORIENTATIONAL PROPERTIES OF THE SELECTED AMPHIPHILIC PORPHYRINS AT THE AIR / WATER INTERFACE AND IN LB FILMS

Amphiphilic Porphyrins	Molecular Area (at 35 mN/m)	Visible Bands in LB Films (nm)	Tilt Angle θ	d_{001} (Å)
H_2P_2	162	427, 521, 550, 592	30°	48
CuP_2	158	432, 554	28°	48
CoP_2[a]	158	450, 562	27°	46
CoP_2Im	212	442, 555	34°	44
$(FeP_2)_2O$	330	420, 568	31°	46
H_2P_1	166	425, 515	27°	41

[a]Oxidized form.

monolayers are easily transferred onto solid substrates with a transfer ratio close to 1.

Linear dichroism[11] and Electron Spin Resonance (ESR) measurements[12–14] undoubtedly prove that the planar macrocycles are lying almost flat on solid substrates, the tilt angle lying around 30 deg (Table 4.1). It is noteworthy that the semiamphiphilic films, where the tetra cationic porphyrins are connected via Coulomb interactions with anionic amphiphiles, exhibit the same structure as the purely amphiphilic ones.

ESR studies indicate the P_1 and P_2 are dimerized within the LB matrix.[8] This dimerization certainly occurs at the air/water interface in order to increase the packing of the amphiphiles. Indeed, the final homodimers bear eight aliphatic chains and fill an area about 330 Å^2 at the air/water interface while the corresponding flat lying monomers should fill about 250 Å^2 for only four chains. Thus, the dimerization leads to more densely packed monolayers. The resulting monolayer tends to fulfill the intuitive rule that the aliphatic chains have to fill almost the same area as the polar heads in order to result in well-ordered monolayers.

Well-defined heterodimers are spontaneously formed[8] at the air/water interface when mixing complementary amphiphilic porphyrins such as P_1 and P_2. In this case, the dimerization process observed on the pure monolayers is reinforced by a salt formation occurring between the peripheral functions of the two macrocycles. The distance between the two planar cores within the heterodimer has been estimated from the ESR spectra to 4.5 Å and P_1 is always located on top of P_2. Figure 4.3 shows the expected structure of the heterodimer.

The homodimerization does not occur when the central metallic ion of the porphyrin is axially complexed. For instance, CoP_2 axially complexed

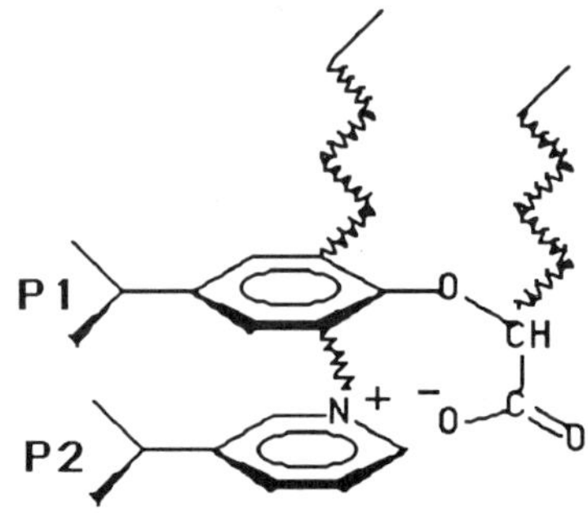

FIG. 4.3. Expected structure of the heterodimer P_1–P_2. Reprinted with permission from F. Porteu *et al.*, *The Journal of Physical Chemistry* **95,** 7438, Copyright © 1991 American Chemical Society.

with octadecylimidazole (CoP_2Im)[15] appears as a flat lying monomer within its LB film (Table 4.1).

4.3. Ionic Network

This section is devoted to the building of a 2-D ionic network, i.e., a macromolecular salt formed via lateral association between two amphiphilic ions. For this purpose, the amphiphilic porphyrins P_1 and P_2 will be used since they are chemically complementary: P_2 is bearing four peripheral cations (the pyridinium rings) while P_1 is bearing four ionizable carboxylic functions. The spontaneous formation of a pyridinium carboxylate at the air-water interface is supposed to link the two molecules.[16,17]

As mentioned earlier, direct mixing of the amphiphilic porphyrins P_1 and P_2 gives rise to well-defined heterodimers where the two macrocycles are chemically bound via salt formation on four points at their periphery. The chemical linkage between P_1 and P_2 occurs perpendicularly to the water surface because this leads to the most stable system. This shows that the complementary functions located on the periphery of the two macrocycles are suitable for linking the two amphiphiles, but it is necessary to force this chemical recognition to occur within the plane instead of perpendicularly to it in order to form a planar ionic network.

As shown for the case of CoP_2, the homodimerization can be precluded when the volume located just on top of the macro cycle is filled by a bulky axial ligand. This result is turned into account in the case of the mixed monolayer. Thus, filling the volume located on top of P_2 prevents the molecule P_1 to climb onto P_2 and allows the chemical linkage between P_1

and P_2 to occur parallel to the water surface. The axial ligand has to be amphiphilic to be located on the "hydrophobic side" of the macro cycle (on the same side as the aliphatic chains), and to be strongly linked and bulky enough to preclude the heterodimerization and lead to a so-called "ionic polymer" where each macro cycle is chemically linked to four of its neighbors, as shown in Fig. 4.4. The following results show that this strategy has been successfully applied to two different axial ligands: octadecylimidazole, linked to the cobaltous ion, and the amphiphilic porphyrin P_2 itself, which may play the role of a bulky axial ligand when used in the stable μ-oxo dimer.

4.3.1. Formation of the Hindered Amphiphilic Porphyrins

As already shown,[5] the cobaltous-imidazole complex can be spontaneously formed at the air/water interface by spreading together a cobaltous porphyrin and octadecylimidazole. The formation of the axial complex can then be checked by studying the molecular area of the porphyrin at the air/water interface (if the complexation occurs, the porphyrin cannot dimerize and the molecular area is higher than 200 $Å^2$) and the kinetics of oxidation of the cobaltous ion by molecular dioxygen. Indeed, cobaltous porphyrins axially complexed by electron-donating ligands such as imidazole are easily oxidized by molecular oxygen into copper(III) complexes, whereas free cobaltous porphyrins are more stable in air.[18,19] It is thus easy to check that mixing one equivalent of CoP_1, one equivalent of CoP_2, and two equivalents of octadecylimidazole in the spreading solution leads to a mixed LB film containing both axially complexed cobaltous porphyrins.

The formation of the μ-oxo dimer of P_2, $(FeP_2)_2O$, is easier than the imidazole complex since it is synthesized by oxidation of the ferrous

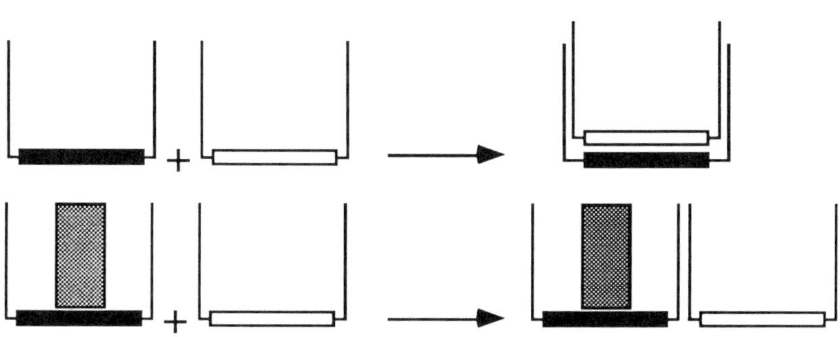

FIG. 4.4. Strategy to preclude the heterodimerization.

porphyrin. The μ-oxo bond can be clearly characterized by infrared spectroscopy (stretching band at 880 cm^{-1}) and visible spectroscopy (Soret band at 420 nm and single Q band at 570 nm). These features are found again on the resulting LB films built from pure $(FeP_2)_2O$, ensuring that the μ-oxo dimer is perfectly stable at the air/water interface. Thus, it is possible to build stable mixed LB films from the mixture $P_1 + P_2$ where both porphyrins (in the case of the cobaltous-imidazole complex) or only P_2 (in the case of the μ-oxo dimer) is sterically hindered on the hydrophobic side of the flat lying macrocycles.

4.3.2. Lateral Linking Between Adjacent Porphyrins

The infrared spectra of the multilayers built up from the $[CoP_1Im + CoP_2Im]$ and $[(FeP_2)_2O + P_1]$ mixtures indicate that all of the carboxylic functions of P_1 have been ionized into carboxylates. This shows that the peripheral functions have met. Since P_1 cannot be vertically linked to P_2 as in the heterodimer P_1-P_2 (Im or FeP_2 is filling the place previously devoted to P_1), the chemical recognition between the two complementary macrocycles must occur parallel to the water surface. The only possible structure is then an "ionic polymer" where each amphiphilic macro cycle is bound via a salt formation to four of its neighbors.

This assumption is confirmed when studying the solubility of these mixed LB films. Indeed, the $[CoP_1Im + CoP_2Im]$ and the $[(FeP_2)_2O + P_1]$ multilayers are totally insoluble in chloroform (which is the spreading solvent), while the $[MP_1 + MP_2]$ LB films, whatever the metallic ion M, appear perfectly soluble in that solvent. The latter result is consistent with the claimed heterodimeric structure of the $[MP_1 + MP_2]$ films: Each heterodimer P_1-P_2 is a neutral supermolecule independent of its neighbors. On the contrary, if each macro cycle is chemically linked to its closest neighbors, a collective consistency is created and insolubility is likely to occur. So, the insolubility of the $[CoP_1Im + CoP_2Im]$ and $[(FeP_2)_2O + P_1]$ films in chloroform is strongly in favor of a "polymeric" structure, built up at the air/water interface by lateral recognition between coplanar porphyrins.

4.3.3. Structure of the Ionic Networks

Table 4.2 gathers the orientational information given by the linear dichroism in the infrared and visible ranges on the $[CoP_1Im + CoP_2Im]$ and $[(FeP_2)_2O + P_1]$ films. It shows that the two ionic networks are slightly different: Although the $[(FeP_2)_2O + P_1]$ LB film exhibits the same geo-

TABLE 4.2

CHARACTERISTIC FEATURES OF THE MIXED MULTILAYERS

Mixtures	Molecular Area (at 35 mN/m)	Solubility in Chloroform	Tilt Angle θ	Vertical Period d_{001} (Å)
$MP_1 + MP_2$	140	yes	25°	46
$[CoP_1Im + CoP_2Im]$	185	no	39°	53 ± 5
$[(FeP_2)_2O + P_1]$	210	no	25°	47

metrical parameters as the heterodimeric film, the $[CoP_1Im + CoP_2Im]$ films are slightly disorganized.

In the case of the $[(FeP_2)_2O + P_1]$ LB films, it is possible to visualize directly the planar network: The periodicity within the monolayer can be checked by normal incidence transmission x-ray diffraction. We observed a lateral period of about 17.5 Å, which is probably the shortest distance between neighboring ferric ions. No in-plane periodicity is observed on the heterodimeric films.

Thus, it has been shown that insoluble ionic networks can be built via lateral linking between adjacent amphiphilic porphyrins. The slight differences observed between the two examples (tilt angles, vertical periods) result from the low stability of the cobaltous-imidazole complex. Indeed, the axial complexation of the cobaltous ion by octadecylimidazole is supposed to occur spontaneously at the air/water interface. Thus, some competition between this process and the heterodimerization of the two complementary porphyrins is likely to occur, leading to some local heterogeneity within the mixed monolayers. This problem is resolved by using very stable axially complexed porphyrins, which are thermodynamically more stable than the heterodimer and not supposed to compete with it within a kinetic formation process at the air/water interface, as is the case with the μ-oxo dimer $(FeP_2)_2O$.

Of course, the system has to be improved, for example, by the use of a mixture of two preformed complementary dimers, which will probably strengthen the mixed monolayer via the increase in the packing of the aliphatic chains. Indeed, the described ionic networks clearly suffer from the lack of crystallinity within the aliphatic matrix (only 10 chains for 370 Å^2 in the $[CoP_1Im + CoP_2Im]$ monolayer, 12 chains for 420 Å^2 in the $[(FeP_2)_2O + P_1]$ monolayer). Some improvement should certainly be found in the increase of the number of aliphatic chains per square angstrom, provided this better packing does not hinder the lateral molecular recognition between the complementary amphiphilic macrocycles. However, this ionic network can easily be broken since the lateral linkages are purely ionic: Any acidic treatment, either in the gas phase or in aqueous solution, can chemically separate the two partners and restore the solubility of the assembly in organic solvents. The next section shows that it is possible to replace the ionic bonds with covalent ones and give rise to a true 2-D polymer.

4.4. Two-Dimensional Covalent Polymer

As in the case of the 2-D ionic network, the building of the first 2-D polymer is governed by the topochemical rule: the monomers have to be previously organized in a regular structure fully compatible with the final polymer. This is easily achieved via a proper design for the monomer. However, building covalent bonds cannot result from spontaneous processes and requires a coupling step. Of course, this coupling reaction must not disturb the geometry of the monomers.

4.4.1. The Monomers and the Coupling Reaction

As shown earlier, porphyrins can be organized as a regular paving by the LB technique either as amphiphilic or semiamphiphilic derivatives, provided strong polar heads are properly located on their periphery. Thus, water-soluble porphyrins, such as P_3, can form well-ordered LB films when associated with suitable amphiphilic counter ions and give rise to the expected regular paving: P_3 is lying almost flat on the water surface and on the substrates. This ensures that the four methyl groups of each porphyrin in the same monolayer are actually located within the same polar plane. Replacing the methyl group by any polymerizable function allows a 2-D polymerization process to occur within each monolayer, since all the reacting sites are located within the same plane.

Performing a solid-state polymerization requires the coupling step to be safe for the precursor structure. This means that the reacting species have to be brought into the monomeric matrix via nondestructive ways. In the case of LB films, the coupling reactions have to proceed only via photochemical processes[2,3] or involve water-soluble reagents.[5,6] Because porphyrins are somewhat fragile under irradiation, the Glaser reaction has been chosen. This oxidative coupling[20] appears to be quite well suited to this purpose since the coupling reagent is an ammoniacal cuprous chloride solution, which can link two neighboring terminal triple bonds into a diacetylenic bond, according to Fig. 4.5. Thus, the monomer P_4 is bearing

$$-C{\equiv}C{-}H + H{-}C{\equiv}C- \xrightarrow[CuCl/NH_3]{} -C{\equiv}C{-}C{\equiv}C-$$

FIG. 4.5. Scheme of the Glaser oxidative coupling between terminal triple bonds.

four terminal triple bonds on its periphery , as shown in Fig. 4.2. This molecule is water soluble and can be easily organized by the LB technique as P_3 when associated with amphiphilic acidic molecules such as dihexadecylphosphoric acid (HDHP). The oxidative Glaser coupling may be performed in two distinct ways: (1) by diffusion of the reactive aqueous solution through the precursor LB film or (2) by direct reaction between the same reactant dissolved in the subphase and the floating semi-amphiphilic monolayer. Both cases are described.

4.4.2. Structural Study of the Precursor LB Film

The semiamphiphilic precursor LB films have been thoroughly characterized as described in the following paragraphs.

A quartz microbalance measurement indicates that the water-soluble porphyrins are densely packed within the semiamphiphilic monolayer, as is the case for purely amphiphilic porphyrins.[21]

The infrared spectrum recorded on the semiamphiphilic LB film exhibits the vibration bands of both the DHP molecule (P=O bond at 1237 cm^{-1}) and the porphyrin [$-C{\equiv}C{-}H$ group at 3300 cm^{-1} ($\nu \equiv C{-}H$) and 2120 cm^{-1} (ν $C{\equiv}CH$)].

The aliphatic chains of the DHP anions are found to be tilted at 30 deg versus the normal to the substrate, whereas the average tilt angle between the planar macrocycles and the substrate is 30 deg.

The vertical period of the semiamphiphilic LB films, measured by x-ray diffraction and by ellipsometry, is 57 Å. The Kiessig fringes, observed on a 29-layer sample, confirm the high regularity of the deposition. The vertical period is only 47 Å for the pure HDHP LB film, so one can estimate as 14 Å the average thickness of the "porphyrin bilayer" located in the polar planes of the semiamphiphilic LB film. This is consistent with a polar plane containing two superimposed porphyrins, each one coming from the deposition of successive monolayers.

Even though the macrocycles appear to be well ordered within the semiamphiphilic LB film, no in-plane crystalline structure can be derived from our transmission x-ray studies. This indicates that the distance between two adjacent porphyrins within each monolayer is not perfectly constant.

As expected, the precursor LB film is totally soluble in chloroform and other organic solvents. This is consistent with the monomeric structure of the film at this stage.

4.4.3. Chemical and Structural Study of the LB Film Polymerized by Diffusion

The chemical coupling between the neighboring acetylenic groups is performed on the precursor LB films by diffusion of the coupling reagents in aqueous solution in order to realize the oxidative coupling. Features were observed as explained in the following paragraphs.

The infrared spectrum of the resulting multilayers shows the complete vanishing of the characteristic bands of the triple bond at 3300 and 2120 cm^{-1}. At the same time, a band located at 2245 cm^{-1} appears after the action of the ammoniacal cuprous chloride solution in the SERS spectrum of the multilayers. This fully reproducible band can be attributed to the symmetrical stretch of the diacetylenic bond, which appears in this region for symmetrically substituted diacetylenes.[22]

The orientational order of the lamellar structure is only slightly affected by the action of the ammoniacal cuprous chloride solution. Indeed, the tilt angle of the aliphatic chains remains the same (30 deg versus the normal to the substrate) and the average angle between the flat macrocycles and the substrate decreases from 30 to 20 deg $\pm$ 5 deg: The diffusion of the ammoniacal cuprous chloride solution within the precursor LB film tends to improve the parallel orientation of the planar macrocycles with respect to the substrate.

The x-ray diffraction and ellipsometry experiments performed after exposure to the ammoniacal cuprous chloride solution give the vertical periodicity of the polymerized film $d_{001} = 53$ Å.

Transmission x-ray diffraction experiments give two diffraction rings at $d = 18.6$ and $d = 27$ Å. These values are consistent with the expected average distances between the neighboring copper ions within a square lattice containing one porphyrin per unit cell.

Unlike the precursor film, the polymerized semiamphiphilic film is totally insoluble in chloroform. This indicates that the oxidative coupling has given rise to some collective consistency within the lamellar structure. Moreover, it is possible to benefit from the semiamphiphilic character of the film: On treatment with hydrochloric acid vapors, the ionic bonds between the DHP anions and the porphyrin rings are destroyed. Then, the aliphatic chains can be removed from the ultrathin film via extraction by chloroform. This removal of the chains does not affect the porphyrin part of the ultrathin film since the Soret band of the film loses less than 10% of its intensity, as indicated in Fig. 4.6. Hence, the resulting ultrathin film is totally free of aliphatic chains and totally insoluble in water and in all the available solvents. The strength of this "polymerized porphyrin" ultrathin film gives evidence of the covalent character of the bonds formed between the adjacent macrocycles within each monolayer. It is noteworthy that, even though no diffraction order could be observed on this chain-free film, ESR spectra recorded at 300 K indicate that the porphyrin rings are still lying almost parallel to the substrate. This last result is indirect evidence of the high value of the polymerization index: Indeed, one would expect small polymerized areas to "fall" and collapse without any preferential orientation on removal of the aliphatic chains, while large and rigid polymerized areas are more or less forced to keep their previous orientation because they do not have enough room to rotate during the solubilization of the aliphatic chains.

4.4.4. Oxidative Coupling Performed on the Trough

To study the influence of the mobility of the porphyrin monomers on polymerization, the coupling reaction can also be performed on the trough, i.e., on the floating monolayer, as has already been described for 1-D amphiphilic polymers.[23-25] For this purpose, the solution under the semi-amphiphilic monolayer has to be exchanged for the coupling reactant solution by simultaneous slow draining and automatic filling of the trough. The resulting floating monolayer is quite stable and can be transferred onto solid substrates by the horizontal dipping method. Thus, the polymerized film can be deposited onto electron microscope grids with different hole sizes.

Figure 4.7 shows an optical reflection micrograph of the two-layer membrane deposited on a grid following this process, i.e., reaction between ammoniacal cuprous chloride dissolved in the subphase and the floating monolayer at the air/water interface. The folds and cracks that can be seen on this photograph can be attributed to the manual downstroke and upstroke used in this particular case. These macroscopic defects help us to visualize the ultrathin membrane. The number of defects is very low, although it is impossible from only this result to derive any significant value for the polymerization index. Thus, the strengthening expected from the 2-D nature of the polymerization actually occurs, since the maximum bridged area is found to be 100 times higher than the one observed with polydiacetylene LB films.

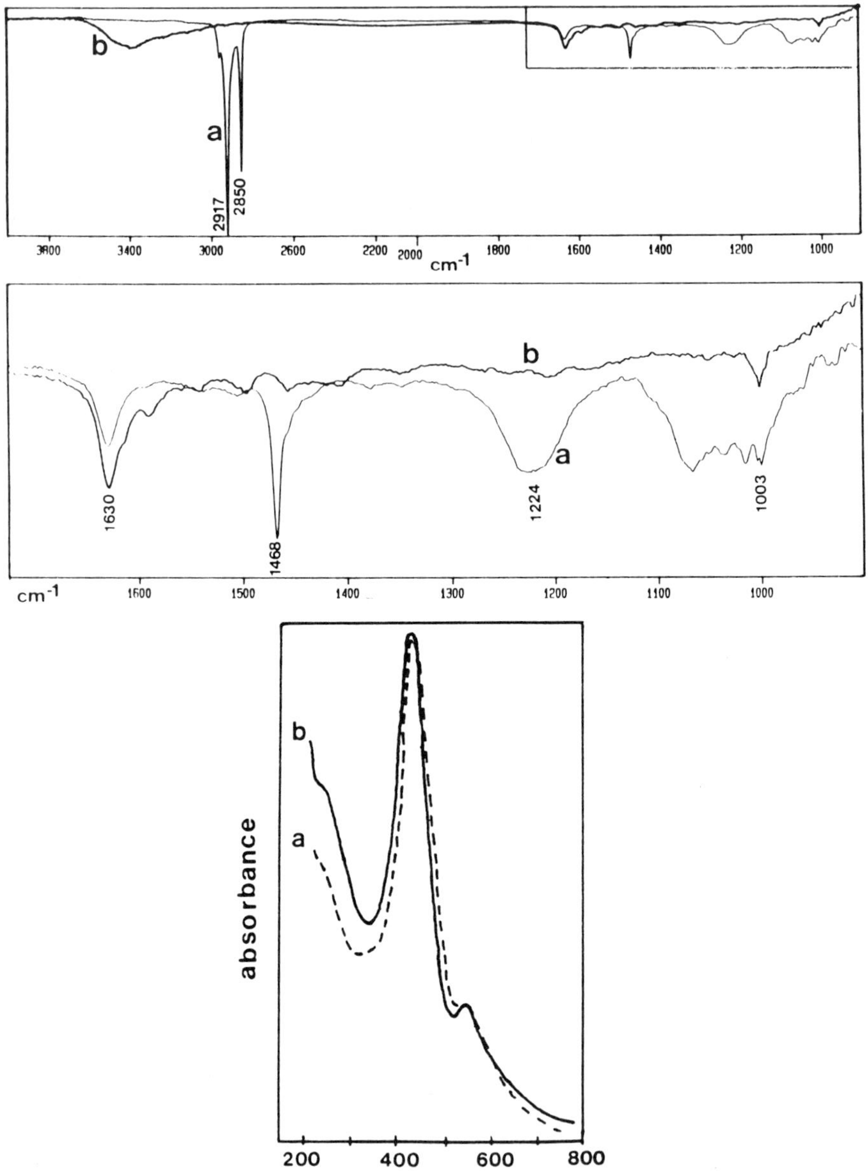

FIG. 4.6. Infrared spectra (top) and UV-visible spectrum (bottom) recorded on the polymerized [DHP, H_2Pr] semiamphiphilic LB film (17-layer sample) (a) before and (b) after removal of the aliphatic chains by exposure to hydrochloric acid vapors and then extraction with chloroform. Reprinted with permission from D. Lefevre *et al. Langmuir* **9,** 150, Copyright © 1993 American Chemical Society.

FIG. 4.7. Optical reflection microscope view of the two-layer membrane formed by the semi-amphiphilic process. By means of several types of microscope square grids, it has been shown that even a monolayer-thick membrane can actually cover areas up to 0.01 mm^2 (200-mesh Agar grids). This means that about 5×10^9 molecules are involved in the bridging being transferred onto the microscope grid via a horizontal lifting method. The size of the hole is about 0.003 mm^2.

4.5. Conclusion

The two examples described here undoubtedly show that following the topochemical rule within an LB matrix can provide truly 2-D networks. Taking advantage of the elective affinity between complementary functions or performing highly efficient coupling reactions enables us to build well-defined 2-D polymers, either of the ionic or covalent type. The former provides an elegant example of supramolecular engineering where the chemical, geometrical, and crystallographic properties of the system have to be taken into account to reach the desired goal. The latter polymer shows that mechanical properties can be tremendously improved at the molecular scale via judicious design of the dimensionality and the energy of the building blocks. It opens a fascinating field of research in the field of highly stable and clever molecular materials.

Acknowledgments

The authors thank Pierre-Antoine Albouy, Peter Balog, Patrick Berthault, Didier Lefèvre, Christophe Mingotaud, Louis Patard, Monique Roulliay, and Gilbert Zalczer for different spectroscopic studies and André Barraud for fruitful discussions.

References

1. Cohen, M. D., and Schmidt, G. M. J. (1964). *J. Chem. Soc.* 1996.
2. Barraud, A., Rosilio, C., and Ruaudel-Teixier, A. (1977). *J. Coll. Interface Sci.* **62,** 509.
3. Tieke, B., Lieser, G., and Wegner, G. (1979). *J. Polym. Sci. Polym. Chem. Ed.* **17,** 1631.
4. Baker, S., Seki, A., and Seto, J. (1989). *Thin Solid Films* **180,** 263.
5. Rosilio, C., and Ruaudel-Teixier, A. (1975). *J. Polym. Sci. Polym. Chem. Ed.* **13,** 2459.
6. Hong, K., and Rubner, M. F. (1988). *Thin Solid Films* **160,** 187.
7. Ruaudel-Teixier, A., Barraud, A., Belbeoch, B., and Roulliay, M. (1983). *Thin Solid Films* **99,** 33.
8. Porteu, F., Palacin, S., Ruaudel-Teixier, A., and Barraud, A. (1991). *J. Phys. Chem.* **95,** 7438.
9. Porteu, F., Palacin, S., Ruaudel-Teixier, A., and Barraud, A. (1991). *Makromol. Chem. Macromol. Symp.* **46,** 37.
10. Palacin, S., and Barraud, A. (1989). *J. Chem. Soc. Chem. Commun.* 45.
11. Vandevyver, M., Barraud, A., Ruaudel-Teixier, A., Maillard, P., and Gianotti, C. (1982). *J. Colloid Interface Sci.* **85,** 571.
12. Palacin, S., Ruaudel-Teixier, A., and Barraud, A. (1986). *J. Phys. Chem.* **90,** 6237.
13. Palacin, S., Lesieur, P., Stefanelli, I., and Barraud, A. (1988). *Thin Solid Films* **159,** 83.
14. Palacin, S., Ruaudel-Teixier, A., and Barraud, A. (1989). *J. Phys. Chem.* **93,** 7195.
15. Lecomte, C., Baudin, C., Berleur, F., Ruaudel-Teixier, A., Barraud, A., and Momenteau, M. (1985). *Thin Solid Films* **133,** 103.
16. Porteu, F., Palacin, S., Ruaudel-Teixier, A., and Barraud, A. (1992). *Mol. Cryst. Liq. Cryst.* **211,** 193.
17. Porteu, F., Palacin, S., Ruaudel-Teixier, A., and Barraud, A. (1992). *Thin Solid Films* **211,** 769.
18. Hoffmann, B. M., Diemente, D. L., and Basolo, F. (1970). *J. Am. Chem. Soc.* **92,** 61.
19. Stynes, D. V., Stynes, H. C., James, B. R., and Ibers, J. A. (1973). *J. Am. Chem. Soc.* **95,** 1796.
20. Viehe, H. G. (1979). In: *Chemistry of Acetylenes.* Marcel Dekker, New York.
21. Lefevre, D., Porteu, F., Balog, P., Roulliay, M., Zalczer, G., and Palacin, S. (1993). *Langmuir* **9,** 150.
22. Meister, A. G., and Cleveland, F. F. (1944). *J. Chem. Phys.* **12,** 393.
23. Day, D. R., and Ringsdorf, H. (1978). *J. Polymer Sci. Polymer. Lett. Ed.* **16,** 205.
24. Duran, R. S., and Zhou, H. C. (1992). *Thin Solid Films* **210,** 356.
25. Peltonen, J. P., Pingsheng, H., and Rosenholm, J. B. (1992). *Thin Solid Films* **210,** 372.

Surface Modification

Polymer Surface Modification

THOMAS J. MCCARTHY, TIMOTHY G. BEE, JOAN V. BRENNAN, ELISA M. CROSS, ANTHONY J. DIAS, NICOLE L. FRANCHINA, KANG-WOOK LEE, and MOLLY S. SHOICHET

University of Massachusetts
Polymer Science and Engineering Department
Amherst, Massachusetts 01003

5.1. Introduction: Polymer Surfaces and Dynamic Interfaces

The surface properties of solid organic polymers affect the majority of their materials applications. Examples of properties affected by polymer surface structure are wettability, friction, cohesive strength of polymer blends, mechanical properties of adhesive joints, adhesive strength of composite materials, permeability, dyeability, gloss, corrosion, surface electostatic charging, cellular recognition, and biocompatibility. All solids (and liquids) have surfaces, and the fact that they must communicate with their environment through their surfaces (interfaces) emphasizes the importance of surface structure and properties in materials performance. Organic polymer surfaces are special: Unlike surfaces of more rigid materials such as metals, metal oxides, and ceramics they can be mobile and can resemble a viscous fluid. This property, unique to polymer surfaces, complicates their study and also has limited our understanding of them; at the same time it invites many opportunities for research and indeed, practical application.

Because of the unique mobility and flexibility of polymer chains at a solid polymer surface there can be no specific definition of the term "polymer surface." We can define the *surface* as *that part of the object that affects or can be affected by its environment*, but what we mean by *that part* is situation dependent and because there are so many situations, this definition is so general that it is not useful.

At a given temperature and in a given environment, the magnitude of surface chain mobility depends on the chemical structure of the polymer

ISBN 0-12-523485-6

chain and can be, to a first order, assessed by the difference between the temperature of interest and the glass transition temperature of the polymer. This applies to amorphous polymers and the amorphous regions of semicrystalline polymers. Common polymers have T_g's that span more than 400°C; therefore, at any one temperature and in any one environment, surface mobility varies greatly from polymer to polymer. For example, polyethylene has a T_g of -125°C and polystyrene has a T_g of 100°C. At room temperature, the mobility of surface chains of these two materials is significantly different: The polystyrene chains are rigid and the polyethylene chains in the amorphous regions are rotating and restructuring rapidly. For semicrystalline polymers, the percent crystallinity in the surface region contributes to the degree of mobility. Polymer molecular weight and molecular weight dispersity are also important; in certain molecular weight regimes, mobility is decreased with increasing molecular weight.

For a given polymer the mobility is temperature dependent, and it is not at present clear whether or not surface chain mobility has the same or a similar temperature dependence as bulk polymer mobility. It is likely that the surface mobility characteristics of some polymers can be determined from bulk properties, but other polymers may have surface characteristics that vary greatly from the bulk. Consider a polymer chain at the surface of an amorphous polymer in contact with vacuum or a noninteracting gas. We might assume that it is more mobile than a bulk (subsurface) chain because it has fewer nearest neighbors and conclude that the "surface" T_g should be lower than the bulk T_g and that at a given temperature a surface chain is more mobile than subsurface chains. On the other hand, the surface chain may adopt a constrained conformation to minimize surface free energy (critical surface tension), and we can envision a situation where the low surface energy conformation is in a deep enough potential well that the surface chain is less mobile than bulk chains. The extent to which each of these opposing effects manifests itself in a real polymer sample certainly varies from polymer to polymer.

To understand the structure of a polymer surface, the environment of the polymer sample must also be considered and in many cases it is the most important consideration. Polymers are used in every conceivable environment and thus a range of structures for a polymer surface exists. A polymer surface in contact with vacuum or a noninteracting gas behaves differently than an identical one in the presence of an interacting gas. Gases can adsorb, penetrate to varying depths, and have variable solubility in the polymer. Likewise the structure of a polymer surface in contact with a liquid has different structures that depend on the liquid/solid interactions. Liquids can wet the surface to varying extents, not wet the surface or swell the sample to various extents (the extreme is dissolution). The structure exhibited by a polymer surface in contact with a solid depends on the solid. Chains can adsorb with a variable number and variable location of attachment sites to make interface structures with chains parallel or perpendicular or any degree of orientation in between. The polymer can adhere, not adhere, or crystallize in contact with the solid.

It is clear that there is no generic picture of a polymer surface that can be applied to all situations; there will never be a "model polymer surface." The mobility of polymer chains allows them to respond to their environment and lower interfacial free energy, thus a *polymer surface* should always be considered a *polymer interface*. A further complexity of polymer surfaces (interfaces!) is a complexity inherent in solid polymers. Because of their high molecular weight they exhibit nonequilibrium structures. Surface structure thus depends on the thermal, processing, and environmental history of the sample.

It is often useful to determine a "thickness" of a polymer surface, that is, in terms of the preceding definition, how deep into a polymer sample a polymer segment has an effect on a surface property or measurement. This type of thickness is usually useful only in describing one specific property or measurement; this gives rise to a number of different thicknesses as well as nomenclature difficulties. The *wettability surface* describes the region of the sample (outermost 5 to 10 Å) that affects wettability. This is a different region than the *adhesion surface* or *coefficient of friction surface*. The *contact angle surface* is that part of the sample assessed by a contact angle measurement and is dependent on the probe fluid and in particular the degree to which the probe fluid and polymer sample interact. Advancing and receding contact angles can assess different depths. The *XPS surface* is the region of the sample measured by x-ray photoelectron spectroscopy. The depth assessed by this technique depends on the sample identity (the mean free paths of electrons ejected from specific elements in the sample being analyzed), the x-ray source, and the geometry of the experiment.

To summarize this introductory section, the term "polymer surface" is extremely and necessarily nonsuccinct. When using the term we describe a region of material that in many cases is mobile, at the interface between a complex nonequilibrium solid and its environment.

5.2. Polymer Surface Modification

In light of the diffuse definition of polymer surface, a discussion of polymer surface modification warrants a narrowing description. Again, however, we cannot be precise. Surface modifications are often carried out in practice to change a surface property without affecting bulk properties, but this cannot be considered a defining statement because whether or not bulk properties are changed depends on the depth of the modification, the thickness of the sample being modified, and the structure and properties of the modified region. Bulk modifications can change surface properties, and surface-selective reactions on thin polymer films can change bulk properties. In practice, researchers in this field describe surface modifications in terms of the depth of reaction or the thickness of the modified layer.

A range of methods has been used to modify the surfaces of polymers. The most common techniques are reactive gas discharge modification (plasma and corona techniques), surface graft polymerization, and chemical reaction modification. Plasma modification[1-3] involves the contact of a solid polymer sample with an electrically excited gas. The plasma consists of reactive ions, electrons, metastables, and UV radiation that cause a range of competing and sequential reactions (homolytic bond cleavage, fragmentation, crosslinking, ablation) in the polymer surface region. Inert, reactive, and polymerizable gases are used to generate plasmas and these have been used to modify a wide array of polymer surfaces.[4-27] The technique is versatile and can be applied to many situations, but it is also brutal and nonselective and chemically well-defined surfaces cannot be designed and prepared. Surface graft modification involves the polymerization of a monomer or monomers from active sites (almost exclusively radicals) generated on the solid polymer surface. Radiation (most common), thermal, mechanical, photochemical, plasma, and wet chemical methods are used to generate sites for initiation. Reactive species can be generated before monomer exposure, after the polymer surface has been infused with monomer or during contact with monomer. Monomers can be used in the gas phase, as neat liquids or in solution. This technique, like plasma modification, is versatile and a wide range of monomers has been grafted to a wide range of solid polymers.[28-44] Controlling the diffuseness of the polymer/polymer interface, the thickness of the grafted layer, the density and location of attachment sites, the molecular weight, molecular weight distribution, branching, and crosslinking of the grafted polymer is difficult and these variables in general have not been addressed analytically. Chemical surface modification is the subject of this chapter and is discussed next. Other methods for surface modification that are not discussed here include adsorption and polymerization of amphiphiles,[45-48] entrapment of terminally functionalized oligomers,[49] deposition of metal oxides,[50] reaction with metal vapors,[51-56] ion beam modification,[57] photoisomerization,[58] and flame treatments.[59]

Surface property control can also be accomplished by taking advantage of the fact that polymer surfaces are not at equilibrium and that they reconstruct to form interfaces with lower energy. Homopolymer, copolymer, and polymer blend surfaces can display different properties by selective orientation of surface functionality or the concentration of one component at the surface.[60-93]

5.3. Chemical Modification

The remainder of this chapter concerns chemical modification of polymer surfaces. The majority of the work discussed involves synthetic methods for introducing organic functional groups to the surface region of relatively unreactive polymers. Several examples of the surface modification of a more reactive polymer (which contains functional groups in the virgin polymer) are included as well. Almost all of the strategies used to effect this type of modification involve a chemical reaction that has precedence in solution, at the interface between the solid organic polymer and a solution in contact with the polymer. A general analysis of the factors that should be considered in carrying out a reaction at a solid/solution interface is warranted and presented first. Several of the issues are unique to this type of reaction and have no counterpart in solution organic chemistry or surface chemistry at more sharply defined interfaces.

A solvent in contact with a solid polymer will interact with the polymer to an extent that ranges from not being wet by the solvent to being swollen by the solvent (ultimately to being dissolved by the solvent). For a given polymer this will depend on both the solvent structure and the temperature. The terms *wet* and *swell* describe regions of this continuum and within these regions polymers are wet or swollen to varying degrees. The solid polymer/solvent interface can vary from being sharply defined to highly diffuse and the solid polymer chains can be oriented at this interface (or in this "interphase") in response to the solvent. The term *interphase* is used to describe this region of solvent/polymer interaction; it can be thin (as in interface) if the interaction is slight or quite thick. A reagent in this solvent will partition between the solution and the interphase (and perhaps the region of the polymer that the solvent does not penetrate). The structure of the solvent at this interface may also be important.[60] The depth of the modification reaction and the thickness of the modified layer of the product will depend on the structure of this interphase. If solvent mixtures are used the interphase may be more complex; one solvent may partion differently than another.

If a reaction occurs between the polymer and a reagent, the structure of the interphase changes and a range of scenarios can be envisioned. If the polymer product of the reaction interacts with the solution to less of an extent than the virgin polymer, then the product will collapse and form a barrier to further penetration. If the original interphase was rather sharp, a thin modified layer will result. This type of surface modification reaction exhibits autoinhibitive kinetics, that is, the reaction proceeds rapidly at initial times and then stops. If the original interphase was diffuse, one can imagine a resulting stratified interphase of the same dimensions that could continue to react as long as reagent was present, or also imagine that the interphase could reconstruct, "burying" the product and exposing more unreacted polymer to the solution. If the solution interacts to a greater extent with the polymer product than the unreacted polymer, then the solid can be swollen more greatly and a thick modified layer can result, or the product can be dissolved by the solution, resulting in a thin modified layer on a corrosively degraded solid.

This discussion addresses the interphase structure of the unreacted polymer and product polymer in contact with the reactive solution. It should be appreciated, however, that the interphase region changes throughout the reaction from unreacted polymer to modified product, so the partition coefficient of a reagent and the interaction of the solvent with the polymer may be dynamic. One can envision scenarios where a partially modified surface (unreacted polymer-modified polymer copolymer) interacts with the solution to a greater or lesser extent than does either the unreacted polymer or the modified product.

Other effects of the chemistry that occur in the modification reaction affect the structure of the modified surface. If crosslinking occurs, then the product will not dissolve even if the solvent interacts extensively with the modified product; a thick modified layer will result. If chains are cleaved in the reaction, then low molecular weight fragments will dissolve and chain ends will be important surface-chemical features.

Characterization of the products of surface modification is also a major concern. The techniques of contact angle analysis, x-ray photoelectron spectroscopy (XPS), and attenuated total reflectance infrared (ATR IR) spectroscopy are currently the standard tools that are used in analysis. They are complementary techniques in many respects and together can often be used to describe a modification reaction product in detail. Each of these analytical techniques, however, is performed on an isolated (from the reaction solution) product and must be used with the knowledge that what is observed may not be the same as the product that was formed at the polymer/solution interface. Reconstruction of the surface in response to the new environments that the sample experiences during rinsing, drying, and analysis can occur.

A wide variety of chemical modifications of polymer surfaces has been reported. For relatively unreactive polymers these include oxidations of hydrocarbon polymers,[94-99] reductions of fluoropolymers,[100-103] ozonations,[104-106] sulfonations,[107-109] halogenations,[110-115] and insertions of carbenes, nitrenes, and sulfur atoms.[116-119] For polymers containing functionality these include hydrolyses,[120,121] dehydrohalogenations,[122-124] urethanations,[125] amidations,[126,127] epoxidations,[128] ene reactions,[129] and electrophilic aromatic substitution.[130] In general, the reports of these reactions do not contain adequate surface analytical data and do not address the issues discussed earlier. We have chosen to discuss here only several systems in enough detail to indicate that polymer surface modification is a complex undertaking and that the issues just discussed are important.

5.4. Hydroxylation of Polypropylene[131]

The oxidation of polypropylene with CrO_3 in acetic acid/acetic anhydride was carried out using conditions reported to give hydroxylated hydrocarbons.[132-134] Retention of configuration is observed on low molecular weight hydrocarbons, and the relative rates of oxidation[133] at tertiary, secondary, and primary carbons is 3500 : 65 : 1. Based on the reported solution chemistry, the reaction depicted in Eq. (5.1) was expected in high yield.

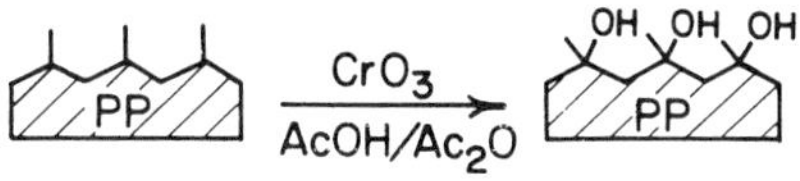

(5.1)

Polypropylene film samples were submerged in a solution of chromium(VI) oxide (1.0 g) in acetic acid (10 mL)/acetic anhydride (10 mL), and the reaction was allowed to proceed at room temperature for 30 min to 16 h. The film samples became slightly yellow on reaction and this is indicated in their UV-visible spectra: a broad absorbance from low wavelength tailing out to 330 nm that increases rapidly over the first four hours of reaction and levels off thereafter. These data are plotted in Fig. 5.1. The XPS spectrum indicates the incorporation of oxygen on reaction (none was present in virgin polypropylene). The reactions that incorporate oxygen exhibit kinetics similar to those observed by UV-visible spectroscopy. Figure 5.2 plots the oxygen : carbon ratios for film samples versus reaction time. The ratios were determined from spectra recorded at takeoff angles of 15 and 75 deg (between the film plane and the detector) that assess the composition of the outer ~ 10 and ~ 40 Å, respectively (~ 95% of the measured photoelectrons originate in these regions[135]). As was observed by UV-visible spectroscopy, the reaction is rapid over the first ~ 4 h and then levels. The take-off angle dependence indicates a faster reaction in the outer 10 Å than in the region between 10 and 40 Å deep. Water contact angle data are consistent: polypropylene exhibits advancing and receding contact angles (θ_A/θ_R) of 111 ± 3 deg/88 ± 3 deg that decrease rapidly over the first 4 h of reaction to θ_A/θ_R = 103 ± 3 deg/49 ± 3 deg. The ATR IR spectra of oxidized polypropylene exhibit, in addition to strong PP bands, weak absorbances at 1742, 1711, 1620, and 1551 cm^{-1}, which are assigned to ester carbonyl, ketone, trisubstituted olefin, and disubstituted olefin, respectively. The infrared spectrum gives no indication of hydroxyl absorbances but other analyses (described later) indicate that they are present. The hydroxyl band is difficult to observe because it is broad and has a relatively low extinction coefficient.

The combination of the UV-visible and XPS data indicates a surface-selective, autoinhibitive reaction. The kinetics suggest that the reactions

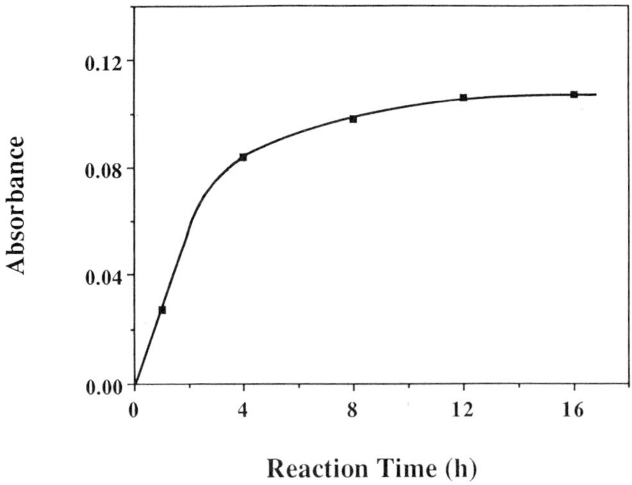

FIG. 5.1. Absorbance (260 nm) of PP samples versus reaction time. Reprinted with permission from Lee, K. W., and McCarthy, T. J., *Macromolecules* **21,** 309, Copyright © 1988 American Chemical Society.

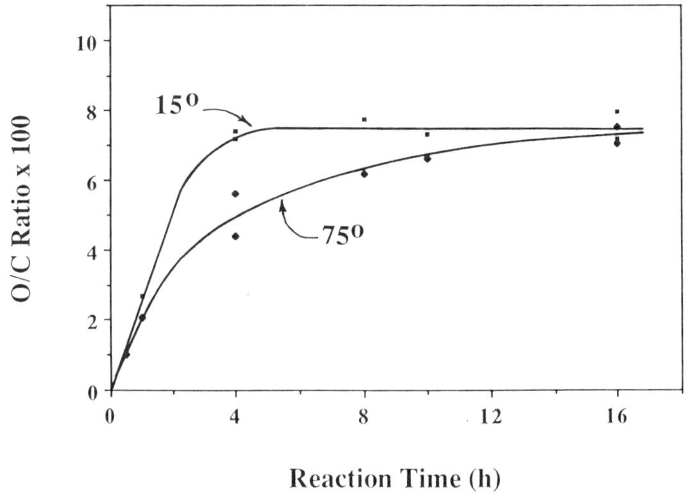

FIG. 5.2. Oxygen : carbon ratio in the XPS sampling depth versus reaction time. The 75° and 15° indicate the angle between the film surface and the analyzer. Reprinted with permission from Lee, K. W., and McCarthy, T. J., *Macromolecules* **21,** 309, Copyright © 1988 American Chemical Society.

selectively occur at sites accessible to the oxidizing solution and then cease. Angle-resolved XPS analysis indicates that reaction in the outer 10 Å is complete before the reaction in the outer 40 Å, and the fact that the increase in the amount of oxidized product in the outer 40 Å levels off at the same point that the increase in the total amount of oxidized product (in the entire film) levels off (UV-visible data) indicates that the thickness of the modified layer is on the order of the XPS sampling depth. The analytical data also indicate that the simple and selective hydroxylation depicted in Eq. (5.1) does not occur. UV-visible and ATR IR spectra indicate the presence of carbonyls and olefins, conjugated to varying extents. It was reported that olefins and carbonyl-containing compounds are produced in prolonged oxidations of hydrocarbons.[133] Acid-catalyzed dehydration [Eq. (5.2)] of tertiary and tertiary allylic alcohols is the likely source of olefins. Carbonyls can form by the oxidation of this unsaturation and esterification of the hydroxyl groups. The carbonyl bands are assigned to a mixture of ketones and esters and specifically not carboxylic acids. Brief treatment with aqueous base did not shift a portion of the peak at 1711 cm^{-1} to a lower frequency as it would have done if carboxylic acids were present. Long exposure to base removes the 1742-cm^{-1} band due to ester hydrolysis. PP-$OCOCH_3$ is formed by the reaction of PP-OH with acetic anhydride. The XPS spectra indicate a maximum O : C ratio of 1 : 12, which corresponds to one oxygen per four polypropylene repeat units. A portion of this oxygen (~ 25%) is present as hydroxyl groups. This is verified by their reaction with heptafluorobutyryl chloride.

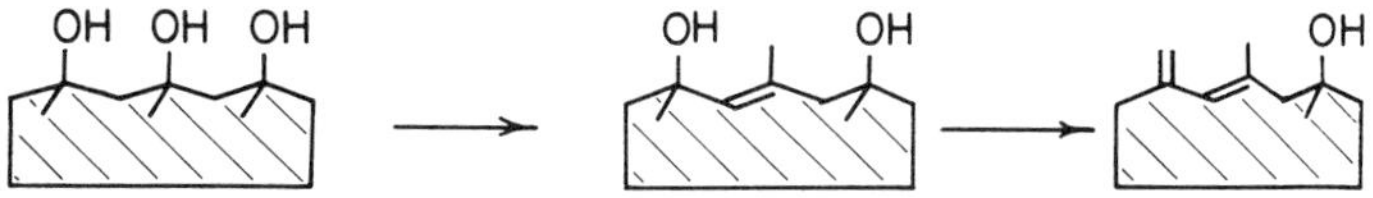

(5.2)

Gravimetric analysis indicates that the reaction is more complex than indicated by the preceding analytical data. Figure 5.3 shows that an increase in mass occurs over the first 4 h of reaction, followed by a decrease in mass thereafter. Two explanations for this mass loss are (1) that the oxidation reaction cleaves chains and low molecular weight fragments are dissolved and (2) that the modified polymer dissolves into the oxidizing solution. The second explanation is more likely because of the absence of carboxylic acids, which would be present if chains were cleaved. Storing the oxidized film in reaction solvent causes the O : C ratio, observed by XPS, to decrease by 30% to 40%; this supports the second explanation. Figure 5.4 shows a graphic representation of the oxidation reaction.

It was possible to increase the surface concentration of hydroxyl groups by hydroboration [Eq. (5.3)]. This reduced the carbonyl functionality and a portion of the olefins as evidenced by ATR IR, UV-visible (the yellow tint

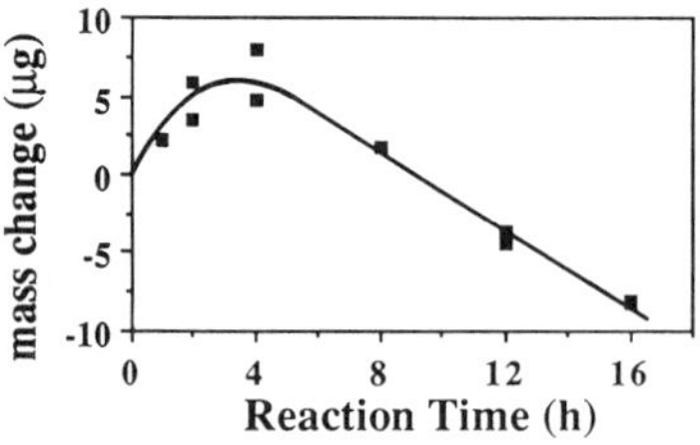

FIG. 5.3. Mass change versus reaction time for polypropylene oxidation Film samples had 6 cm^2 surface area. Reprinted with permission from Lee, K. W. and McCarthy, T. J., *Macromolecules* **21,** 309, Copyright © 1988 American Chemical Society.

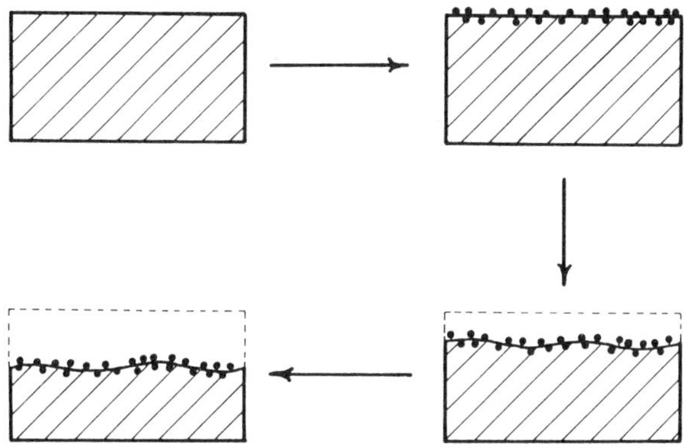

FIG. 5.4. Graphic representation of the oxidation reaction of polypropylene. Reprinted with permission from Lee, K. W., and McCarthy, T. J., *Macromolecules* **21,** 309, Copyright © 1988 American Chemical Society.

disappears), and the yield of the heptafluorobutyryl chloride labelling reaction.

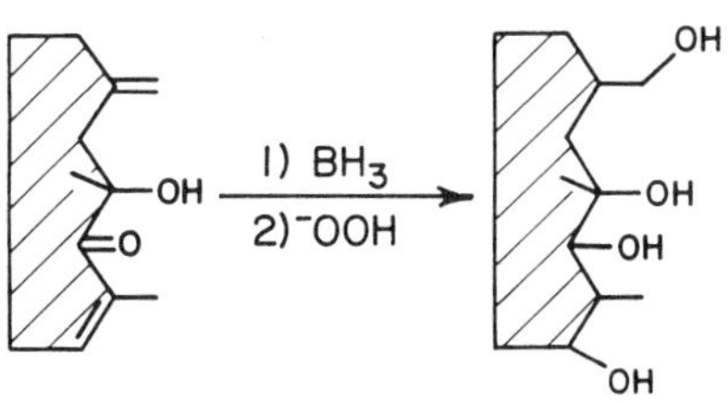

(5.3)

5.5. Functionalization of Poly(chlorotrifluoroethylene)

Much of the surface modification chemistry that has been carried out in our laboratory has been with fluoropolymer substrates and we have developed surface modification techniques for poly(vinylidene fluoride) (PVF_2), poly(tetrafluoroethylene) (PTFE), poly(tetrafluoroethylene-co-hexafluoropropylene) (FEP), and poly(chlorotrifluoroethylene) (PCTFE). We have chosen these materials for study because of their chemical resistant qualities. Our primary reason for polymer surface modification is to study the subsequent chemistry of the functionalized surfaces (this review does not present this work) and we want substrates with reactive surfaces and inert interiors. Further advantages of these fluoropolymers as substrates for the study of surface chemistry include the facts that the surfaces remain clean under normal laboratory conditions (they do not spontaneously oxidize, hydrate, or adsorb impurities, as do metal and inorganic surfaces), the bulk of these materials is mostly transparent in the infrared, UV, and visible regions of the spectrum, thus surface functionality can be observed, and the carbon 1s region of the XPS spectrum is lucid with regard to changes because fluorinated carbon has an especially high binding energy.

Alkyllithium reagents in THF/heptane mixtures react with PCTFE film [Eq. (5.4)] to incorporate the alkyl groups of the lithium reagent on polymer chains near the surface of the film.

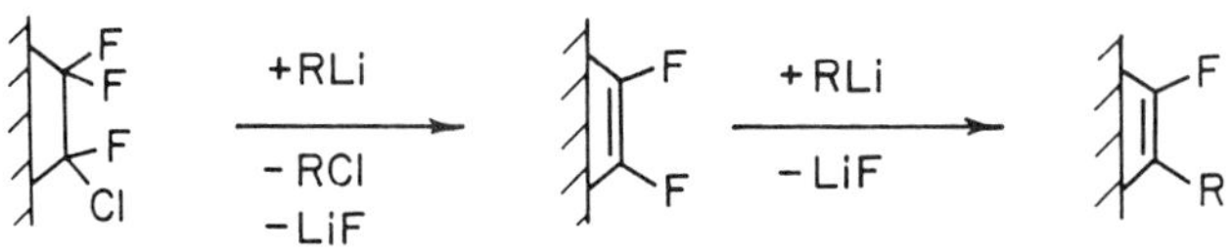

(5.4)

Reprinted with permission from Dias, A. J. and McCarthy, T. J., *Macromolecules* **20,** 2068, Copyright © 1987 American Chemical Society.

The mechanism of this reaction was studied using PCTFE film, oil, and high surface area powder.[136] Metal-halogen exchange forms a chloroalkane and an unstable lithiated polymer that eliminates lithium fluoride to form a difluoroolefin. A second equivalent of alkyllithium adds to the difluoroolefin and lithium fluoride is eliminated to yield the alkyl-substituted olefin. Lithium reagents containing protected functional groups (Scheme 5.1) have been used to introduce carboxylic acids, alcohols, and aldehydes to the PCTFE surface.[137,138] The deprotected alcohol and one of the carboxylic acid surfaces (the one with the 3-methylene spacer) proved to be versatile starting materials for further chemistry.[138–140] Two of the modification reactions are discussed here: the reaction of PCTFE with

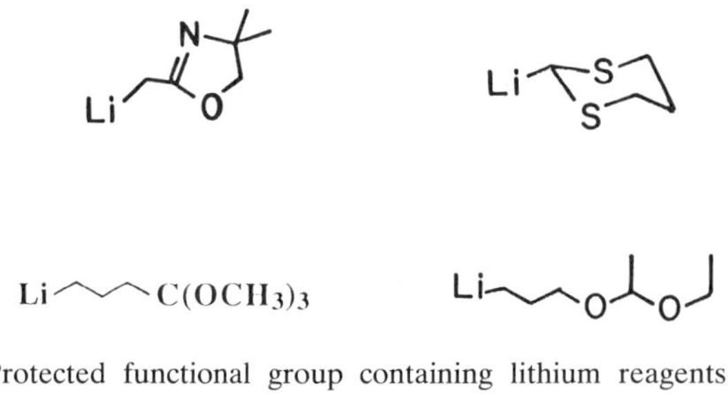

SCHEME 5.1. Protected functional group containing lithium reagents used for PCTFE modification.

2-lithiomethyl-4,4-dimethyl-oxazoline (Li-MDO, a protected carboxylic acid), and acetaldehyde 3-lithiopropyl ethyl acetal (Li-PEAA, a protected alcohol). These two reactions exhibit very different kinetics and product structure.

The reaction of PCTFE film with Li-MDO in THF/heptane was studied at three temperatures with a battery of analytical techniques. The kinetics were determined using UV-visible spectroscopy and the results are shown in Fig. 5.5. The reactions display autoinhibition and are essentially complete after 5 min. These results show that the product film surface does not interact extensively with the solution and that the modified layer inhibits further reaction. The XPS spectrum of PCTFE-MDO exhibits the expected nitrogen and oxygen photoelectron lines and the expected decreased chlorine and fluorine intensities. The C_{1s} regions of the XPS spectra (recorded at take-off angles of 15 and 75 deg) of PCTFE-MDO prepared at −78°C for 60 min are shown in Fig. 5.6. These spectra indicate the composition of the outer ~ 10 and ~ 40 Å of the sample, respectively, and when observed along with the kinetics suggest a very

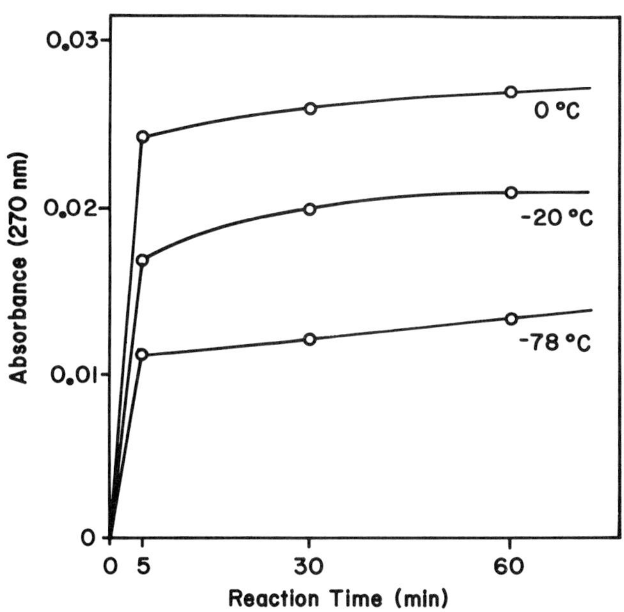

FIG. 5.5. Plots of absorbance (270 mm) versus reaction time for the reactions of PCTFE film with Li-MDO in 50:50 heptane/THF at 0, −20, and −78°C. Reprinted with permission from Dias, A. J., and McCarthy, T. J., *Macromolecules* **20,** 2068, Copyright © 1987 American Chemical Society.

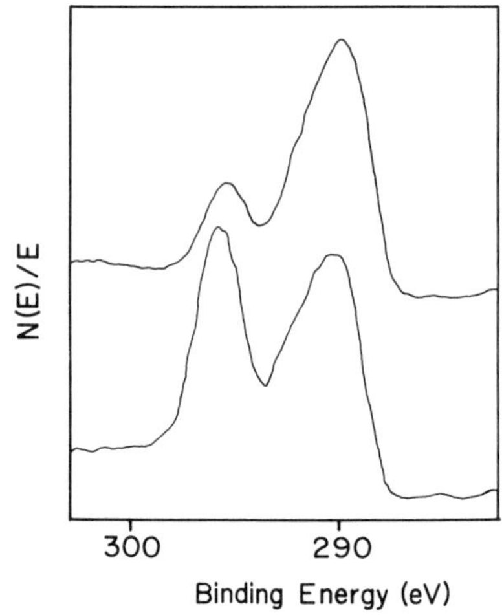

FIG. 5.6. XPS spectra (expanded C_{1s} region) for PCTFE-MDO prepared at −78°C in 50 : 50 heptane/THF (reaction time: 60 min). The take-off angles used to obtain the spectra were 15 deg for the upper spectrum and 75 deg for the lower spectrum. Reprinted with permission from Dias, A. J., and McCarthy, T. J., *Macromolecules* **20,** 2068, Copyright © 1987 American Chemical Society.

surface-selective reaction. The majority of the MDO functionality is in the outer 10 Å. When the reaction is carried out at −20°C for 60 min, the outer 10 Å contains no unreacted PCTFE and a small amount remains in the outer 40 Å. At 0°C, the outer 40 Å is completely reacted in 5 min. The depth of this reaction and the thickness of the modified layer can be controlled with temperature. At higher temperatures the chains are more mobile, probably because of increased solvent interaction and more reaction sites being exposed to reagents. The mobility and the thickness of the modified layer also depend on the composition of the reaction solvent; higher THF content increases chain mobility and causes deeper modifications. THF wets both PCTFE and PCTFE-MDO to a greater extent than does heptane.

The reactions of PCTFE with Li-PEAA are very different than those for Li-MDO. The corresponding kinetics are displayed in Fig. 5.7. The reactions proceed to much greater depths and no autoinhibition occurs. We estimate (using a combination of XPS and UV spectroscopies) that the modified layer thicknesses for PCTFE-PEAA after 60-min reaction time are 50, 1150, and 1600 Å at −78, −20, and 0°C, respectively. This compares with estimated depths of reaction of 20, 35, and 50 Å for PCTFE-MDO prepared under identical conditions. The pronounced dif-

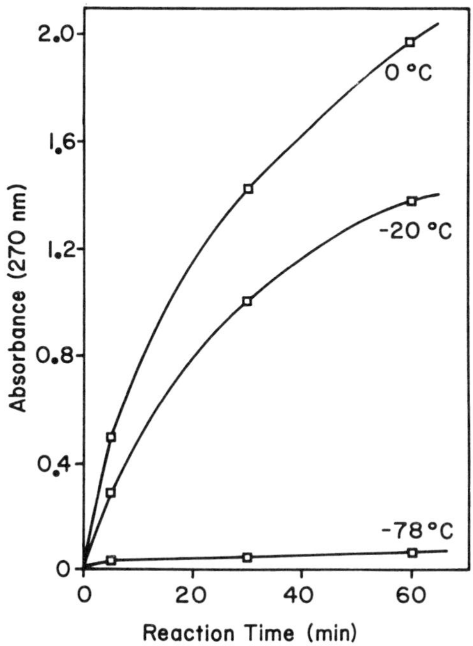

FIG. 5.7. Plots of absorbance (270 nm) versus reaction time for the reaction of PCTFE with Li-PEAA in 50:50 THF/heptane at −78, −20, and 0°C. Reprinted with permission from Dias, A. J., and McCarthy, T. J., *Macromolecules* **20,** 2068, Copyright © 1987 American Chemical Society.

ferences in reaction depth are due to the differences in solution-modified polymer surface interactions. The dangling acetal in PCTFE-PEAA promotes swelling in the film.

5.6. Dehydrofluorination of Poly(vinylidene fluoride)[141–143]

PVF_2 film is conveniently modified by reaction with bases to form unsaturation (fluoroolefins) at the surface and the olefins are versatile reactive handles with which further chemistry can be effected. The reaction with aqueous sodium hydroxide at room temperature provides an interesting example of how interfacial free energy controls the chemistry. When PVF_2 film is immersed in 50% aqueous NaOH at room temperature, no reaction takes place. A sharp interface exists between the solution and the film and hydroxide cannot approach a PVF_2 repeat unit (cross the interface) and effect dehydrofluorination. The surface is not wet by the solution. The addition of tetrabutylammonium ion promotes the elimination reaction. Tetrabutylammonium functions as a phase transfer catalyst, transporting hydroxide from bulk solution to the interface. Tetrabutylammonium ion can also be termed a wetting agent in this situation. The reaction shows pronounced autoinhibition and stops when the accessible polymer repeat units have been dehydrofluorinated. A very thin modified layer (< 10 Å) results. The modified polymer does not interact with the solution and functions as a barrier to further reaction. In contrast, the use of 1,8-diazabicyclo[5.4.0]undec-7-ene (DBU) in heptane as the base produces a deep reaction with significant amounts of unreacted PVF_2 in the outer 40 Å. This reaction proceeds corrosively; the modified polymer interacts with the solution and a diffuse interphase results.

5.7. Indirect Syntheses of Carboxylic Acid-Functionalized Fluoropolymers[144]

The surfaces of hydrocarbon polymers can be conveniently functionalized with carboxylic acids by reaction with strong oxidizing agents. The polymer chains are cleaved, carbon dioxide is produced and the resulting chain ends remain functionalized with carboxylic acids. Whitesides' group has studied in detail[95–99] and reviewed[145] the chemistry and wettability of oxidized polyethylene and its derivatives prepared by this type of reaction. The fluoropolymers PTFE, PVF_2, FEP, and PCTFE are inert to strong oxidants and cannot be functionalized by this method. We have used two-step modification procedures to introduce carboxylic acid groups to these surfaces using the strategy depicted in Eq. (5.5).

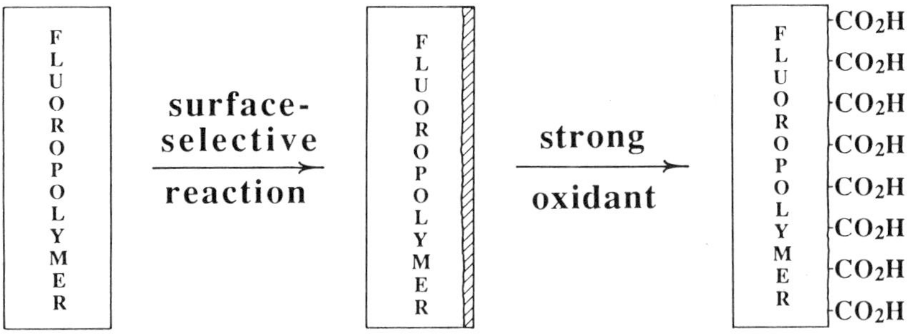

(5.5)

Fluoropolymer functionalization reactions are first carried out to modify a thin layer of fluoropolymer and the modified product is then completely removed by oxidation. At sites where virgin polymer and the modified layer were attached (copolymer junction points), carboxylic acids (carbon in its highest attached oxidation state) are produced as chain termini. They are present on the surface in submonolayer level quantity. Butyllithium-modified PCTFE, sodium naphthalide-reduced FEP,[146] and dehydrofluorinated PVF_2 were modified using this methodology. The PVF_2 modification is described in some detail here.

PVF_2 was allowed to react with aqueous sodium hydroxide and tetrabutylammonium bromide as described earlier (at 40°C). The eliminated layer was then oxidized with potassium chlorate/sulfuric acid to yield the carboxylic acid-functionalized surface (PVF_2-CO_2H). The C_{1s} regions of the XPS and the UV-visible spectra of PVF_2, eliminated PVF_2, and PVF_2-CO_2H are displayed in Fig. 5.8. The UV-visible and C_{1s} XPS spectra of PVF_2-CO_2H are indistinguishable from those of PVF_2. The conjugated fluoroolefin bonds observed in the UV and in the XPS by a decrease in the intensity of the high binding energy CF_2 peak are removed on oxidation. The XPS spectra of PVF_2-CO_2H indicate the presence of an O_{1s} line at

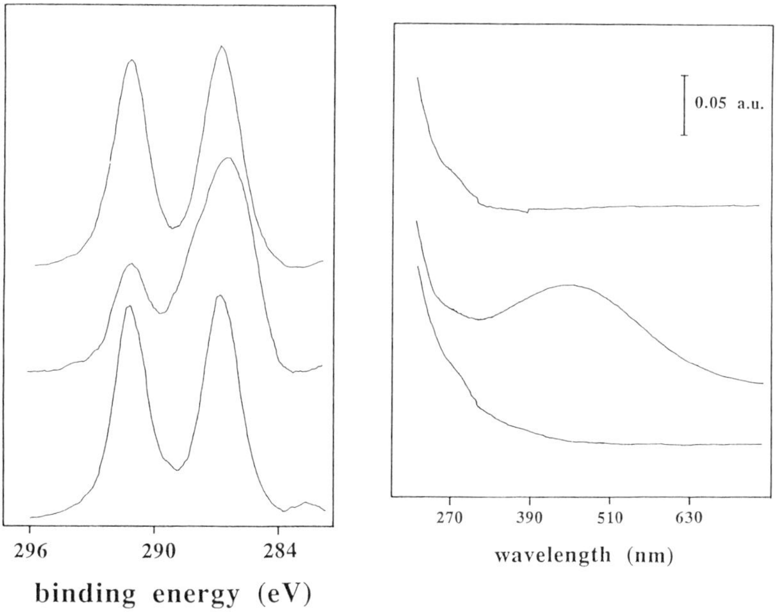

FIG. 5.8. XPS (C_{1s} region, 15 deg take-off angle) and UV-visible spectra of (top) PVF_2, (middle) eliminated PVF_2, and (bottom) PVF_2-CO_2H. Reprinted with permission from Shoichet, M. S., and McCarthy, T. J., *Macromolecules* **24,** 982, Copyright © 1991 American Chemical Society.

536 eV with an intensity that corresponds to a structure of $(CH_2CF_2)_{13}CO_2H$ (one carboxylic acid per 13 repeat groups). The carboxylic acids were identified by labeling with thallium ethoxide, measuring the pH dependence of the water contact angles (θ_A/θ_R = 77 deg/39 deg at low pH and θ_A/θ_R = 68 deg/25 deg at high pH), and subsequent chemical transformations.

5.8. Gas Phase Chlorination of Polyethylene[147]

The gas phase chlorination of polyethylene has been studied extensively, but not with regard to the issues discussed in the introductory sections. This reaction is obviously a different type than the others discussed here; no solvent is used. The chlorinated product is reported to be less permeable to gases than is polyethylene so we sought to control conditions so that chlorination would be more rapid than diffusion of chlorine into polyethylene or the reverse. The objective of our work was to demonstrate, as the principles described previously indicate, that simultaneous and independent control of the depth of chlorination and the chain density of chlorination could be effected. We prepared four different types of chlorinated polyethylene to demonstrate this control: a shallow layer of lightly chlorinated chains, a shallow layer of highly chlorinated chains, a deep layer of lightly chlorinated chains, and a deep layer of highly chlorinated chains. The diffusion rate was controlled by the partial pressure of chlorine, and the chlorination rate was controlled by the photointensity. It was necessary to control the reaction time to prepare the four target materials.

A shallow layer of highly chlorinated chains was prepared with a 5-min reaction time at 5 mm Cl_2 using UV irradiation. The composition of the outer 10 Å of this surface (determined using XPS) was $C_{100}Cl_{107}$. Take-off angle dependence was observed and the outer 40 Å (75-deg take-off angle) had a composition of $C_{100}Cl_{75}$, indicating the surface selectivity of the reaction. The sample gained 2.2 $\mu g/cm^2$ mass on reaction, corresponding to a depth of reaction of ~ 100 Å. A shallow layer of lightly chlorinated chains was prepared with a 30-min reaction time with 2 mm Cl_2 and ambient light. XPS indicated a composition of $C_{100}Cl_{10}$ (15-deg take-off angle) and $C_{100}Cl_7$ (75-deg take-off angle). No mass gain was detected, which indicates a very shallow reaction depth. A deep layer of lightly chlorinated chains was prepared with a 22-h reaction time with 1 atm Cl_2 and near total darkness. XPS indicated a take-off angle independent composition of $C_{100}Cl_9$ and gravimetric analysis indicated a modified layer thickness of greater than 10 μm. A deep layer of highly chlorinated chains was prepared with a 22-h reaction time with 1 atm Cl_2 and ambient light.

XPS indicated a take-off angle independent composition of $C_{100}Cl_{107}$ and gravimetric analysis a mass gain that corresponded to a reaction depth of $\sim 3.5\ \mu m$.

5.9. Carbonyl Surface Chemistry of Poly(ether ether ketone)[148]

The surface chemistry of poly(ether ether ketone) (PEEK) differs from that of the other polymers discussed in this chapter in that the virgin polymer contains a potentially reactive functional group (the diaryl ketone), which is present in known concentration. PEEK is a thermoplastic with excellent mechanical properties and for a reactive polymer it has unusually good resistance to solvents and chemicals. Learning the extent to which this ketone reacts as a functional group at PEEK/solvent interfaces was the objective of our study.

Both semicrystalline and amorphous PEEK film samples (ICI) were studied initially. Amorphous PEEK was eliminated as a substrate because it was found that a range of solvents (benzene, toluene, dichloromethane, or pyridine) induces crystallization and plasticization, causing the films to become opaque and roll up into scrolls within 30 min at room temperature. Semicrystalline film is considerably more solvent resistant, exhibiting no visible changes after 24 h in refluxing THF, methanol, ethanol, benzene, toluene, water, hexane, or heptane.

A series of carbonyl-selective reactions was carried out on PEEK using modifications of literature conditions for high-yield reactions of benzophenone. PEEK reacts with 2,4-dinitrophenyl-hydrazine to yield

$$\text{PEEK} \sim \text{C}{=}\text{N–NH–}\phi(\text{NO}_2)_2$$

and with hydroxylamine to yield PEEK ~ C=N–OH. Reaction with methylenetriphenylphosphorane yields $\text{PEEK} \sim \text{C} = \text{CH}_2$. Dimsylsodium adds to the PEEK carbonyl rendering $\text{PEEK} \sim \text{C(OH)CH}_2\text{S(O)CH}_3$, and sodium bis(2-methoxyethoxy)-aluminum hydride reduces the PEEK carbonyl to PEEK ~ CH–OH. Cyanosilylation of PEEK with trimethylsilyl cyanide yields $\text{PEEK} \sim \text{C(CN)OSiMe}_3$. Reactions were monitored and products were characterized by contact angle, XPS, and ATR IR. Modification depths were uniformly ~ 100 Å. The yields, based on the ketone, were assessed by quantitative XPS analysis of the outer 40 Å and were about 50% for all of the reactions except the reduction with sodium bis(2-methoxyethoxy)aluminum hydride. This reaction proceeded in 75% to 90% yield. We believe that the reaction yields are limited by the crystallinity of the film; only the amorphous regions react in most cases. We propose that the reaction yield for the reduction is high due to disruption of the crystalline phase.

5.10. Conclusion

We conclude by restating that polymer surface modification is a complex field. The gross differences in the examples shown here emphasize that model systems will not be available that can be drawn on to predict reactivity. This field must be considered to be still in its infancy. Key questions remain to be addressed and answered: What properties can be controlled by surface modification and what is the extent of control? Under what conditions do surface functional groups in contact with reagents in solution react in the same manner as functional groups in solution and when do interfacial conditions (surface energy, solvation, wetting, swelling) and geometric constraints alter their reactivity? What is the effect of the location (both distance from the surface and pattern on the surface) of functional groups on reactivity and properties?

Acknowledgments

We thank the Office of Naval Research and the University of Massachusetts Materials Research Laboratory for financial support and our coworkers for help and guidance.

References

1. Boegin, H. V., Ed. (1982). Plasma Science and Technology, Cornell University Press, Ithaca, NY.
2. Clark, D. T., Dilks, A., and Shuttleworth, D. (1978). In: *Polymer Surfaces* (D. T. Clark, and W. J. Feast, eds.). Wiley, New York.
3. Yasuda, H. (1976). *J. Macromol. Sci. Chem.* **A10,** 383.
4. Anand, M., Cohen, R. E., and Baddour, R. F. (1982). In: *Photon, Electron and Ion Probe of Polymer Structure and Properties* (D. W. Dwight, T. J. Fabish, and H. R. Thomas, eds.). American Chemical Society, Washington DC.
5. Hansen, R. H., and Schonhorn, H. (1966). *J. Polym. Sci.*, **B4,** 203.
6. Schonhorn, H., and Hansen, R. H. (1967). *J. Appl. Polym. Sci.* **11,** 1461.
7. Westerdahl, C. A. L., Hall, J. R., Schramm, E. C., and Levi, D. W., (1974). *J. Coll. Interface Sci.* **47,** 610.
8. Moshonov, A., and Avny, Y. (1980). *J. Appl. Polym. Sci.* **25,** 89.
9. Nuzzo, R. G., and Smolinsky, G. (1984). *Macromolecules* **17,** 1013.

10. Anand, M., Cohen, R. E., and Baddour, R. F. (1981). *Polymer* **22,** 361.
11. Corbin, G. A., Cohen, R. E., and Baddour, R. F. (1982). *Polymer* **23,** 1546.
12. Loh, I. H., Cohen, R. E., and Baddour, R. F. (1986). *J. Appl. Polym. Sci.* **31,** 901.
13. Corbin, G. A., Cohen, R. E., and Baddour, R. F. (1985). *Macromolecules* **18,** 98.
14. Stradal, M., and Goring, D. A. I. (1975). *Can. J. Chem. Eng.* **53,** 427.
15. Owens, D. K. (1975). *J. Appl. Polym. Sci.* **19,** 265.
16. Blythe, A. R., Briggs, D., Kendall, C. R., Rance, D. G., and Zichy, V. J. I. (1978). *Polymer* **19,** 1273.
17. Briggs, D., and Kendall, C. R. (1979). *Polymer* **20,** 1053.
18. Yamakawa, S., Yamamoto, F., and Kato, Y. (1976). *Macromolecules* **9,** 754.
19. Clark, D. T., and Dilks, A. (1977). *J. Polym. Sci. Polym. Chem. Ed.* **15,** 2321.
20. Clark, D. T., and Dilks, A. (1978). *J. Polym. Sci. Polym. Chem. Ed.* **16,** 911.
21. Wertheimer, M. R., and Schreiber, H. P. (1981). *J. Appl. Polym. Sci.* **26,** 2087.
22. Bosisio, R. G., Tang, T. W. C., and Lamontagne, P. B. (1980). *J. Appl. Polym. Sci.* **45,** 711.
23. Everhart, D. S., and Reilley, C. N. (1981). *Surf. Interface Anal.* **3,** 258.
24. Bosisio, R. G., Tang, T. W. C., and Lamontagne, P. B. (1980). *J. Appl. Polym. Sci.* **45,** 711.
25. Clark, D. T., and Dilks, A. (1979). *J. Polym. Sci. Polym. Chem. Ed.* **17,** 957.
26. Clark, D. T., and Wilson, R. (1983). *J. Polym. Sci. Polym. Chem. Ed.* **21,** 837.
27. Inagaki, N., and Yasuda, H. (1981). *J. Appl. Polym. Sci.* **26,** 3333.
28. Yamakawa, S. (1979). *Macromolecules* **12,** 1222.
29. Feng, X. D., Sun, Y. H., and Qin, K. Y. (1985). *Macromolecules* **18,** 2105.
30. Yamakawa, S., Yamamoto, F., and Kato, Y. (1976). *Macromolecules* **9,** 754.
31. Hegazy, E.-S. A. (1984). *J. Polym. Sci. Polym. Chem. Ed.* **22,** 493.
32. Barkalov, I. M., Kiryukhin, D. P., and Muydinov, M. R. (1981). *J. Polym. Sci. Polym. Chem. Ed.* **19,** 2331.
33. Rao, K. N., Rao, M. H., Moorthy, P. N., and Charlesby, A. (1972). *J. Polym. Sci. Polym. Lett. Ed.* **10,** 893.
34. Yamakita, H, and Hayakawa, K. (1972). *J. Polym. Sci. Polym. Lett. Ed.* **10,** 823.
35. Pugh, C., and Hsieh, Y.-L. (1984). *J. Appl. Polym. Sci.* **29,** 3641.
36. Morgan, P. W., and Corelli, J. C. (1983). *J. Appl. Polym. Sci.* **28,** 1879.
37. Decker, C. (1983). *J. Appl. Polym. Sci.* **28,** 97.
38. Misra, B. N., Sood, D. S., and Mehta, I. K. (1985). *J. Polym. Sci. Polym. Chem. Ed.* **23,** 1749.
39. Cohn, D., Hoffman, A. S., and Ratner, B. D. (1987). *J. Appl. Polym. Sci.* **33,** 1.
40. Suzuki, M., Kishida, A., Iwata, H., and Ikada, Y. (1986). *Macromolecules* **19,** 1804.
41. Bellobono, I. R., Selli, E., and Marcandalli, B. (1986). *J. Appl. Polym. Sci.* **32,** 4323.
42. Lai, J. Y., Chang, T. C., Wu, Z. J., and Hsieh, T. S. (1986). *J. Appl. Polym. Sci.* **32,** 4709.
43. Rao, M. H., Rao, K. N., Lokhande, H. T., and Teli, M. D. (1987). *J. Appl. Polym. Sci.* **33,** 2707.
44. Hamil, H. F., Adams, L. M., Harlowe, W. W., Jr., and Martin, E. C. (1971). *J. Polym. Sci. A1* **9,** 363.
45. Regen, S. L., Kirszensztejn, P., and Singh, A. (1983). *Macromolecules* **16,** 335.
46. Albrecht, O., Laschewsky, A., and Ringsdorf, H. (1984). *Macromolecules* **17,** 937.
47. Albrecht, O., and Laschewsky, A. (1984). *Macromolecules* **17,** 1292.
48. Foltynowicz, Z., Yamaguchi, K., Czajka, B., and Regen, S. L. (1985). *Macromolecules* **18,** 1394.
49. Bergbreiter, D. E., Chen, Z., and Hu, H. (1984). *Macromolecules* **17,** 2111.
50. Galenbeck, F. (1970). *J. Polym. Sci. Polym. Lett. Ed.* **15,** 107.

51. Roberts, R. F., and Schonhorn, H. (1975). *Polym. Prepr.* **16**(2), 146.
52. Bodö, P., and Sundgren, J. E. (1986). *Surf. Interface Anal.* **9,** 437.
53. Kovac, C. A., Jordan, J. L., and Pollak, R. A. (1986). *Mater. Res. Soc. Symp. Proc.* **72,** 247.
54. Vogel, S. L., and Schonhorn, H. (1979). *J. Appl. Polym. Sci.* **23,** 495.
55. Francis, C. G., Lipera, S., Morand, P. D., Morton, P. R., Nash, J., and Radford, P. P. (1986). *Polym. Prepr. (Am. Chem. Soc. Div. Polym. Chem.)* **27**(2), 78.
56. Roberts, R. F., Ryan, F. W., Schonhorn, H., Sessler, G. M., and West, J. E. (1976). *J. Appl. Polym. Sci.* **20,** 225.
57. Rabalais, J. W. (1980). *Polym. Prepr. (Am. Chem. Soc. Div. Polym. Chem.)* **21**(1), 142.
58. Ishihara, K., Okazaki, A., Negishi, N., Shinohara, I., Kataoka, K., Okano, T., and Sakurai, Y. (1982). *J. Appl. Polym. Sci.* **27,** 239.
59. Briggs, D., Brewis, B. M., and Konieczko, M. B. (1979). *J. Mater. Sci.* **14,** 1344.
60. Yasuda, H., Charlson, E. J., Charlson, E. M., Yasuda, T., Miyama, M., and Okuna, T. (1991). *Langmuir* **7,** 2394.
61. Andrade, J. D., Gregonis, D. E., and Smith, L. M. (1986). In: *Surface and Interfacial Aspects of Biomedical Polymers* (J. D. Andrade, ed.). Plenum, New York, Vol. 1, Chapters 2 and 7.
62. Carre, A., and Schreiber, H. P. (1982). *J. Coat. Tech.* **54,** 31.
63. Good, R. J., Kvikstad, J. A., and Bailey, W. O. (1971). *J. Coll. Interface Sci.* **35,** 314.
64. Baier, R. E., and Zisman, W. A. (1970). *Macromolecules* **3,** 70.
65. Baier, R. E., and Zisman, W. A. (1970). *Macromolecules* **3,** 462.
66. Good, R. J., and Kotsidas, E. D. (1978). *J. Coll. Interface Sci.* **66,** 360.
67. Schonhorn, H. (1968). *Macromolecules* **1,** 145.
68. Schonhorn, H., and Ryan, F. W. (1966). *J. Phys. Chem.* **70,** 3811.
69. Wang, L.-H., and Porter, R. S. (1983). *J. Appl. Polym. Sci.* **28,** 1439.
70. Reardon, J. P., and Zisman, W. A. (1974). *Macromolecules* **7,** 920.
71. O'Malley, J. J., and Thomas, H. R. (1979). *Macromolecules* **12,** 323.
72. Thomas, H. R., O'Malley, J. J., and Lee, G. M. (1979). *Macromolecules* **12,** 996.
73. Gagnon, D. R., and McCarthy, T. J. (1984). *J. Appl. Polym. Sci.* **29,** 4335.
74. Kumler, P. L., Matteson, H. L., and Gardella, J. A., Jr. (1991). *Langmuir* **7,** 2479.
75. Phillips, R. W., and Dettre, R. H. (1976). *J. Coll. Interface Sci.* **56,** 251.
76. Sohn, J., and Ree, T. (1984). *J. Appl. Polym. Sci.* **29,** 4237.
77. Schmitt, R. L., Gardella, J. A., Jr., Magill, J. H., Salvatti, L., Jr., and Chin, R. L. (1985). *Macromolecules* **18,** 2675.
78. Holly, F. J., and Refojo, M. F. (1975). *J. Biomed. Mat. Res.* **9,** 315.
79. Lavielle, L., and Schultz, J. (1985). *J. Coll. Interface Sci.* **106,** 438.
80. Hogt, A. H., Gregonis, D. E., Andrade, J. D., Kim, S. W., Dankert, J., and Feijen, J. (1985). *J. Coll. Interface Sci.* **106,** 289.
81. Andrade, J. D., Ma, S. M., King, R. N., and Gregonis, D. E. (1979). *J. Coll. Interface Sci.* **72,** 488.
82. Pennings, J. F. M., and Bosman, B. (1979). *Coll. Polym. Sci.* **257,** 720.
83. Pekala, R. W., and Merrill, E. W. (1984). *J. Coll. Interface Sci.* **101,** 120.
84. Ruckenstein, E., and Gowisankar, S. V. (1985). *J. Coll. Interface Sci.* **107,** 488.
85. Ruckenstein, E., and Gowisankar, S. V. (1986). *J. Coll. Interface Sci.* **109,** 557.
86. Pennings, J. F. M. (1978). *Coll. Polym. Sci.* **256,** 1155.
87. Schwarcz, A. (1974). *J. Appl. Polym. Sci. Polym. Phys. Ed.* **12,** 1195.
88. Le Grand, D. G., and Garnes, G. L., Jr. (1970). *Polym. Prepr. (Am. Chem. Soc. Div. Polym. Chem.)* **11**(2), 442.
89. Owen, M. J. (1980). *Ind. Eng. Chem. Prod. Res. Dev.* **19,** 97.

90. Tse, J., and Adamson, A. W. (1979). *J. Coll. Interface Sci.* **72,** 515.
91. Yoon, S. C., and Ratner, B. D. (1986). *Macromolecules* **19,** 1068.
92. Widmaier, J. M., and Meyer, G. C. (1983). *J. Appl. Polym. Sci.* **28,** 1429.
93. Takahara, A., Tashita, J., Kajiyama, T., Takayanagi, M., and MacKnight, W. J. (1985). *Polymer* **26,** 987.
94. Baszkin, A., Deyme, M., Nishino, L., and Ter-Minassian-Saraga, L. (1979). *Prog. Colloid Polym. Sci.* **61,** 97.
95. Rasmussen, J. R., Stedronsky, E. R., and Whitesides, G. M. (1977). *J. Am. Chem. Soc.* **99,** 4736.
96. Rasmussen, J. R., Bergbreiter, D. E., and Whitesides, G. M. (1977). *J. Am. Chem. Soc.* **99,** 4746.
97. Holmes-Farley, S. R., Reamey, R. H., McCarthy, T. J., Deutch, J., and Whitesides, G. M. (1985). *Langmuir* **1,** 725.
98. Holmes-Farley, S. R., and Whitesides, G. M. (1986). *Langmuir* **2,** 266.
99. Holmes-Farley, S. R., Bain, C. D., and Whitesides, G. M. (1988). *Langmuir* **4,** 921.
100. Dwight, D. W., and Riggs, W. M. (1974). *J. Coll. Interface Sci.* **47,** 650.
101. Barker, D. J., Brewis, D. M., Dahm, R. H., and Hoy, L. R. (1978). *Polymer* **19,** 856.
102. Jansta, J., and Dousek, F. P. (1981). *Electrochim. Acta* **26,** 233.
103. Costello, C. A., and McCarthy, T. J. (1987). *Macromolecules* **20,** 2819.
104. Clark, D. T., and Munro, H. S. (1984). *Polymer* **25,** 826.
105. Peeling, J., and Clark, D. T. (1983). *J. Polym. Sci. Polym. Chem. Ed.* **21,** 2047.
106. Andries, J. C., and Diem, H. E. (1974). *J. Polym. Sci. Polym. Lett. Ed.* **12,** 281.
107. Bergbreiter, D. E., and Kabza, K. (1991). *J. Am. Chem. Soc.* **113,** 1447.
108. Masuda, T., and Litt, M. H. (1974). *J. Polym. Sci. Polym. Chem. Ed.* **12,** 489.
109. Gibson, H. W., and Bailey, F. C. (1980). *Macromolecules* **13,** 34.
110. Luttinger, J., and Cooper, C. (1968). *J. Polym. Sci. C.* **24,** 257.
111. Rouchon, J.-P., Sage, D., Romand, M., Bador, R., and Berticat, P. (1977). *Die Angew. Makro. Chem.* **65,** 223.
112. Sage, D., Berticat, P., and Vallet, G. (1976). *Die Angew. Makro. Chem.* **54,** 151.
113. Nakagawa, T., and Yamada S. (1972). *J. Appl. Polym. Soc.* **16,** 1997.
114. Elman, J. F., Gerenser, L. J., Goppert-Bararducci, K. E., and Pochan, J. M. (1990). *Macromolecules* **23,** 3922.
115. Cross, E. M., and McCarthy, T. J. (1992). *Macromolecules* **25,** 2603.
116. Osteraas, A. J., and Olsen, D. A. (1969). *J. Appl. Polym. Sci.* **13,** 1537.
117. Olsen, D. A, and Osteraas, A. J. (1969). *J. Appl. Polym. Sci.* **13,** 1523.
118. Osteraas, A. J., and Olsen, D. A. (1969). *Nature* **221,** 1140.
119. Olsen, D., and Osteraas, A. J. (1969). *J. Polym. Sci. Part A-1* **7,** 1913.
120. Erbil, H. Y. (1987). *J. Appl. Polym. Sci.* **33,** 1397.
121. Chatzi, E. G., Tidvick, S. L., and Koenig, J. L. (1987). *Polym. Prepr. (Am. Chem. Soc. Div. Polym. Chem.)* **28**(1), 13.
122. Dias, A. J., and McCarthy, T. J. (1984). *Macromolecules* **17,** 2529.
123. Kuhn, K. J., Hahn, B., Percec, V., and Urban, M. W. (1987). *Appl. Spectroscopy* **41,** 843.
124. Urban, M. W., and Salazar-Rojas, E. (1988). *Macromolecules* **21,** 372.
125. Varma, D. S., and Nedunyadi, C. (1976). *J. Appl. Polym. Sci.* **20,** 681.
126. Avny, Y., and Rebenfeld, L. (1986). *J. Appl. Polym. Sci.* **32,** 4009.
127. Kim, K. J., and Ko, S. W. (1986). *J. Appl. Polym. Sci.* **32,** 6017.
128. Schilling, F. C., Bovey, F. A., Tseng, S., and Woodward, A. E. (1983). *Macromolecules* **16,** 808.
129. Schectman, L. A. (1986). *Polym. Prepr. (Am. Chem. Soc. Div. Polym. Chem.)* **27**(2), 48.
130. Trott, G. F. (1974). *J. Appl. Polym. Sci.* **18,** 1411.

131. Lee, K.-W., and McCarthy, T. J. (1988). *Macromolecules* **21,** 309.
132. Shilov, A. E. (1984). *Activation of Saturated Hydrocarbons by Transition Metal Complexes*. D. Ridel Publishing, Holland.
133. Wiberg, K. B., and Foster, G. (1961). *J. Am. Chem. Soc.* **83,** 423.
134. Rocek, J. (1962). *Tetrahedron Lett.* 135.
135. These values assume a photoelectron mean free path of 14 Å. This was measured for C_{1s} photoelectrons in poly(*p*-xylylene): Clark, D. T., and Thomas, H. R. (1977). *J. Polym. Sci. Polym. Chem. Ed.* **15,** 2843.
136. Dias, A. J., and McCarthy, T. J. (1985). *Macromolecules* **18,** 1826.
137. Dias, A. J., and McCarthy, T. J. (1987). *Macromolecules* **20,** 2068.
138. Lee, K.-W., and McCarthy, T. J. (1988). *Macromolecules* **21,** 2318.
139. Lee, K.-W., and McCarthy, T. J. (1987). *Macromolecules* **21,** 3353.
140. Bee, T. G., and McCarthy, T. J. (1992). *Macromolecules* **25,** 2093.
141. Dias, A. J., and McCarthy, T. J. (1984). *Macromolecules* **17,** 2529.
142. Brennan, J. V., and McCarthy, T. J. (1987). *Polym. Prepr. (Am. Chem. Soc., Div. Polym. Chem.)* **29**(2), 338.
143. Brennan, J. V., and McCarthy, T. J. (1989). *Polym. Prepr. (Am. Chem. Soc., Div. Polym. Chem.)* **30**(2), 152.
144. Shoichet, M. S., and McCarthy, T. J. (1991). *Macromolecules* **24,** 982.
145. Whitesides, G. M., and Ferguson, G. S. (1988). *Chemtracts: Org. Chem.* **1,** 171.
146. Bening, R. C., and McCarthy, T. J. (1990). *Macromolecules* **23,** 2648.
147. Cross, E. M., and McCarthy, T. J. (1992). *Macromolecules* **25,** 2603.
148. Franchina, N. L., and McCarthy, T. J. (1991). *Macromolecules* **24,** 3045.

Lithographically Patterned Self-Assembled Films

JEFFREY M. CALVERT

Naval Research Laboratory
Center for Bio/Molecular Science & Engineering (Code 6900)
Washington, DC 20375-5348

6.1. Introduction

The objective of this chapter is to review recent (through early 1993) research conducted at the Naval Research Laboratory (NRL) on patterned self-assembled films. The NRL has a rich history of work in the general area of ultrathin films on surfaces that derives from the pioneering studies of Zisman on wetting and adhesion at interfaces.[1] In the Center for Bio/Molecular Science and Engineering at NRL the theme of molecular self-assembly underlies, or is a component of, almost all of the ongoing research programs.

Our work on patterned thin films was an offshoot of a research program on polymerizable diacetylenic phospholipid microstructures. The discovery in early 1986 of electroless (EL) deposition as a method for metallizing the lipid structures[2,3] led to the lithographically attractive concept of selectively metallizing patterned planar lipid films deposited by the Langmuir-Blodgett (LB) technique. Later that year, initial experiments to metallize UV patterned LB films failed due to desorption of the film from the substrate during EL plating. Chemisorbed organosilane films, originally intended only as adhesion promoters for the LB film, were soon found to be effective by themselves as ultrathin imaging layers to produce metal patterned surfaces, especially for lithographic applications.[4,5]

Since the initial demonstration of selective metal buildup in mid-1986, our interest in patterned chemisorbed organosilane films has spanned a wide variety of research areas and applications, including high-resolution microlithographic processes, patterned wetting, controlled adhesion of metals, proteins, and cells, and fabrication of coplanar molecular assemblies.[4–28]

ISBN 0-12-523485-6

In our work with self-assembled (SA) films,[29,30] we have focused on the use of chemisorbed organosilane films. These are a class of materials of the general form R_nSiX_{4-n}, where $n = 1, 2$, or 3 and X = Cl, OCH_3, or OC_2H_5. Organosilanes react with surface hydroxyl groups on a wide variety of substrates to produce covalent Si-O-substrate (siloxane) linkages.[31] The ability to functionalize a great diversity of substrates through siloxane bond formation also enables a multitude of technological applications that are not readily accessible with other SA material systems such as organothiol adsorption on gold surfaces. The covalent bonding of organosilane SA films to the surface, compared to coordinatively bound systems such as the thiols, leads to improved thermal and chemical stability for the organosilane films.

At the NRL, we have emphasized *lithographic* techniques for directly patterning SA films, although a number of other methods have been reported. These include mechanical scraping ("machining") to remove an SA film from the substrate,[32] or squirting a stream of organothiol SA film precursor "ink" from the aperture of a needle onto a gold substrate to produce features.[32] However, exposure tools such as a UV lamp-based contact aligner or an excimer laser projection system provide the best means of reproducibly fabricating high-resolution features of controlled linewidth with arbitrary geometries over large areas. Also, the use of such industrially accepted tools is the only way to evaluate materials and processes under the conditions that will ultimately be required for practical applications. Furthermore, as will be discussed in Section 6.2, ultrathin films are well suited as imaging layers for high-resolution patterning and lithography. For areas such as microelectronics, it is therefore essential to be able to adapt the surface modification and patterning chemistry to relevant substrates *as required by the particular application*, and not be restricted by the surface attachment chemistry to selected substrates.

6.2. Lithographic Considerations for Ultrathin Films as Imaging and Resist Layers

Our initial interest in patterned monolayer films was based on the use of SA films as imaging layers for high-resolution optical lithography.[5–10] Lithography is a critical step in the fabrication of microelectronic devices such as integrated circuits (ICs). A lithographic process can be thought of as combining the elements of photography (image formation *in the film*) and sculpture (feature definition *in the substrate*). New approaches for optical lithography must therefore meet the necessary criteria for *both* aspects of the overall process.

Some of the most important factors that must be considered for optical imaging include the resolution of the imaging layer, compatibility of the imaging film with the depth of focus of the exposure system, and sufficient transparency of the imaging layer to the exposing radiation. The Rayleigh criterion for the maximum resolution (minimum feature width, R) of a projected image is directly proportional to the exposure wavelength λ and a process and material-specific k factor, whereas it is inversely proportional to the numerical aperture (NA) of the imaging optics:

$$R = k\lambda/\mathrm{NA} \tag{6.1}$$

The continuing drive for producing features of higher resolution has therefore resulted in exposure tools evolving to shorter wavelength sources, i.e., from *g*-line (436 nm) and *i*-line (365 nm) to deep UV ($\leq$ 254 nm).

A second important imaging parameter, the depth of focus (DOF), is directly proportional to wavelength and inversely proportional to $(\mathrm{NA})^2$.

$$\mathrm{DOF} = \pm k\lambda/(\mathrm{NA})^2 \tag{6.2}$$

The effects of using shorter wavelengths on resolution and DOF are illustrated by the following example. Using the Rayleigh criterion, a photoresist process with a k factor of 0.6, exposed to 248-nm radiation from an imaging system with a lens having an NA of 0.5, yields a resolution of 0.30 μm; the DOF for this process would be ± 0.6 μm. If the same process could operate at 193 nm, values for R and DOF would be 0.23 μm and ± 0.46 μm, respectively. Thus, although decreasing the exposure wavelength and increasing the NA of the imaging system lenses results in improved resolution, this combination also leads to DOF values that become too small to produce a focused image throughout a standard 1.0- to 1.5-μm-thick photoresist.

Another imaging parameter that must be considered is photoresist transparency. To expose a typical photoresist, it is necessary for the radiation to penetrate the layer completely. However, conventional micron-thick polymer-based photoresists that are moderately transmissive at near-UV wavelengths such as 365 nm are essentially opaque in the deep UV due to the intense adsorptions of aromatic chromophores in the photoresist formulations.

One solution to the various limitations to high-resolution optical imaging discussed earlier is to replace the thick photoresist layer with an ultrathin imaging film, thereby eliminating DOF restrictions, resist opacity problems, and other deleterious effects such as standing waves.

After creation of an image, the patterned layer must then be able to withstand the harsh reactive ion etch plasmas that are used to transfer the pattern into an underlying substrate such as silicon. Organic materials and Si etch at comparable rates under typical plasma etching conditions. To remove ~ 0.5 μm of Si, as required for definition of microelectronic devices such as MOSFET transistors (see Section 6.5), the etch resistant layer must be on the order of 1 μm thick. Also, polymeric photoresists are usually formed by spin-coating techniques, and it is necessary to produce micron-thick layers to minimize pinhole densities to acceptable levels. Therefore, the use of conventional photoresists in which an organic polymer serves as *both* the imaging and etch resistant layer, leads to a trade-off in optimizing these two key functions for high-resolution lithography.

We and others (see Refs. 8 and 21, and references therein) have attacked this problem by developing processes in which different components of the overall photoresist were responsible for the imaging and etching functions. In our case, we chose organosilane SA films as the ultimate imaging layers due to their monolayer thickness and the thermodynamic driving force of the chemisorption process, which minimizes the pinhole density of the film. For the pattern transfer function, the high etch resistance of metals compared to Si allows a very thin EL metal layer to suffice as the plasma etch barrier.

Thus, the strategy of using an ultrathin SA imaging film in combination with a thin metal etch mask is well matched to the problems presented in advanced photon-based lithographies. These include deep-UV (193-nm) and soft x-ray (14-nm) projection exposure tools, which are being developed for commercial scale production of ICs with minimum feature sizes to ~ 0.1 μm. However, radiation with a considerably smaller beam diameter (to < 10 nm) can be obtained with sources such as focused ion beams, electron beam writers, and scanning tunneling microscopes. For such nanolithographic applications, the SA precursors can potentially provide monomeric surface adlayers whose only inherent limitation to image formation in the SA film layer is the occupied surface area of the individual molecules, which is of the order of < 1 nm.

Another essential attribute of a lithographic process is sensitivity. Our initial patterning studies with SA films[4,5] used organosilanes with low absorbance in the UV that required large amounts of incident energy (high doses) to give selective metallization. It was therefore necessary to develop imaging layers with improved absorbance at the wavelengths used in deep-UV exposure sources, specifically 193 nm (ArF excimer laser), 248 nm (KrF excimer laser), and 254 nm (Hg lamp).

6.3. Deep-UV Photochemistry of Aromatic Hydrocarbon Organosilane Films

Organic molecules with one or more aromatic ring systems typically have large molar absorptivities (10^4 to 10^5 M^{-1} cm^{-1}) in the deep UV.

We have investigated the deep-UV photochemistry of organosilane films with aromatic R groups[7,24] to determine the effect of molecular structure on photospeed. Structures of the molecules are shown in Fig. 6.1. The simplest aromatic organosilane molecule, phenyltrichlorosilane (PTCS), was studied as a paradigm for this class of materials.

Figure 6.2 shows[24] a UV absorption spectrum of a PTCS film[33] on fused silica. This film exhibits a peak maximum at 208 nm and has even greater absorptivity at 193 nm ($\varepsilon_{soln} = \sim 5 \times 10^4$ M^{-1} cm^{-1}). Exposure of the PTCS film to increasing amounts of ArF laser radiation[34] caused a monotonic decrease in absorbance, until the spectrum was reduced to baseline noise at ~ 300 mJ/cm^2. As shown in the inset of Fig. 6.2, the initially hydrophobic PTCS surface (water contact angle $\sim 60°$) was converted to a hydrophilic surface upon irradiation (water contact angle $\sim 15°$). We will return to this point in Section 6.7.

Laser-induced desorption Fourier transform mass spectrometry (FTMS) was used to determine the species removed from the surface by irradiation. The FTMS spectrum in Fig. 6.3 shows[24] an intact benzene molecule ($m/z = 78$) as the primary photoproduct of 193-nm irradiation. No evidence for Si or any organosilicon fragments was found.

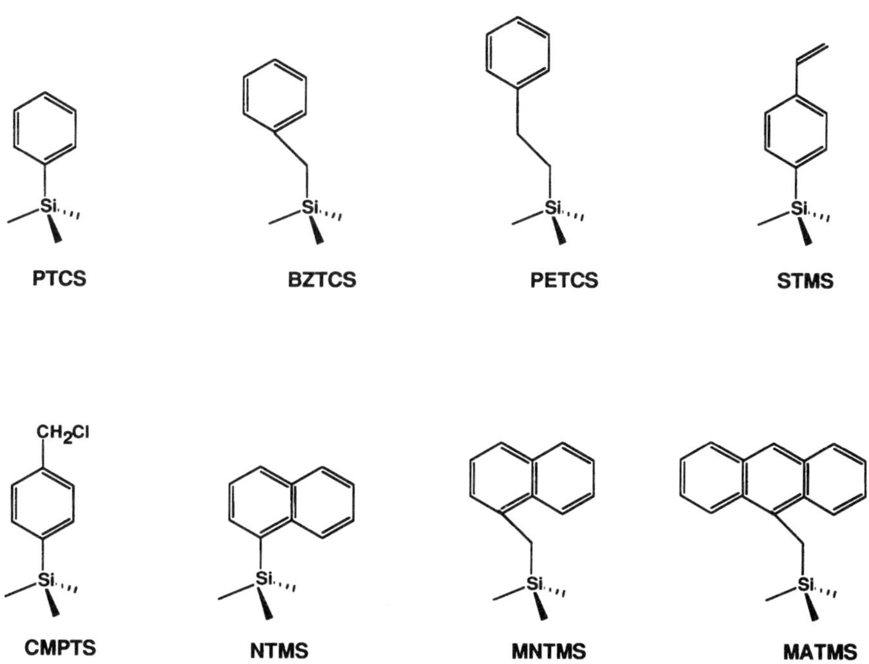

FIG. 6.1. Structures of aromatic hydrocarbon organosilanes.

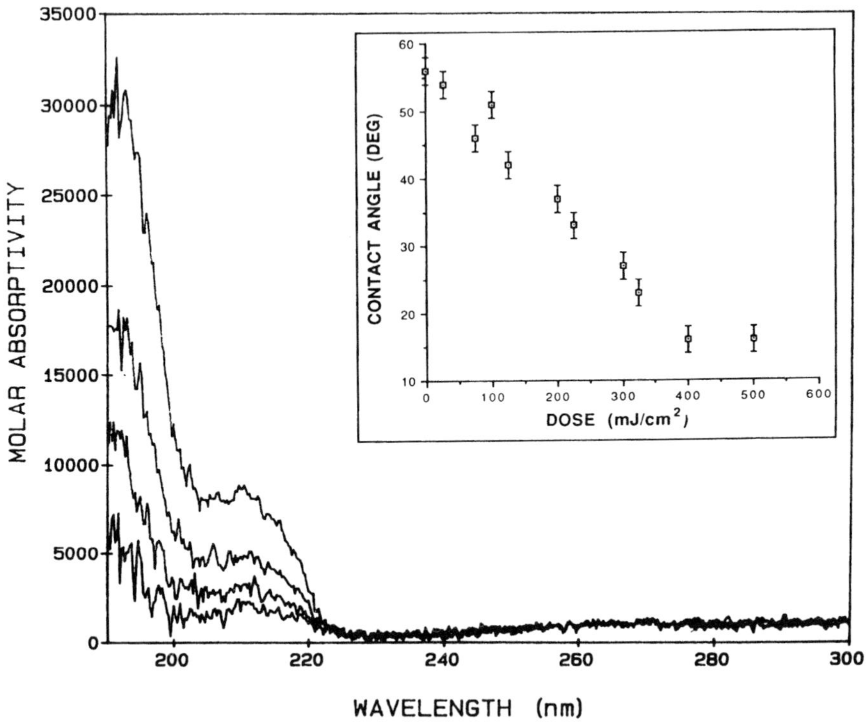

FIG. 6.2. UV spectrum of a PTCS film on fused silica with increasing exposure to 193-nm radiation. From top to bottom, traces correspond to unirradiated film and 100, 200, and 300 mJ/cm^2 exposure. The inset shows the water contact angle of a PTCS film exposed to varying amounts of 193-nm radiation.

X-ray photoelectron spectroscopy (XPS) was used to identify the surface bound species remaining after irradiation of a PTCS film on a Pt substrate. The intensity of the Si(2p) signal, centered at 103 eV, was unchanged by irradiation but is shifted to higher binding energy (~ 104 eV) after ArF exposure. This result is consistent with the Si atom in a more oxidized state after irradiation.

The combination of results from UV spectroscopy, wetting measurements, FTMS, and XPS led us to propose a mechanism for deep-UV photochemistry of PTCS that involves photochemical cleavage of the aromatic ring at the Si–C bond, followed by hydroxylation of the reactive exposed region to form surface silanol (Si–OH) groups (Fig. 6.4).[7,24] The observation that photocleavage occurs on deep-UV exposure of the SA film is not surprising because the energy in a 193-nm photon (147 kcal/mol)

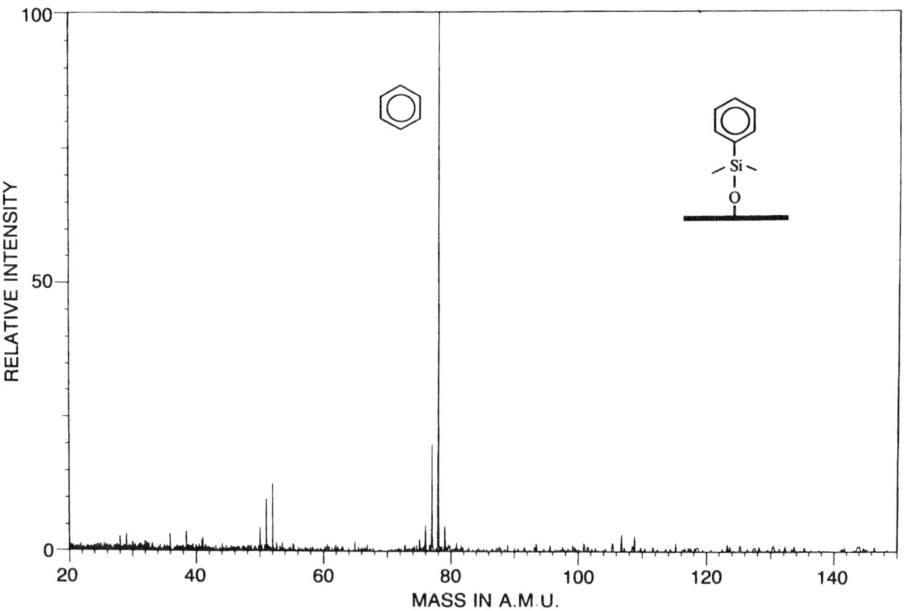

FIG. 6.3. Laser desorption FTMS of PTCS film irradiated with 193-nm light.

considerably exceeds the dissociation energies for single chemical bonds. Breakage occurs at Si–C because it is the weakest bond in the molecule (88 kcal/mol).[35] The measured dose of ~ 300 mJ/cm^2 for PTCS at 193 nm corresponds to a quantum yield of approximately 0.06. This moderately high efficiency likely results from constraining the excitation energy in a small ring system that does not have other predominant channels for energy dissipation.

Photochemical mechanism studies have also been performed on other aromatic organosilane films. The observed dose values, quantum yields (Table 6.1),[24] and spectroscopic behavior upon irradiation for single-ring systems such as benzyltrichlorosilane (BZTCS) and phenethyltrichlorosilane (PETCS) were similar to those observed for PTCS (with the exception that FTMS shows toluene and ethylbenzene, respectively, as the primary desorbed photoproducts). These results indicate that the excitation energy absorbed by the $\Pi \rightarrow \Pi^*$ transition of the aromatic ring is delocalized over the entire molecule, and ultimately breaks the Si–C bond.

Our interest in deep-UV photochemistry also includes processes that occur upon excitation with 248-nm radiation (115 kcal/mol photon energy).[24] The molar absorptivity of single-ring systems such as PTCS at 248 nm is ~ 4 to 5 × 10^2 M^{-1} cm^{-1}. The dose for a PTCS film at 248 nm

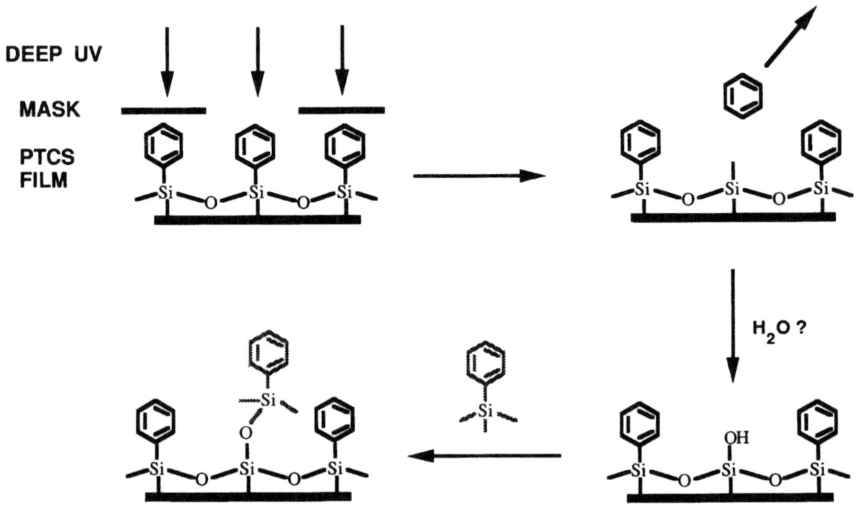

FIG. 6.4. Schematic of PTCS deep-UV photochemistry mechanism and formation of a coplanar molecular assembly. The first step shows the photochemical cleavage of the organic group at the Si–C bond. The second step shows formation of a silanol group at the reactive (exposed) film. The third step illustrates resilanization of the exposed regions by reaction with a solution of PTCS.

has been measured to be of the order of $\sim 50\ \mathrm{J/cm^2}$, corresponding to a quantum yield of ~ 0.01. Although the photochemical efficiency is still rather high at the longer wavelength, the extremely low absorptivity is manifested in large dose values.

To increase the molecular absorptivity at 248 nm, a number of aromatic organosilanes with extended Π conjugation were synthesized and the film photochemistry was investigated as described earlier. These materials

TABLE 6.1

PHOTOCHEMICAL PROPERTIES OF AROMATIC HYDROCARBON ORGANOSILANE FILMS AT 193 NM

Film	Dose ($\mathrm{J/cm^2}$)	ε_{193} ($\mathrm{M^{-1}\ cm^{-1}}$)[a]	Φ_{193}
PTCS	0.3	5×10^4	0.04
BZTCS	0.3	6×10^4	0.03
PETCS	0.3	6×10^4	0.03

[a] Molar absorptivity values are for the organosilanes in CH_3CN solution.

exhibit peak maxima with high molar absorptivity in the spectral region near 250 nm. Upon irradiation, they exhibit a photochemical mechanism similar to the single-ring systems. In each case, FTMS shows the primary photoproduct to be RH; i.e., the entire organic functionality plus a hydrogen atom. However, the 248 nm doses for the multiple-ring system films range from 4 to > 20 J/cm^2, corresponding to quantum yields of 4×10^{-2} to 8×10^{-4} (Table 6.2).[24] The lower photocleavage efficiencies of the multiple-ring organosilane films are attributed to dissipation of the excitation energy into other photophysical pathways (e.g., rotation about the C–C bond between the rings of the biphenyl silane, fluorescence or oxygen quenching of the naphthyl and anthracenyl derivatives). In this group of extended conjugation systems, the styryl compound had the lowest 248-nm dose of ~ 1 J/cm^2. Although the absorptivity of STMS is somewhat lower than the multiple-ring systems at 248 nm, the photocleavage pathway in the smaller molecule is much more favorable ($\Phi = \sim 0.02$); therefore, the dose—which is related to the product of absorptivity and quantum yield—is lower.

6.4. Selective Electroless Metallization of Patterned SA Films

6.4.1. EL Metallization with a Colloidal Pd/Sn Catalyst

Selective metallization of substrates is of great importance for a variety of microelectronics applications. As mentioned, one use of a patterned metal layer is as an etch mask for IC definition. Other uses include the

TABLE 6.2

Photochemical Properties of Aromatic Hydrocarbon Organosilane Films at 248 nm

Film	Dose (J/cm^2)	ε_{248} (M^{-1} cm^{-1})[a]	Φ_{248}
PTCS	$\sim$ 50	2×10^2	0.05
BPTMS	21	2×10^4	0.001
NTMS	24	3×10^3	0.007
MNTMS	4	3×10^3	0.04
MATMS	6	1×10^5	0.0008
STMS	1	1.8×10^4	0.03

[a] Molar absorptivity values are for the organosilanes in CH_3CN solution.

fabrication of conductive paths (interconnects, vias, contacts) for various levels of circuitry in both printed circuit (PC) and IC technology. High-resolution metal patterns are also used as the opaque regions in optical and x-ray lithographic masks.

At NRL we have developed several lithographic processes based on selective EL metallization of patterned ultrathin films.[4–10,14–17,21–23,26–28] In EL deposition a catalyst, typically Pd-based, is adhered to the substrate to be metallized. The catalyzed surface is immersed into an EL metallization bath containing a metal ion, reducing agent, buffers, and stabilizers, resulting in autocatalytic deposition of metal onto the substrate.

Our initial work[4–10] used a commercial EL catalyst, which is a colloidal dispersion of ~ 5- to 20-nm-diameter Pd particles that are surrounded by a protective sphere of hydrolyzed tin. The Sn layer serves to adhere the Pd particle to the substrate, and to maintain the Pd in a low valent oxidation state. With this catalyst, a sufficient amount of the protective layer must be removed to allow initiation of EL deposition in the plating bath. This step is referred to as *acceleration.* EL deposition could be initiated on SA films with alkyl, olefinic, aromatic, heteroaromatic, amine, thiol, and halogen-terminated surfaces. The water contact angle of these surfaces ranges from ~ 10° to > 85°, so catalyst adhesion is not determined solely by surface free energy considerations. One of the surfaces on which plating could *not* be initiated with the Pd/Sn catalyst is clean (hydroxylated) SiO_2. Since deep-UV photochemistry of aromatic organosilane films produces silanol groups, the nonspecific adsorption of the catalyst could be prevented in the exposed regions of the film. Therefore, patterned irradiation confines adsorption of the catalyst to the unexposed regions of the film, producing selective EL metallization in the same areas. This process, shown[8–10] schematically in Fig. 6.5, produces a positive tone image because metal deposition (formation of the plasma resistant etch barrier) occurs in the unirradiated regions. Figure 6.6 shows selective EL metal deposition on an organosilane SA film that was patterned to subhalf-micron resolution with a deep-UV exposure tool.

6.4.2. Selective EL Metallization by Ligand-Metal Complexation

The strategy of preventing nonspecific adsorption of the Pd/Sn catalyst for patterned metallization places important limitations on the overall process chemistry. Since silanol groups must be generated to prevent catalyst binding, the organosilane imaging films were limited to those classes of materials whose deep-UV photochemistry removed the entire R group above the Si atom, such as the aromatic films described in Section 6.3.

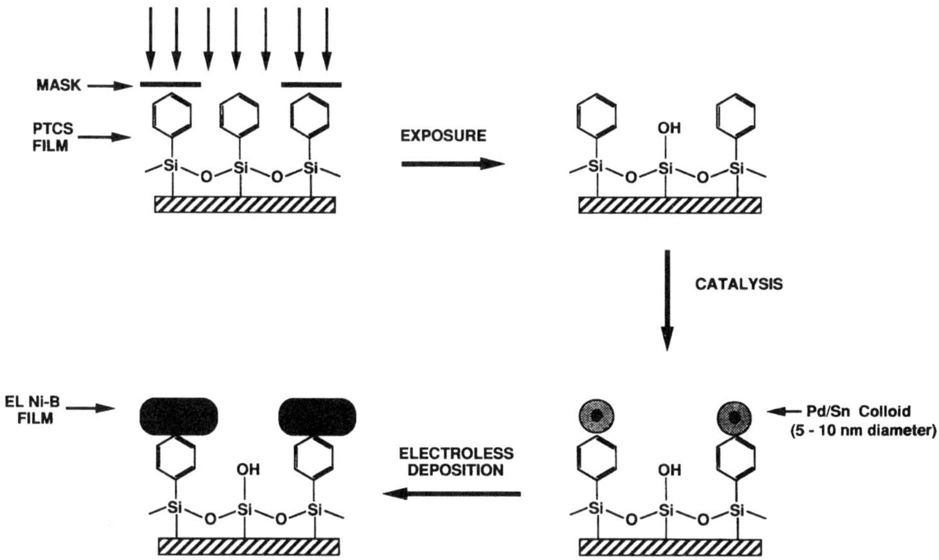

FIG. 6.5. Schematic of selective EL metallization using a patterned PTCS film and the colloidal Pd/Sn catalyst.

We turned to metal-ligand complexation chemistry to improve the selectivity of catalyst binding to patterned, functionalized surfaces. This is an attractive approach because metal ion binding to ligands is highly sensitive to the nature of the ligand. Factors such as the number and type of donor atom, the availability of Π backbonding interactions, and the bite angle of the donor atoms can affect metal ion equilibria by many orders of magnitude.[36] Ligand complexation offers a degree of control, or tunability, in the strength of the interaction of a metal ion with a modified surface. This approach also enhances flexibility in molecular design or deep-UV patterning, because the photochemistry is no longer required to remove the entire R group. Instead, it must only impair the ligating ability of the film sufficiently to prevent metal complexation. This approach is shown[15,16] schematically in Fig. 6.7.

Ligands with nitrogen donor atoms such as amines and pyridines are particularly effective in forming stable complexes of Group VIII transition metals such as palladium. We have investigated the deep-UV photochemistry and selective metallization of a variety of ligating organosilane SA films, the chemical structures of which are shown in Fig. 6.8.

To utilize effectively the ligating ability of these films to complex Pd directly, it was necessary to develop new catalyst solutions that eliminated the protective Sn layer and were still sufficiently stable to be used over a

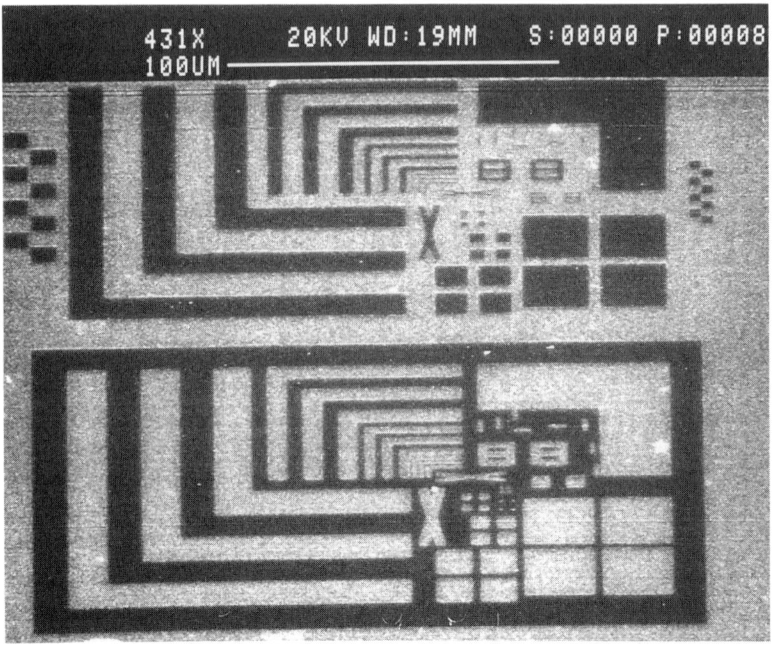

FIG. 6.6. Electron micrograph of microstructures fabricated by selective EL metallization of a deep-UV patterned SA film. The substrate was an Si wafer with 400 nm of poly-Si and a top layer of 12 nm of silicon dioxide. The film was patterned with a GCA 10 × reduction projection stepper system with a KrF (248-nm) laser source. The patterned film was catalyzed with the Pd/Sn colloid, accelerated, and subsequently plated with ~ 20 nm of EL copper. (From Calvert et al., *J. Electrochem. Soc.* **139,** 1677, copyright 1992, The Electrochemical Society.)

period of weeks. Catalyst formulations based on aqueous $PdCl_4^{2-}$ were buffered to pH ~ 5 to facilitate Pd complexation to amine surfaces (e.g., the pK_a of pyridine in solution is ~ 4.5), and stabilized with Cl^- to retard hydrolytic decomposition.[14,15,22,37]

Substrates modified with ligating films, such as 2-(trimethoxysilyl)-ethyl-2-pyridine (PYTMS), were treated with the Pd catalyst solution and immersed into an EL bath, resulting in homogeneous metal deposition across the surface.[14–16,21,22] Note here that metallization with the Sn-free Pd catalyst is obtained without the use of an acceleration step, since the protective layer is absent. The selectivity of the catalyst binding to the ligand surface is demonstrated in several ways. Clean substrates, or those with nonligating films such as PETCS (the structural analog of PYTMS, without the N atom) exhibit no metallization when treated with the Pd catalyst. Both UV spectroscopy[14,22] and XPS[37] give evidence for Pd surface

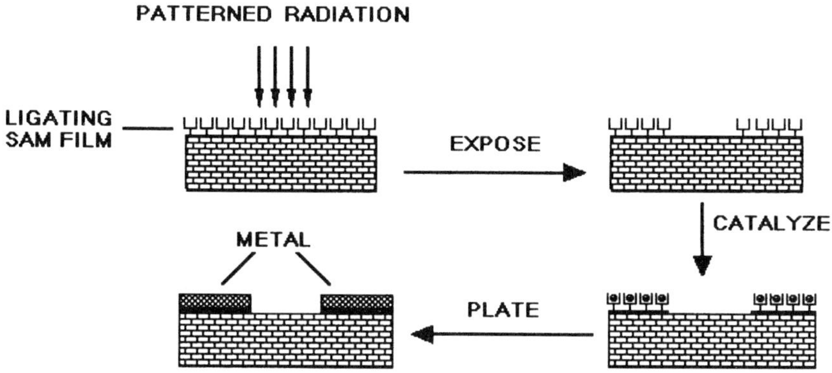

FIG. 6.7. Schematic of selective EL metallization using the metal-ligand complexation approach. The goalpost structures represent ligating groups of the UTF layer. Removal of the goalpost indicates loss of ligating ability, and not necessarily complete removal of the ligand or organofunctional groups in the film.

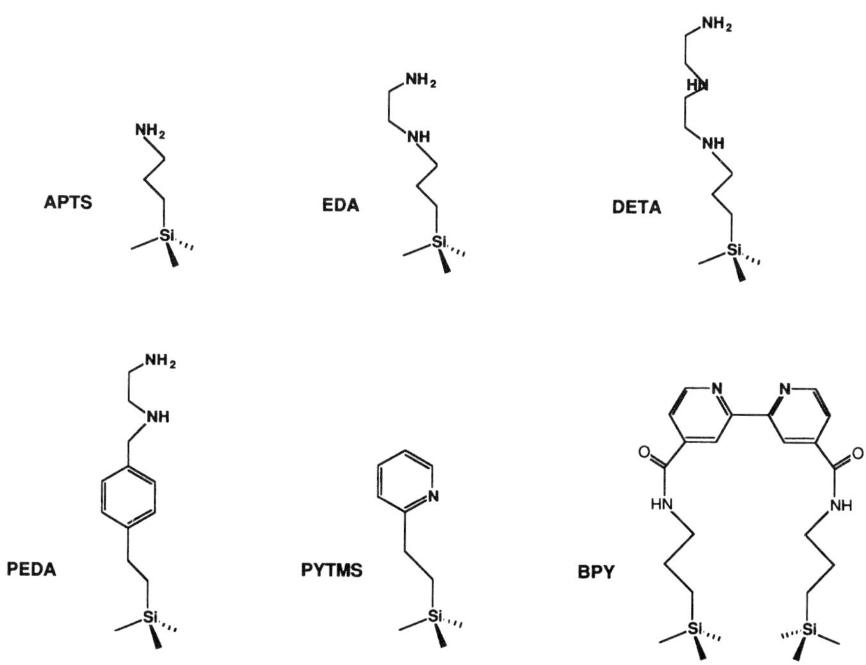

FIG. 6.8. Structures of ligating organosilanes.

complexation that involves a bis(pyridyl)palladium dichloride coordination environment.

The deep-UV photochemistry of PYTMS[14,22] exhibits similar characteristics to the aromatic hydrocarbon silane films, with a photochemical dose at 193 nm of 1.5 J/cm^2 ($\Phi \sim 0.03$). UV spectroscopy, XPS, and wettability measurements indicate that the mechanism is consistent with Si–C bond cleavage; FTMS shows ehtylpyridine as the primary desorbed photoproduct. Exposure of a PYTMS-treated silicon wafer surface to patterned 193-nm radiation through a lithographic line/space mask, followed by treatment with the Pd catalyst, yields selective attachment of the catalyst only to the unexposed regions of the film. An Auger electron spectroscopy linescan (Fig. 6.9)[24] clearly shows the modulation in Pd concentration on the patterned surface. A similarly patterned, catalyzed surface was immersed in an EL Ni bath to produce selective metal deposition. In this case, a higher resolution mask was employed and the resulting metal patterns with linewidths to ~ 0.4 μm are shown[21] in Fig. 6.10.

The photochemistry and selective metallization of other ligand SA films has also been investigated.[12–17,21,22,26–28] The EDA silane, due to its weak adsorption at 193 nm, has a photochemical dose[13] (elimination of the N signal by XPS analysis) of ~ 13 to 15 J/cm^2. However, the related PEDA silane, which has a large adsorption at 195 nm due to the phenyl chromophore, exhibits a dose for selective metallization of ~ 300 mJ/cm^2 at 193 nm on an Si thermal oxide substrate.[21] Recently, we have used a different approach to produce patterned ligating surfaces with a much

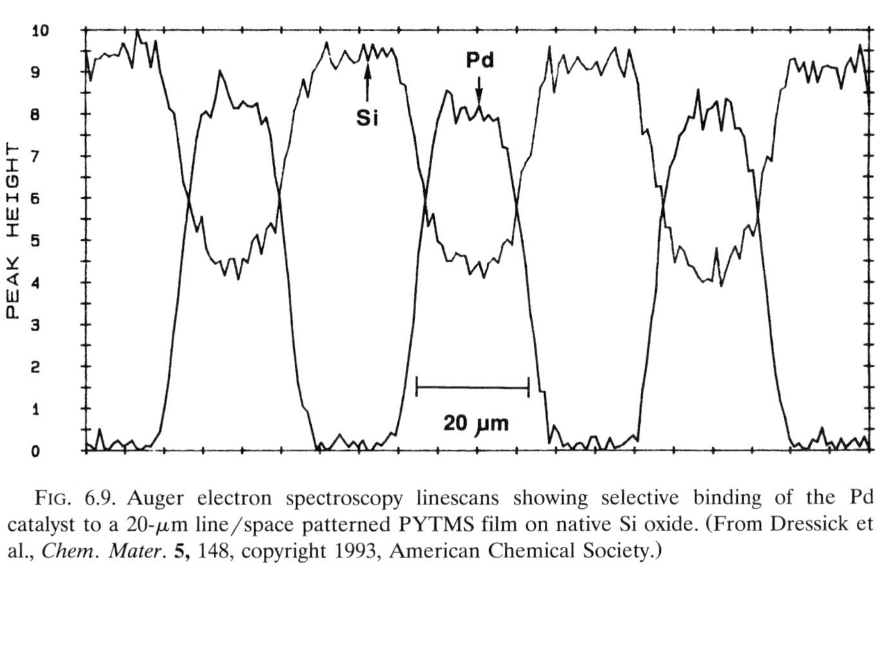

FIG. 6.9. Auger electron spectroscopy linescans showing selective binding of the Pd catalyst to a 20-μm line/space patterned PYTMS film on native Si oxide. (From Dressick et al., *Chem. Mater.* **5**, 148, copyright 1993, American Chemical Society.)

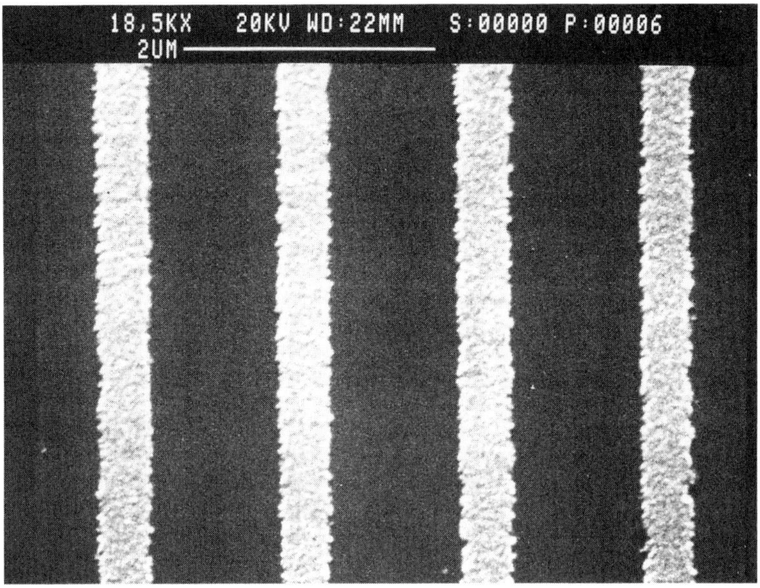

FIG. 6.10. Electron micrograph of 0.4-μm-wide EL Ni patterns produced using the ligand-based metallization process. A Si native oxide wafer was coated with the PYTMS film, exposed to $\sim 4\ J/cm^2$ of 193-nm radiation through a high-resolution grating mask in hard contact with the wafer. The wafer was treated with the Pd catalyst and then metallized with EL Ni. (From Calvert *et al.*, in: *Polymers for Microelectronics* (C. G. Willson *et al.*, eds.). American Chemical Society, Washington, DC, copyright 1993, American Chemical Society.)

more efficient photochemical process. A benzyl chloride silane film (CMPTCS, see Fig. 6.1) exposed to a dose of $< 50\ mJ/cm^2$ at 193 nm results in the elimination of HCl from the molecule. A subsequent nucleophilic substitution reaction is then used to attach ligands covalently to the intact benzyl chloride groups of the film, leading to selective metallization in the unexposed regions of the film.[26–28] This two-step approach is a more flexible way to produce patterned SA films because the initial SA film can be optimized for efficient photochemistry, and essentially any desired ligating (or other) functionality can be attached to the surface by appropriate choice of the subsequent chemical reaction.

6.4.3. Controlled Adhesion of EL Metal Deposits

In addition to selectivity, another attribute of the ligand-based metallization process is that the *strength* of the catalyst binding to the surface should be "tunable" to some extent based on the equilibrium formation constant K_f of the Pd-ligand complexation reaction,[36] which in turn is affected by factors such as the type of ligand donor atoms and the number of binding sites (denticity). We have observed that EL metal deposits formed on monodentate ligand films such as PYTMS exhibit lower adhesion than with films of corresponding multidentate ligands.[14,15,21,22] For example, ~ 40-nm-thick EL Co films can be deposited on a silicon substrate without flaking using the ligand-based metallization process with films of PYTMS. However, > 70% of the metal film is removed during a tape peel test. In contrast, Co films in excess of 250 nm can be deposited without flaking onto the same Si surface using the BPY silane as the ligating film. Peel tests show that no Co metal is removed from BPY-treated surfaces. A similar trend has been observed with EL Ni-plated Si wafer surfaces: The PYTMS film exhibits only partial metal adhesion during a Scotch tape peel test, whereas all of the metallized bi- and tridentate ligand films in Fig. 6.8 pass the test completely.[21,22] Stud pull measurements of the adhesive strength of Ni-plated surfaces using multidentate ligands have yielded pull strengths in excess of 1000 psi.

6.4.4. Selective Metallization of Nonhydroxylated Substrates

For many applications in microelectronics and other technological areas, it is important to be able to functionalize or selectively metallize substrates that are not intrinsically reactive to organosilane reagents, i.e., those that do not possess a sufficient density of surface −OH groups. Examples include various polymers, plastics, and ceramic materials. Selective metallization of diamond substrates is of particular interest for x-ray mask fabrication and applications in microelectronic packaging. Metal patterns are typically produced on diamond using sputter deposition or evaporation. Acceptable adhesion of the metal to the diamond surface can be obtained by annealing at ~ 800 to 1000°C to form a metal carbide layer at the interface, although such a high-temperature step is undesirable. We therefore sought to use ligand SA films to achieve selective metallization of diamond substrates at room temperature.

As with polymers such as poly(ethylene), the initial diamond surface is usually hydrogen terminated. In contrast to the polymers, relatively little is known about the effects of various oxidation conditions on diamond surfaces. CVD diamond substrates exposed to RF oxygen plasma resulted in surfaces that were highly wettable, with water contact angles of ~ 15–20° (the contact angle of the as-prepared diamond substrates was ~ 65–70°). XPS of oxidized diamond films showed development of a feature on the high-binding energy side of the C_{1s} peak, consistent with a surface layer of oxygen bonded to the diamond surface.

The oxidized diamond surface was then functionalized with a ligating silane film by treatment with a solution of *m,p*-(aminoethylaminomethyl)-phenethyltrimethoxysilane (PEDA). The water contact angle of the silanized surface was ~ 55°, similar to other PEDA-treated surfaces. XPS analysis of the UTF-PEDA–modified surface shows the expected peaks for Si (Fig. 6.11)[16] and N, both constituents of the SA film layer that are absent on the unmodified diamond surface.

The PEDA-treated diamond surface was then treated with the Sn-free Pd catalyst, and plated with EL Ni. No metal was removed from the Ni-plated surface by a tape peel test. This result demonstrates that the selective metallization process can be applied to diamond substrates to produce adhesive metal deposits without the use of high-temperature annealing steps. Analogous results have been obtained for functionalization and adhesive metallization of a number of other nonhydroxylated substrates, including plastics and polymers such as ABS, epoxy, polycarbonate, polysulfone, and fluoropolymers.[17,38]

Patterned metallization on the CVD diamond substrate was obtained by deep-UV irradiation followed by catalysis and EL deposition in the manner described previously for the PYTMS and PEDA films on silicon. A micrograph of patterned metal on diamond is shown[16,17] in Fig. 6.12.

6.5. Pattern Transfer and Microelectronic Device Fabrication by Means of Selectively Metallized SA Films

6.5.1. Pattern Transfer

The ultimate test of a microimaging process is the ability to produce working microelectronic devices. For most applications, this requires defining of high-resolution features of any desired geometry *in* (rather than *on*) Si substrates. Thin (~ 20- to 40-μm-thick) EL metal patterns with linewidths to ~ 0.3 μm were produced on polycrystalline Si (poly-Si) substrates by the selective metallization processes described in Section 6.4. The patterned substrates were etched in standard reactive ion etch plasmas (e.g., SF_6, CF_4, Freon 115) to remove poly-Si from the unprotected regions. Figure 6.13 shows[10] a micrograph of a high-resolution patterned poly-Si layer after etching, demonstrating that the thin EL metal layer is a robust etch mask for pattern transfer to Si.

6.5.2. Device Compatibility

In addition to satisfactory pattern transfer capability, a process must be compatible with the production of microelectronic circuitry; i.e., it must

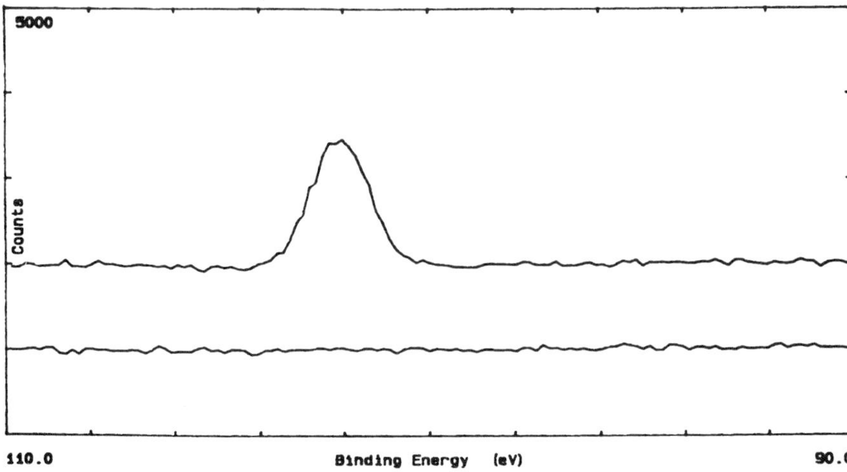

FIG. 6.11. XPS of the Si region of a CVD diamond surface before (lower trace) and after (upper trace) modification with PEDA silane film. (From Calvert et al., *Mater. Res. Soc. Proc.* **260,** 905, copyright 1992, Materials Research Society.)

FIG. 6.12. Optical micrograph of CVD diamond with selectively deposited EL Ni (light regions) showing features with 60-μm linewidths.

FIG. 6.13. Electron micrograph of partially etched patterns (0.4-μm lines and 1.0-μm spaces) produced as described in Fig. 6.6. The wafer was etched with SF_6 to remove the outer SiO_2 layer and ~ 1000 Å of poly-Si. (From Calvert et al., *J. Vac. Sci. Tech.* **B9,** 3447, copyright 1991, American Institute of Physics.)

not introduce defects or contaminants specific to the process chemistry. Metal-oxide semiconductor (MOS) capacitors are particularly useful structures for characterizing the effects of processing on device performance because they are more sensitive than surface spectroscopic techniques to the presence of defects and contaminants. Two types of test structures were employed to assess device compatibility: MOS capacitors and transistors.

In the first test, capacitors were formed on an Si wafer with a 95-nm-thick layer of thermal oxide, and a 400-nm-thick top layer of highly doped poly-Si. The selective metallization process was used to define the top poly-Si plate of the capacitor, using EL Cu as the etch mask. After reactive ion etching of the poly-Si and metal stripping in an acid dip, the electrical characteristics of the capacitors were examined. No detectable changes in oxide breakdown field strength were found compared to oxide wafers not subjected to the process. Mobile ion shifts and carrier storage lifetime degradation also were not observed,[8] even after thermal bias stress testing to 300°C and ± 8 V.

The selective metallization process was also used to fabricate single-level MOS field effect transistor (MOSFET) test structures. The procedure employed was analogous to that described earlier for creating the poly-Si/SiO_2/Si MOS capacitors, except the SA film imaging layer was exposed with a 193-nm excimer laser in a contact printer. After stripping the metal mask, sources and drains were produced by ion implantation. Electrical testing gave the expected MOSFET current-voltage response characteristics for the particular geometries employed (Fig. 6.14).

6.6. Deep-UV Photochemistry of Thiol-Terminal Organosilane Films and Patterned Proteins

The ability to use patterned chemical functional groups as a template for the selective attachment of materials to a substrate led us to investigate

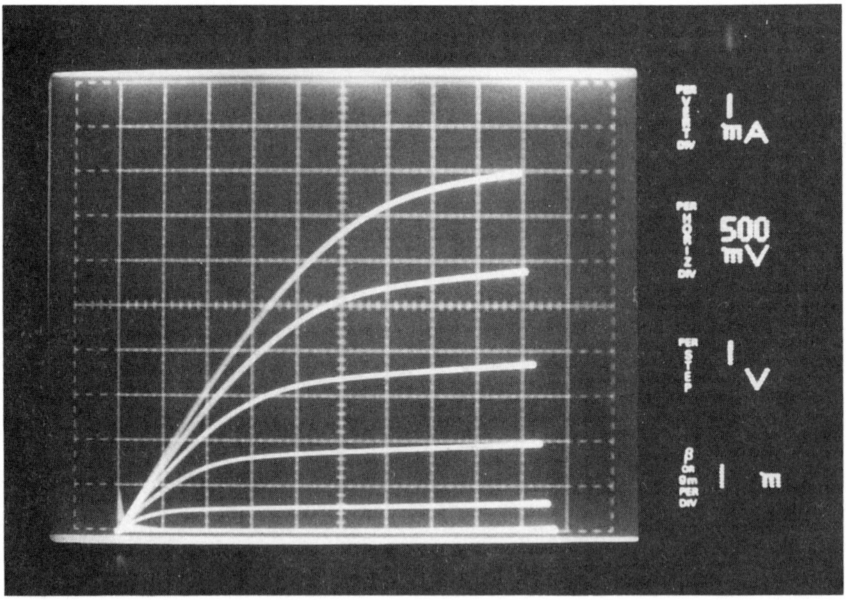

FIG. 6.14. Current-voltage response for a transistor test structure (5-μm gate length) fabricated using the selective metallization process. The SA film was patterned using an ArF excimer laser (193-nm) based contact printer, metallized with EL Cu, and etched down to the underlying oxide. After stripping the metal layer and ion implantation, the transistors were probed with micromanipulators using the three-point probe method. (From Calvert et al., *J. Vac. Sci. Tech.* **B9**, 3447, copyright 1991, American Institute of Physics.)

the attachment of biological moieties to patterned surfaces. The continuing drive for miniaturization to enhance the versatility, specificity, and detection limits of sensors has led to a high degree of interest in biosensor arrays. A key scientific issue in the development of biosensors and many other bioelectronic devices is to understand the factors that control the ability to attach a *functioning* biomolecule to the desired location on a surface while also *preventing* nonspecific adsorption to other areas of the substrate.

We have developed a procedure for covalent immobilization of functional antibodies in which the –SH terminal group of organosilane films such as (3-mercaptopropyl)trimethoxysilane (MTS) were reacted with the heterobifunctional crosslinker N-[(γ-maleimidobutyryl)oxy]succinimide (GMBS). The succinimide terminal group of the GMBS was then available to react with the available amine groups on the surface of a protein to link the biomolecule to the surface.[39,40] Thiol SA film surfaces, after attachment of GMBS, were treated with rabbit antigoat immunoglobulin in phosphate-buffered saline solution to immobilize the antibody. The surface-attached antibody was demonstrated to be functional by subsequent exposure to ^{125}I-labeled specific antigen goat IgG. Scintillation measurements indicated a surface concentration of ~ 1 ng/mm^2 of bound antigen.[19,39,40]

The deep-UV photochemistry of MTS was explored to determine whether proteins could be attached selectively to the SA film surface.[18–20] MTS films were exposed to deep-UV radiation from a low-pressure Hg/Ar lamp (30 min at 4.3 mW/cm^2, measured at 254 nm) or a 193-nm excimer laser (2 J/cm^2). The water contact angle of the MTS surface decreased from $\sim 60°$ to 30° after irradiation. XPS analysis of the S(2p) region showed a shift in binding energy from 164 to 169 eV, but was essentially unchanged in intensity (Fig. 6.15).[18,19] These results are consistent with the photochemical conversion of the thiol to a sulfonate ($R{-}SO_3^-$) group, rather than a photocleavage mechanism as occurs for the aromatic hydrocarbon films.

The thiol and sulfonate surfaces were also characterized with regard to protein adhesion.[18–20] The amount of ^{125}I radiolabeled goat IgG that adsorbed to the photogenerated sulfonate surface was reduced by 86% compared to the unirradiated thiol surface. Irradiation of the MTS-coated surface to produce sulfonate groups also reduced the nonspecific adsorption of other proteins such as bovine serum albumin and glucose oxidase by 74% and 90%, respectively. Although the specific reason(s) for the effectiveness of the photogenerated sulfonate groups in resisting protein adsorption has not been definitely established, the high density of negative charge on the surface may be a significant factor in repelling proteins with

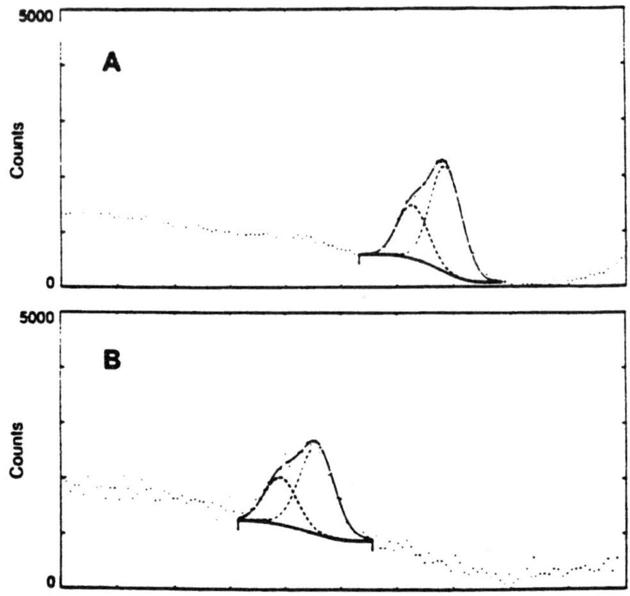

FIG. 6.15. XPS of an MTS film (A) before and (B) after deep-UV irradiation. (From Bhatia et al., *Analyt. Biochem.* **208,** 197, copyright 1993, Academic Press.)

a net negative charge at the physiological pHs used in the attachment chemistry.

Patterns of covalently attached proteins were produced by exposing MTS coated substrates with deep-UV light through a lithographic mask. The irradiated films were treated with GMBS and phycoerythrin, a fluorescent protein. The high selectivity of protein attachment to the unexposed regions of the surface is demonstrated by the fluorescence micrograph shown[18–20] in Fig. 6.16.

6.7. Mixed Monolayers, Coplanar Molecular Assemblies, and Controlled Wetting

The monotonic decrease in contact angle as a function of a deep-UV irradiation dose of PTCS (described in Section 6.3; see Fig. 6.2) has significance in addition to its value as a diagnostic of the photochemical mechanism. It shows that irradiation of chemisorbed SA films can be used to control precisely and directly the wetting characteristics of a single

FIG. 6.16. Confocal fluorescence micrograph of phycoerythrin covalently immobilized on a 20-μm line/space patterned MTS film. The lighter areas are regions to which the fluorescent protein was bound. (From Bhatia et al., *Analyt. Biochem.* **208,** 197, copyright 1993, Academic Press.)

component film.[41] The relationship between contact angle and dose was determined with several other organosilane films that exhibit Si–C photocleavage, and similar results have been obtained.

Manipulation of the wetting characteristics of alkylamine-terminal silane film surfaces has also been investigated.[7,11–13] Films of N-(2-aminoethyl)(3-aminopropyl)trimethoxysilane (EDA) are hydrophilic, exhibiting a water contact angle of ~ 17° when initially formed and ~ 30° after equilibration and conformational rearrangement.[13] XPS analysis of EDA films exposed to ~ 15 J/cm^2 of 193-nm radiation reveals the loss of the N(1s) signal for both protonated and unprotonated amine functional groups. FTMS showed several photoproducts upon 193-nm irradiation of EDA films including the EDA parent ion, $NH_2(CH_2)_2NH(CH_2)_3$, as well as ions at lower mass. These results indicate that deep-UV photolysis of the EDA film causes partial to complete removal of the organic portion of the molecule from the surface. The residual surface species are either silanols or oxidized short chain (< 3 carbon atoms) alkyl derivatives.

The photochemical removal or transformation of the organosilane R group to a silanol or other oxidized species renders the exposed region

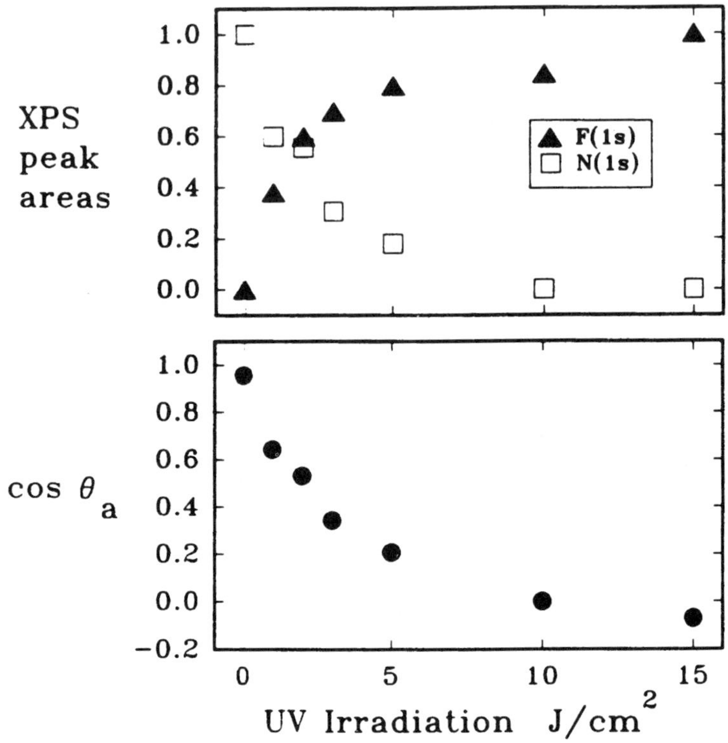

FIG. 6.17. (Top) XPS peak areas of N and F signals for an EDA film exposed to increasing doses of 193-nm radiation, followed by exposure to a 13F solution. (Bottom) cos θ_a values for the surface compositions shown in the top panel show increasing hydrophobicity as the surface concentration of 13F increases. (From Stenger et al., *J. Am. Chem. Soc.* **114,** 8435, copyright 1992, American Chemical Society.)

amenable to resilanization; for example, EDA film surfaces exposed to unpatterned 193-nm radiation in the range of 0 to 15 J/cm^2 and then treated with a solution of a perfluorinated alkylsilane,

$$CF_3(CF_2)_5(CH_2)_2Si(CH_3)_2Cl \text{ (13F)},$$

in toluene. XPS showed that increasing exposure doses decreased the nitrogen surface concentration, which also directly correlated with increased fluorine surface concentration (Fig. 6.17, top)[13] and a corresponding increase in the hydrophobicity of the surface, as represented by the cosine of the advancing contact angle, θ_a (Fig. 6.17, bottom).[13]

Both approaches—UV exposure of a hydrophobic SA film to form hydrophilic silanols, or UV exposure of a hydrophilic film followed by remodification with a hydrophobic material—permit highly controlled fabrication of continuously variable free-energy surfaces.

We have used lithographic patterning techniques and surface modification chemistry to produce assemblies of SA materials that reside in the same molecular plane.[4,5,7,11-13,24] A PTCS film on glass was exposed to patterned 193-nm radiation through a line/space lithographic mask. Faint dark stripes were produced in the glass as a result of color center formation, and served as a convenient fiducial for the exposed regions of the film. A drop of water was placed on the slide to probe the wetting characteristics of the patterned surface. As shown[24] in Fig. 6.18(a), the "scalloped" appearance of the drop graphically illustrates the laterally modulated surface energies due to alternating domains of the hydrophobic PTCS film and photogenerated hydrophilic silanols. The patterned surface was then treated with a solution of PTCS and the wetting behavior was reexamined. A water drop placed on the slide showed[24] essentially uniform wetting across the surface [Fig. 18(b)], indicating that PTCS reacted with silanol groups produced in the initial patterning step, as shown schematically in Fig. 6.4.[7,24]

Fabrication of coplanar assemblies of EDA and 13F offers a method for producing surfaces with laterally modulated wetting and a tonality opposite to that produced by the PTCS approach. An EDA film was exposed to patterned 193-nm radiation and then treated with 13F. The wetting characteristics of this surface are shown[7] in Fig. 6.19. In this case, the dark stripes, which correspond to the UV exposed regions, are hydrophobic (inward curvature) because the perfluorinated silane is attached to the photolytically transformed EDA sites. Spreading of the water drop (outward curvature) on the light stripes is due to the unirradiated, hydrophilic amine groups.

6.8. Patterned Amine Surfaces as Reactivity Templates

Patterned reactive surfaces can also serve as a template for building a variety of organic structures in the direction normal to the substrate. In one example, we have used amine-terminal organosilane films for attaching an organic fluorophore via a condensation reaction. Films of 3-aminopropyltrimethoxysilane on fused silica slides were exposed to 7.5 J/cm^2 of patterned 193-nm radiation. The patterned surface was then reacted with a solution of fluorescein isothiocyanate, and the fluorophore was attached to the amine groups via a thiourea linkage. A fluorescence micrograph of the resulting surface is shown[7] in Fig. 6.20.

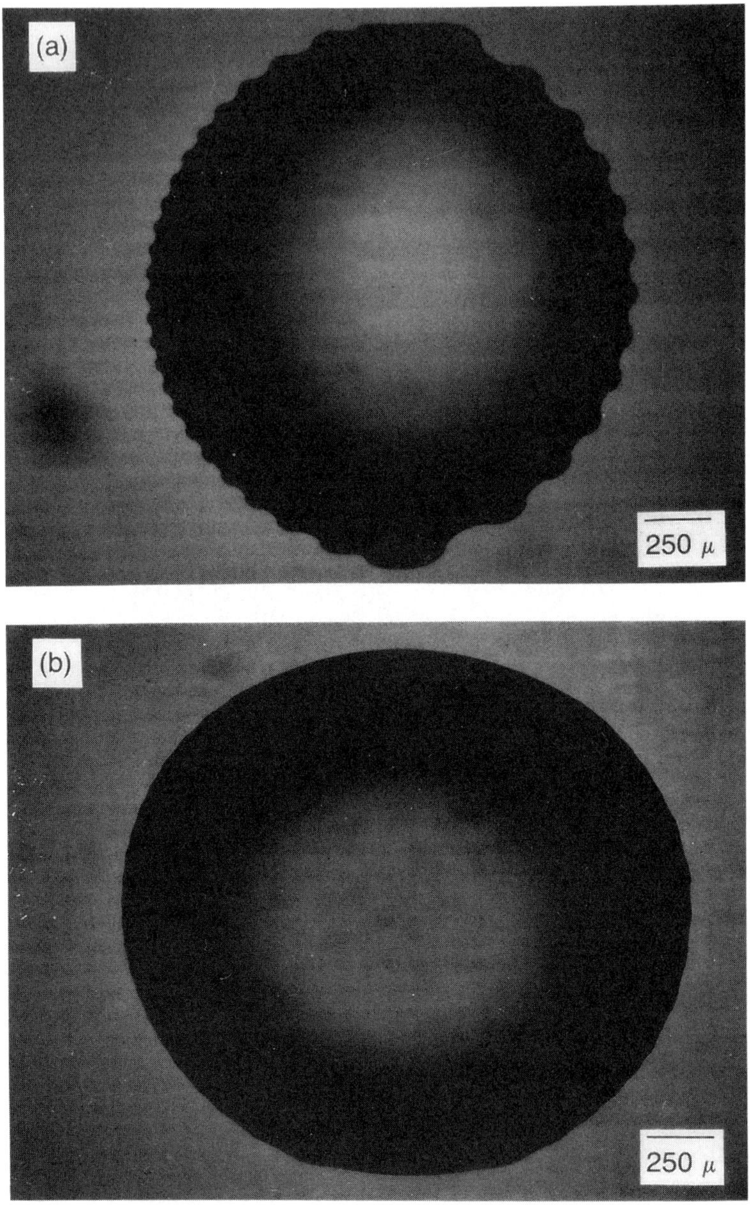

FIG. 6.18. Positional control of wetting: (a) A water drop placed on a PTCS film on glass exposed to 500 mJ/cm^2 of 193-nm radiation exhibits spreading on the exposed (dark) regions. (b) A water drop placed on the surface as described in part (a) after resilanization with a solution of PTCS shows elimination of modulated wetting.

FIG. 6.19. A water drop placed on a coplanar EDA/13F assembly spreads on the hydrophilic EDA film [corresponding to the unexposed (light stripe) regions], and does not spread on the hydrophobic 13F film (dark stripes). The different periodicity of the stripes at the top and bottom of the image corresponds to a change in linewidth of the features on the lithographic mask. (From Dulcey et al., *Science* **252,** 551, copyright 1991, American Association for the Advancement of Science.)

6.9. Selective Cell Adhesion and Outgrowth on Patterned SA Films

Precise geometric control of cell attachment is of interest for providing new approaches to the study of surface-directed growth, intercellular communication, organogenesis, and to control the alignment of individual cells with transducer elements in biosensors and implants. We have used the ability to fabricate coplanar assemblies of molecular monolayers to spatially control the adhesion and outgrowth of several types of biological cells on surfaces.[7,11–13,25]

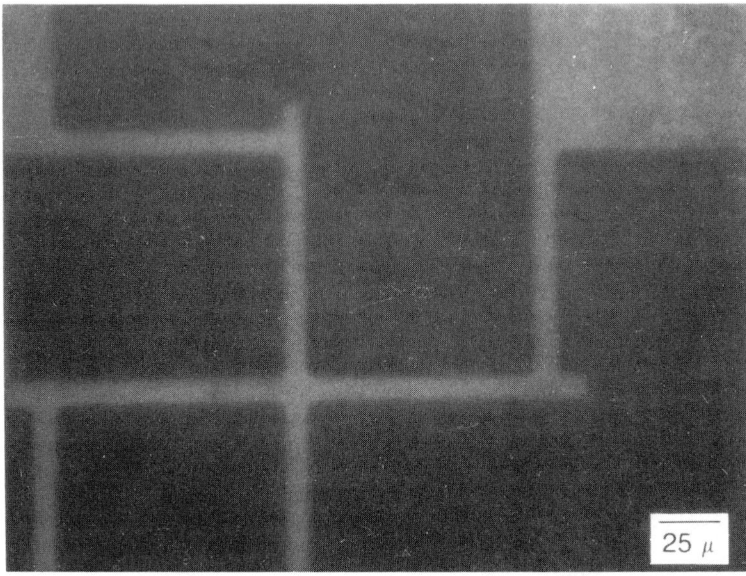

FIG. 6.20. Fluorescence micrograph of fluorescein isothiocyanate coupled to a patterned APTS film. (From Dulcey et al., *Science* **252,** 551, copyright 1991, American Association for the Advancement of Science.)

Initial cell adhesion experiments were performed with cancer cells on glass slides having previous fabricated line/space patterns of EDA and 13F.[7,11,12] The EDA films provide a hydrophilic surface that promotes cell adhesion, and 13F films are hydrophobic and nonadhesive. Within 10 min after plating a culture of SK–N–SH neuroblastoma cells onto the patterned film surface, ~ 98% of the cells adhered preferentially to the EDA regions. After 24 h, the cells exhibited neurite outgrowth that was predominantly limited to the EDA channels. At longer times, increasing numbers of neurites began to spread across the 13F regions. This effect is likely due to the abnormally high growth rates of these cancer cells, and to the fact that they are not particularly selective with regard to the nature of the growth substrate.

Further work was performed with dissociated cells from mammalian tissues, which are generally less adhesive and more dependent on the exact substrate composition for growth.[11,13,25] In a study of the relationship between SA film composition and cell adhesion and outgrowth, optimal results were found for organosilanes such as EDA and DETA (trimethoxysilylpropyldiethylenetriamine) that possessed high surface concentra-

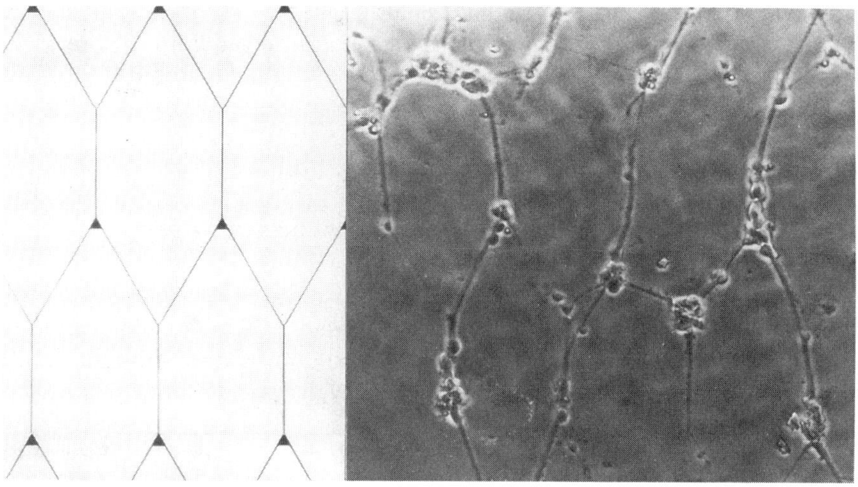

FIG. 6.21. (Left) Lithographic mask design for preferential cell body adhesion and directional definition of neurite outgrowth. (Right) Hippocampal neurons extending processes on EDA pathways at 48 h after plating. (From Stenger et al., *J. Am. Chem. Soc.* **114,** 8435, copyright 1992, American Chemical Society.

tions of protonated amine functional groups.[25] After plating cell cultures onto patterned EDA/13F surfaces, E18 rate hippocampal neurons, human umbilical vein endothelial cells, and porcine aortic endothelial cells were all found to adhere selectively to the EDA regions. Within 90 h, the surface-attached endothelial cells undergo several rounds of cell division and exhibit an increase in cell density, which leads to the formation of cellular multilayers in the direction normal to the substrate surface.[11] These structures may ultimately form the basis for revascularization of damaged tissues.

To achieve precise geometric definition of neurite outgrowth, EDA/13F patterns were fabricated using a lithographic mask with features as shown in Fig. 6.21. The mask was designed to produce triangle-shaped regions with sides ranging from 5 to 20 μm that are large enough to allow adhesion when the hippocampal cells are initially plated. Narrow lines, with widths ranging from 1 to 3 μm, extended from the triangular regions, create EDA channels that do not allow cell body adhesion, but provide pathways for the exploratory growth cones of developing neurons. Figure 6.21 shows[13] the pattern of cells that developed at 48 h after plating. In general, cell adhesion occurred only at the triangular sites and at regions where three narrow lines intersected. Neurite processes were almost exclusively directed along the EDA pathways. The ability to control the geometric outgrowth of neural processes holds considerable promise for formation of synaptic connections and the study of nerve impulse transmission over predefined pathways.

6.10. Conclusions and Future Areas of Research

In our work to date, we have developed a variety of techniques and materials for the controlled placement and manipulation of molecules and larger structures on surfaces. We have also shown how lithographically patterned SA films are of considerable utility for a wide range of practical applications.

Our future work will continue to utilize the principles of self-assembly and surface molecular engineering to provide thin films with useful chemical functionality or physical properties. We will employ characterization techniques with high spatial resolution (x-ray reflection, AFM, ellipsometry[43]) and high surface sensitivity (XPS, scanning auger microscopy) to improve our understanding of the composition, structure, and distribution of molecular SA films.

We are beginning to explore the use of a range of lithographic techniques for pattern definition that have considerably higher resolution capability than deep-UV lithography, including x-ray projection,[26] focused ion beam, and field emission STM[23,42] exposure tools. Recent results show that patterning of organosilane films can be accomplished with *all* of these techniques. Feature sizes of the order of $\sim$ 30 nm have been produced by STM patterning of organosilane films and existing processes for selective metallization. Future work will continue to address the ultimate resolution limits for photon and electron beam patterning of ultrathin layers and further examine the nature of the interaction of energetic radiation with SA films.

We will continue to explore new interfacial properties that result from rational two-dimensional control of chemical functionality, topography (on the molecular scale), and surface energy. In particular, issues such as the wetting and adhesion behavior of solvents and materials such as polymers at surfaces having tunable surface energy characteristics will be investigated, as will the factors that control nonspecific adsorption of biological materials including proteins, nucleic acids,[44] enzymes, and cells.

We also intend to further develop chemistries for attaching a variety of materials at predetermined surface locations. Examples include electroactive and photoredox reagents for studies of directed energy or charge transport in two-dimensional systems, catalytic and nucleating reagents for selective area deposition of materials using solution or vapor phase techniques,[45] and molecular recognition sites that promote specific interactions with, or trigger selected processes in biological moieities such as cells.

Acknowledgments

The work described here incorporates the contributions of many people at the NRL as well as collaborators at other institutions, as evidenced by the many different authors of papers in the Reference section. The interdisciplinary nature of this research requires a group of individuals with diverse skills and technical backgrounds. It is only because of the combination of the talents of such people with the excellent facilities of NRL and those of our collaborators that this work was realized.

I thank all of the people with whom I have had the pleasure of working in this fascinating area of science. I particularly recognize Dr. Suresh Bhatia, Dr. Gary Calabrese, Ms. Mu-San Chen, Dr. Charles Dulcey, Dr. Walter Dressick, Dr. Timothy Koloski, Dr. Frances Ligler, Dr. Christie Marrian, Dr. Marty Peckerar, Dr. Pehr Pehrsson, Dr. Joel Schnur, Dr. Paul Schoen, Dr. Warren Schultz (Captain, USN), and Dr. David Stenger for their many contributions to this work. I would also especially like to thank my colleague Jacque Georger, whose creativity made much of this work possible.

This work was funded by the following sponsors and their support is gratefully acknowledged: the Manufacturing Technology Office of the Assistant Secretary of the Navy, the Defense Advanced Research Projects Agency, the Office of Naval Research (Molecular Engineering Accelerated Research Initiative and Surface Biopatterning Programs), the Office of Naval Technology (Electronics Block Program), and the Shipley Company.

References

1. Zisman, W. (1964). In: *Contact Angles, Wettability and Adhesion, Advances in Chemistry 49* (F. M. Fowkes, ed.). American Chemical Society, Washington, DC, Chapter 1.
2. Schnur, J. M., Schoen, P. E., Yager, P., Calvert, J. M., Georger, J., and Singh, A. (1987). *Thin Solid Films* **152,** 181.
3. Schnur, J. M., Schoen, P. E., Yager, P., Calvert, J. M., Georger, J. H., and Price, R. (1990). U.S. Patent 4,911,981.
4. Calvert, J. M., Georger, J. H., Peckerar, M. C., Pehrsson, P. E., Schnur, J. M., and Schoen, P. E. (1992). *Thin Solid Films* **210 / 211,** 359.

5. Schnur, J. M., Peckerar, M. C., Marrian, C. R. K., Schoen, P. E., Calvert, J. M., and Georger, J. H. (1991). U.S. Patent 5,077,085.
6. Schnur, J. M., Peckerar, M. C., Marrian, C. R. K., Schoen, P. E., Calvert, J. M., and Georger, J. H. (1992). U.S. Patent 5,079,600.
7. Dulcey, C. S., Georger, J. H., Krauthamer, V., Fare, T. L., Stenger, D. A., and Calvert, J. M. (1991). *Science* **252,** 551.
8. Calvert, J. M., Dulcey, C. S., Georger, J. H., Peckerar, M. C., Schnur, J. M., Schoen, P. E., Calabrese, G. S., and Sricharoenchaikit, P. (1991). *Solid State Technol.* **34**(10), 77.
9. Calvert, J. M., Chen, M. S., Dulcey, C. S., Georger, J. H., Peckerar, M. C., Schnur, J. M., and Schoen, P. E. (1991). *J. Vac. Sci. Tech.* **B9,** 3447.
10. Calvert, J. M., Chen, M. S., Dulcey, C. S., Georger, J. H., Peckerar, M. C., Schnur, J. M., and Schoen, P. E. (1992). *J. Electrochem. Soc.* **139,** 1677.
11. Georger, J. H., Stenger, D. A., Rudolph, A. S., Hickman, J. J., Dulcey, C. S., and Fare, T. L. (1992). *Thin Solid Films* **210 / 211,** 716.
12. Stenger, D. A., Georger, J. H., and Fare, T. L. (pending). U.S. Patent Application 07/598,194.
13. Stenger, D. A., Georger, J. H., Dulcey, C. S., Hickman, J. J., Rudolph, A. S., Nielsen, T. B., McCort, S. M., and Calvert, J. M. (1992). *J. Am. Chem. Soc.* **114,** 8435.
14. Dressick, W. J., Calvert, J. M., Dulcey, C. S., Georger, J. H., Calabrese, G. S., Thomas, M. E., and Stever, H. A. (1992). *Mater. Res. Soc. Proc.* **260,** 659.
15. Calvert, J. M., Dressick, W. J., Calabrese, G. S., and Gulla, M. (allowed, but not yet published). U.S. Patent Application 07/691,565.
16. Calvert, J. M., Pehrsson, P. E., Dulcey, C. S., and Peckerar, M. C. (1992). *Mater. Res. Soc. Proc.* **260,** 905.
17. Calvert, J. M., Pehrsson, P. E., and Peckerar, M. C. (pending). U.S. Patent Application 07/933,147.
18. Bhatia, S. K., Hickman, J. J., and Ligler, F. S. (1992). *J. Am. Chem. Soc.* **114,** 4432.
19. Bhatia, S. K., Teixeira, J. L., Anderson, M. A., Shriver-Lake, L. C., Calvert, J. M., Georger, J. H., Hickman, J. J., Dulcey, C. S., Schoen, P. E., and Ligler, F. S. (1993). *Analyt. Biochem.* **208,** 197.
20. Ligler, F. S., Bhatia, S. K., Shriver-Lake, L. C., Georger, J. H., Calvert, J. M., and Dulcey, C. S. (allowed, not yet published). U.S. Patent Application 07/691,491.
21. Calvert, J. M., Dressick, W. J., Dulcey, C. S., Georger, J. H., Stenger, D. A., Koloski, T. S., and Calabrese, G. S. (1993). In: *Polymers for Microelectronics, American Chemical Society Symposium Series* (C. G. Willson, L. F. Thompson, and S. Tagawa, eds.). American Chemical Society, Washington, DC, **537,** 210.
22. Dressick, W. J., Dulcey, C. S., Georger, J. H., and Calvert, J. M. (1993). *Chem. Mater.* **5,** 148.
23. Marrian, C. R. K., Perkins, F. K., Brandow, S. L., Koloski, T. S., Dobisz, E. A., and Calvert, J. M. (1994). *J. Appl. Phys. Lett.*, **64,** 1.
24. Dulcey, C. S., Georger, J. H., McElvaney, S., and Calvert, J. M. (in press). *Langmuir*.
25. Stenger, D. A., Pike, C., Hickman, J. J., and Cotman, C. W. (1993). *Brain Research*, **630,** 136.
26. Calvert, J. M., Koloski, T. S., Dulcey, C. S., Dressick, W. J., Peckerar, M. C., Cerrina, F., Taylor, J., Suh, D., Wood, O., and MacDowell, A. (1993). *Optical Engineering*, **32(10),** 2437.
27. Dulcey, C. S., Koloski, T. S., Dressick, W. J., Chen, M. S., and Calvert, J. M. (1993). *Proc. SPIE* **1925,** 657.
28. Koloski, T. S., Dulcey, C. S., Dressick, W. J., and Calvert, J. M. (1994). *Langmuir*, **10,** 3122.

29. As defined by Ulman: "SA monolayers are molecular assemblies that are formed by the immersion of an appropriate substrate into a solution of an active surfactant in an organic solvent." Organosilane films are one example of an SA film.[30]
30. Ulman, A. (1991). *An Introduction to Ultrathin Organic Films: From Langmuir-Blodgett to Self-Assembly.* Academic Press, San Diego, p. 237.
31. Arkles, B. (1992). In: *1992 Huls Silicon Compounds Register and Review* (R. Anderson, G. L. Larson, and C. Smith, eds.). Huls America Inc., Piscataway, NJ, p. 59.
32. Abbott, N. A., Folkers, J. P., and Whitesides, G. M. (1992). *Science* **257,** 1380.
33. PTCS films were prepared by immersing acid-cleaned substrates into a dilute solution of PTCS in anhydrous toluene for ~ 5 min, rinsed with fresh solvent, and baked on a hot plate in a Class 100 clean room. The monolayer nature of the trifunctional silane film prepared by this process on silicon native oxide wafers was established by ellipsometry, which gave a thickness of ~ 0.5 nm.
34. Pulse energy densities were limited to less than ~ 4 mJ/cm^2 to avoid photoablation.
35. Walsh, R. (1981). *Acc. Chem. Res.* **14,** 246.
36. Martell, A., and Smith, R. (1975). *Critical Stability Constants.* Plenum Press, NY.
37. Dressick, W. J., Dulcey, C. S., Georger, J. H., Calabrese, G. S., and Calvert, J. M. (1994). *J. Electrochem. Soc.*, **141,** 210.
38. Vargo, T. G., Gardella, J. A., Calvert, J. M., and Chen, M. S. (1993). *Science*, **262,** 1711.
39. Bhatia, S., Shriver-Lake, L., Prior, K. J., Georger, J. H., Calvert, J. M., Bredehorst, R., and Ligler, F. S. (1989). *Analyt. Biochem.* **178,** 408.
40. Ligler, F. S., Calvert, J. M., Georger, J. H., Shriver-Lake, L. C., Bhatia, S. K., and Bredehorst, R. (1991). U.S. Patent 5,077,210.
41. Since the PTCS is being transformed into Si–OH groups by irradiation, a partially converted film could be thought of as a two-component mixture of phenyl groups and silanols. The term "two-component" is used in the sense of two *organic* constituents, such as the amine and perfluorinated silane mixtures.
42. Perkins, F. K., Brandow, S. L., Koloski, T. S., Dobisz, E. A., Rhee, K., Marrian, C. R. K., and Calvert, J. M. (1994). *J. Vac. Sci. Tech.*, **B12.**
43. Geer, R. E., Stenger, D. A., Chen, M. S., Calvert, J. M., Shashidhar, R., Jeong, Y. H., and Pershan, P. S. (1994). *Langmuir*, **10,** 1171.
44. Chrisey, L. A., Roberts, P. M., Benezra, V. I., Dressick, W. J., Dulcey, C. S., and Calvert, J. M. (1994). *Mater. Res. Soc. Proc.*, **330,** 179.
45. Potochnik, S. J., Hsu, D. S. Y., Calvert, J. M., and Pehrsson, P. E. (1994). *Mater. Res. Soc. Proc.*, **337,** 429.

Recognition at Surfaces

Langmuir Films of Amphiphilic Alcohols and Surfaces of Polar Crystals as Templates for Ice Nucleation

R. POPOVITZ-BIRO, J. MAJEWSKI, J. L. WANG, L. LEISEROWITZ, and M. LAHAV

The Weizmann Institute of Science, The Department of Materials and Interfaces, Rehovot, 76100, Israel

K. KJAER and J. ALS-NIELSEN

Risø National Laboratory, Department of Physics, DK4000 Roskilde, Denmark

7.1. Introduction

Although ice melts at 0°C, pure water can be supercooled in a clean atmosphere to temperatures below −30°C. To overcome the free-energy barrier to the liquid/solid phase transition, the presence of auxiliaries, which act as catalysts for ice nucleation, is required. Certain inorganic,[1,2] organic,[3–6] and biological[7–9] materials possess the ability to promote ice nucleation. For example, silver iodide seeded in clouds have been used for induced precipitation of rain by their promotion of ice nucleation.[10,11] On the other hand, large-scale damage to crops is caused by particular proteins present in frost bacteria.[12–14] Although widely studied, the detailed mechanisms by which these auxiliaries act are not clear. It has been suggested that ice nucleating substrates bear surfaces that match the crystal structure of ice and thus serve as templates that induce nucleation by epitaxy.[1,6,15] Another mechanism that invokes the presence of local

ISBN 0-12-523485-6

electric fields capable of aligning water molecules into ice-like embryos has been proposed.[16,17] To unravel the complexity of this transformation, we undertook two approaches where these two suggested mechanisms were studied independently. In one, we designed experiments, using Langmuir layers, aimed at a demonstration of the validity of the first-mentioned mechanism.[18] In the second approach, we used single crystals of racemic and chiral-resolved α-amino acids to demonstrate that an electric field, present in the crevices on surfaces of polar crystals, helps stabilize ice-like structures.[19]

7.2. Self-Aggregates of Amphiphilic Alcohols as Templates for Ice Nucleation

It has been previously demonstrated that Langmuir monolayers composed of amphiphilic molecules may be designed that, in their compressed or uncompressed state, act as templates for induced crystal nucleation at the air/solution interface. The idea is to design the monolayer such that the hydrophilic headgroups of the amphiphilic molecules will create a surface similar or complementary to a layer of the to-be-nucleated crystal. In this way, amphiphilic molecules bearing chiral-resolved α-amino acid headgroups promoted the oriented nucleation of α-glycine crystals by creating the first layer of the crystal. Slight variation in the packing of the headgroups affected both the rate of crystal nucleation and the degree of orientation.[20] Furthermore, monolayers of ionic or zwitterionic amphiphiles may bind and organize a counter-ionic layer of atomic or molecular ions in the subphase by electrostatic interactions and trigger the nucleation of ionic crystals at the interface. For example, the first layers of silver propionate crystals were created by monolayers of aliphatic carboxylic acids as a result of electrostatic interactions.[21] Similarly, highly oriented lead sulfide crystals have been generated under a monolayer of arachidic acid, due to a good lattice match and electrostatic forces.[22] Moreover, depending on the headgroup lattice matching, stereochemistry, or ionic state, preferential nucleation from different faces was obtained in the case of sodium chloride,[23] calcium carbonate,[24] and barium sulphate.[25] The reverse process can be taken advantage of, by which this transfer of structural information from two-dimensional (2-D) to three-dimensional (3-D) systems could be used to probe the structure of surface aggregates at the air/solution interface.[26] We have used this method for the design of a monolayer system for induced ice nucleation via epitaxy. As described later in detail, we made use of the general observation that long-chain alkanes form a layer structure in which the chains are separated by about 4.6 Å in an approximate hexagonal arrangement. This lattice is similar in dimensions to the *ab* lattice of hexagonal ice in which $a = b = 4.5$ Å. Direct information on the packing arrangements of the monolayer can be obtained by grazing incidence x-ray diffraction measurements (GID) using synchrotron radiation.[27] Such a strategy also provides a structure-function correlation, on a molecular level, as to the mechanism responsible for ice nucleation, because it has been proven possible to monitor the ice nucleation process under the monolayer by GID.[28]

7.2.1. Hexagonal Ice

The stable form of ice (ice I) crystallized under atmospheric pressure, is hexagonal[29,30] (Fig. 7.1), space group $P6_3/mmc$, with axes $a = b = 4.5$ Å, $c = 7.3$ Å, and $\gamma = 120$ deg. The crystal structure may be described, in terms of the oxygen atoms, as composed of 0.9-Å-thick bilayers parallel to the *ab* plane, and separated by 2.75 Å along the hexagonal *c* axis. The axial length of 4.5 Å in the *ab* plane is the separation distance between two water molecules, both hydrogen bonded to the same water molecule within the bilayer. Each oxygen atom participates in three O–H···O hydrogen bonds within the bilayer and a fourth O–H···O bond of 2.75 Å

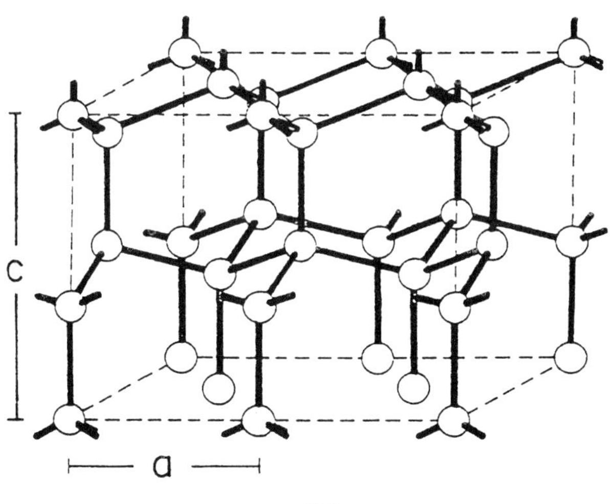

FIG. 7.1. Crystal structure of hexagonal ice.[29,30] The oxygen atoms are denoted by spheres and the disordered O–H ··· O bonds by rods.

interlinking the neighboring bilayers along the c axis. The area per molecule of water within an ab layer is $a^2 \sin 120° = 17.5$ Å^2.

A GID pattern of a *compressed* monolayer of alcohol $C_{21}H_{43}OH$ on water at 0°C yielded[29] a unit cell that, to a first approximation, may be regarded as c-centered rectangular with dimensions $a = 4.98$ Å and $b = 7.42$ Å [Fig. 7.2(a)]. For the purpose of comparison with the unit cell of ice, we may represent this cell in terms of a distorted hexagonal $a_h = b_h = 0.5(a \pm b) = 4.5$ Å, $\gamma = 113$ deg [Fig. 7.2(b)]. It had already been demonstrated in a variety of systems that amphiphilic molecules tend to self-aggregate into 2-D crystalline clusters.[26,31] We therefore anticipated that long-chain aliphatic alcohols may also form crystalline self-assemblies, which would function as promoters of ice nucleation[18,32] as already alluded to at the beginning of this section. Moreover, we had anticipated that the chains would tend to be vertically aligned in the uncompressed state since in this way the chains would be separated by ~ 4.5 Å and so could be easily interlinked by O–H ··· O hydrogen bonds by bound water molecules at the interface.

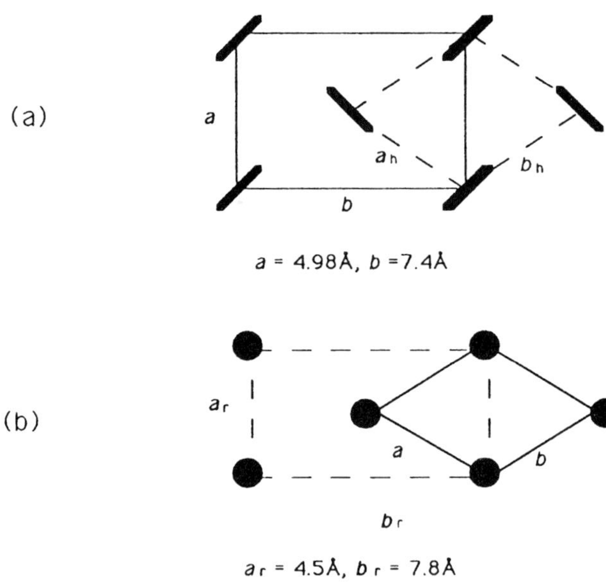

FIG. 7.2. (a) Schematic representation of the orthogonal packing of hydrocarbon chains of an alcohol monolayer in a rectangular cell (a, b), viewed along the chain axis. The distorted hexagonal representation (a_h, b_h) is also depicted. (b) Schematic representation of hexagonal ice in terms of the hexagonal cell (a, b) and the c-centered rectangular cell (a_r, b_r).

7.3. Nucleation of Ice Under Monolayers of Aliphatic Alcohols

The activity of the alcohol monolayers as ice nucleators was evaluated by measuring the threshold freezing temperatures of supercooled water drops (10 μL) covered by the monolayers and compared to drops covered by monolayers of the corresponding carboxylic acids, for which we do not expect a structural match. Freezing point measurements were carried out on drops of water covered by aliphatic alcohols $C_nH_{2n+1}OH$ with $n = 14$ to 31. The results (Fig. 7.3) show that the aliphatic alcohols nucleate ice at higher temperatures and more reproducibly than the corresponding carboxylic acids. Moreover, freezing points appear to be strikingly sensitive to the length and parity of the aliphatic chain. The freezing point curve for the n = odd series increases with chain length, reaching a value of -1°C for $n = 31$, whereas in the n = even series a plateau at -8°C is reached for $n = 22$. Freezing temperatures obtained for drops covered by monolayers of the analogous aliphatic carboxylic acids ($C_{n-1}H_{2n-1}CO_2H$) were scattered in the range of -12 to -18°C.

7.3.1. Effect of Molecular Area (and Lattice Match) on Ice Nucleation

The area per water molecule in the *ab* layer of hexagonal ice is 17.5 Å^2. Monolayers of the alcohols such as perfluorododecanol and cholesterol, which have cross-sectional areas of 28 and 38 Å^2, respectively, do not induce nucleation of ice. The bulky hydrophobic parts of these molecules cause separation of the hydroxyl groups to a distance larger than 4.5 Å. As already mentioned, such a distance would be favorable for hydrogen bonding of a water molecule to two neighboring hydroxyl groups in a way that is complementary to the structure of ice. In the monolayers containing bulky chains, a lattice or structural match cannot be obtained between the monolayer and ice. The highest freezing points were observed with aliphatic alcohols having molecular areas of 18.5 to 20 Å^2, which are reasonably close to the *ab* unit cell area of water in ice. Among this group of aliphatic alcohols we obtained a large range of ice nucleating activities that cannot result from the minor changes in molecular area, thus requiring that other structural parameters be invoked.

X-Ray Structure of Aliphatic Alcohol Monolayers

GID measurements were performed on uncompressed monolayers of $C_nH_{2n+1}OH$, for $n = 16$, 20, 23, 30, and 31 over water at 5°C. The GID

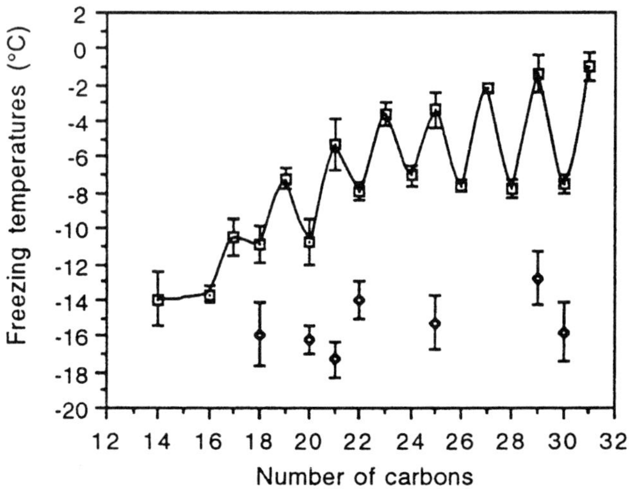

FIG. 7.3. Freezing temperatures of supercooled water drops covered by monolayers of aliphatic alcohols $C_nH_{2n+1}OH$ (□) and carboxylic acids $C_{n-1}H_{2n-1}CO_2H$ (◇) versus the number of carbon atoms in the chain.

patterns of the three monolayers[27,33] $n = 23$, 30, and 31 are very similar, indicating that their 2-D crystal structures are also very similar (Table 7.1) as shown in Fig. 7.4. Monolayers of $C_{30}H_{61}OH$ and $C_{31}H_{63}OH$ are highly crystalline; their coherence lengths as derived from the widths of the observed Bragg reflections are anisotropic, in the range of 300 to 1000 Å. Each crystal structure appears in a primitive rectangular unit cell (see Table 7.1), which contains two molecules related essentially by glide symmetry in the 2-D plane group $p1g1$. (The symbol $p1g1$ notes a primitive (p) unit cell with a glide (g) plane perpendicular to the a axis, the glide being directed along b.) The $C_{30}H_{61}OH$ and $C_{31}H_{63}OH$ molecules are tilted from the vertical by about the same angle of 9 ± 1 deg. The molecules tilt along the b axis, between nearest neighbors. The area per molecule, as projected along the molecular axis A_p (Table 7.1), is 18.5 Å^2, which is almost as low as in the densest lamellar crystals. The neighboring hydrocarbon chains are arranged in the herring-bone motif, with a dihedral angle of 85 deg between the planes of the carbon backbone chains, denoted as an orthogonal $O_\perp$ motif.[34] The good lattice match between the aliphatic alcohols and ice is revealed by comparing the oxygen position of OH headgroups in the $C_{30}H_{61}OH$ and $C_{31}H_{63}OH$ monolayers with those in the (001) ab layer of ice. The comparison is easiest when done by expressing the hexagonal lattice ($a = b$, $\gamma = 120$ deg) of ice in terms of a c-centered rectangular cell $a_r = |a + b| = 4.5$ Å, $b_r = |-a + b| = 7.8$ Å,

TABLE 7.1

RESULTS[a] OF THE GID ANALYSIS FOR THE ALCOHOL MONOLAYERS $C_nH_{2n+1}OH$ FOR n = 16, 20, 23, 30, AND 31 AND ESTER-ALCOHOLS *5* AND *6*

Compound	d_{11} (L)	d_{02} (L)	d_{20} (L)	a (a_p)	b (b_p)	A (A_p)	Tilt angle[b]
n = 16[c]	4.21	4.0[d]		4.95[d]	8.0[d]	19.8[d]	19
	180	low[e]		(5.0)[d]	(7.56)[d]	(18.7)[d]	
n = 20[c]	4.27	4.00		5.05	8.0	20.2	19
	700	300		(5.05)	(7.56)	(19.1)	
n = 23	4.18	3.78	2.50	5.00	7.56	18.90	9.5
	($\geq$ 1000)	(360)	($\geq$ 10000	(5.00)	(7.46)	(18.7)	
n = 30	4.15	3.75	2.49	4.99	7.49	18.7	7.7
	($\geq$ 1000)	(230)	($\geq$ 1000)	(4.99)	(7.42)	(18.51)	
n = 31	4.16	3.77	2.49	4.99	7.53	18.79	9.7
	($\geq$ 1000)	(270)	(700)	(4.99)	(7.41)	(18.5)	
n = 31[c]	4.15	3.74		4.98	7.45	18.6	5–11
	850	260		(4.98)	(7.38)	(18.38)	
5	4.57	3.75		5.76	7.50	21.6	30
	(43)	($\geq$ 1000)		(4.99)	(7.50)	(18.7)	
6	4.53	3.74		5.69	7.48	21.3	28
	(450)	($\geq$ 1000)		(5.02)	(7.48)	18.8	

[a] $d_{(h,k)}$ = d spacings (in Å) of the (h, k) planes; L = coherence length (Å) of the crystallites; a, b = lengths (in Å) of the unit cell axes; a_p, b_p = lengths of the unit cell projected along the molecular chain axis. In these systems $a_p = a$ and $b_p = b \cos t$; in the ester-alcohols $a_p = a \cos t$ and $b_p = b$; A = area per molecule = $ab/2$.

[b] Tilt angle t of monolayer molecules from the normal to the water surface in the b direction for the aliphatic alcohols and in the a direction for *5* and *6*.

[c] These data were measured on the synchrotron beamline BW-1; the remaining on beamline D4 at Hasylab, Hamburg. The best Bragg rod fitting for alcohol with n = 31 was obtained by assuming a continuum of chain tilts in the range of 5 to 11 deg, which was modeled by three tilts of 5, 8, and 11 deg with occupancies of 0.25, 0.5, and 0.25, respectively.[32]

[d] The {0, 2} reflection was absent; its d spacing and the axes were derived following a procedure described in Ref. 33.

[e] Probably $\leq$ 50 Å.

γ_r = 90 deg, as shown in Fig. 7.2. The rectangular unit cells of the $C_{30}H_{61}OH$ and $C_{31}H_{63}OH$ crystals are similar in dimension to the corresponding c-centered rectangular representation $(a_r b_r)$ of hexagonal ice, although the plane symmetry of the alcohol cell is not c-centered, rather primitive $p1g1$. Thus we may compare the extent of lateral match between

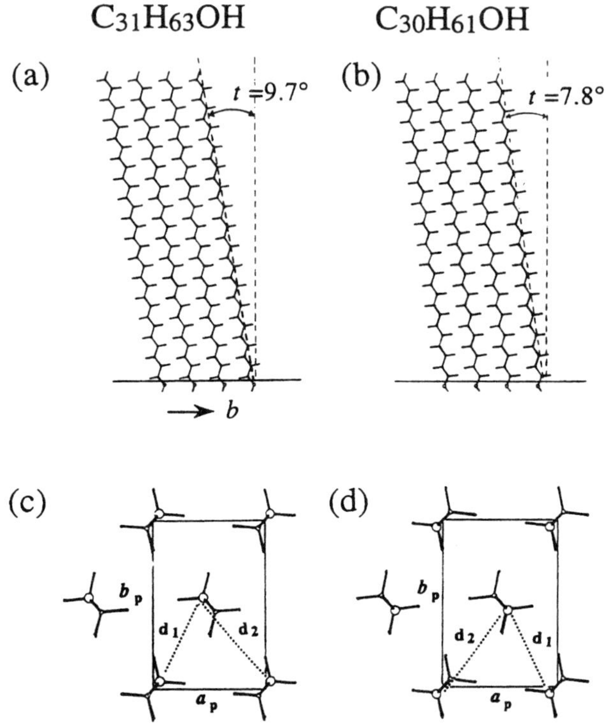

FIG. 7.4. (a) and (b); Structure of alcohol monolayers $C_nH_{2n+1}OH$ (n = 30, 31) view along the a axis. (c) and (d) View along the molecular chain axis. The O ··· O distances between the glide-related molecules in the unit cell are d_1 = 4.2 Å, d_2 = 4.9 Å for both monolayers. Note that a_p and b_p represent the unit cell axes (a and b) projected onto a plane perpendicular to the molecular chain axis.

each of the monolayers of $C_{31}H_{63}OH$ and $C_{30}H_{61}OH$ with the *ab* lattice structure of ice, by superimposing the 2-D nets of their oxygen atom arrangements. There is a reasonable fit over a common area of 600 to 800 $Å^2$, corresponding to about 30 to 40 molecular sites (Fig. 7.5). This common area is much smaller than the area of the crystalline coherence length within either the $C_{30}H_{61}OH$ or $C_{31}H_{63}OH$ monolayers, which is in the range of 1000 × 300 $Å^2$. The GID data of monolayers with n = 20 and 16 were different than those with n = 23, 30, and 31, in that the molecules of the former are tilted from the vertical by 19 deg along the b axis (see Table 7.1). This increase in the tilt angle vis-a-vis n = 23, 30, and 31 leads to an increase in the length of the b axis (Table 7.1) and so to a poorer lattice match with ice resulting in lower freezing points.

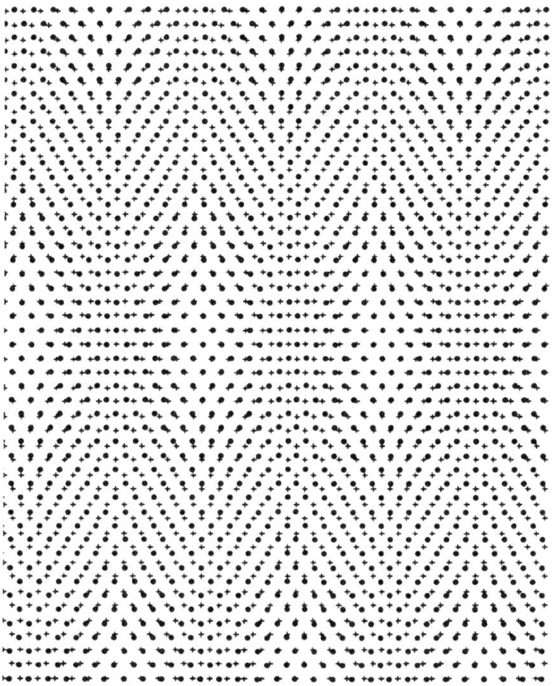

FIG. 7.5. Superimposed diagram of the oxygen positions within the *ab* layer of the hexagonal ice (labeled +) and the oxygen positions in a monolayer of $C_{31}H_{63}OH$ (labeled ●).

Direct Evidence for Epitaxial Relation Between the Monolayer and Ice

To establish which face of ice is nucleated under the alcohol monolayer, x-ray powder diffraction measurements of ice were carried out under various alcohol monolayers. Diffractograms obtained from ice nucleated under a monolayer of the alcohol $C_{31}H_{63}OH$ indicated preferred orientation of the ice crystals with the *c* axis perpendicular to the plane of the sample [Fig. 7.6(b)]. When ice was nucleated in the absence of the monolayer, a randomly oriented ice powder was obtained, which may be regarded as a reference pattern [Fig. 7.6(a)]. Ice created under a monolayer of 1H,1H,2H,2H-perfluorododecanol, which does not have a lattice match with the *ab* plane of ice, did not show any preferred orientation along the *c* axis.

Independent support for the structure of the $C_{31}H_{63}OH$ monolayer on hexagonal ice has been obtained by cryo-electron microscopy (EM). The EM studies of the $C_{31}H_{63}OH$ monolayer with hexagonal ice formed by

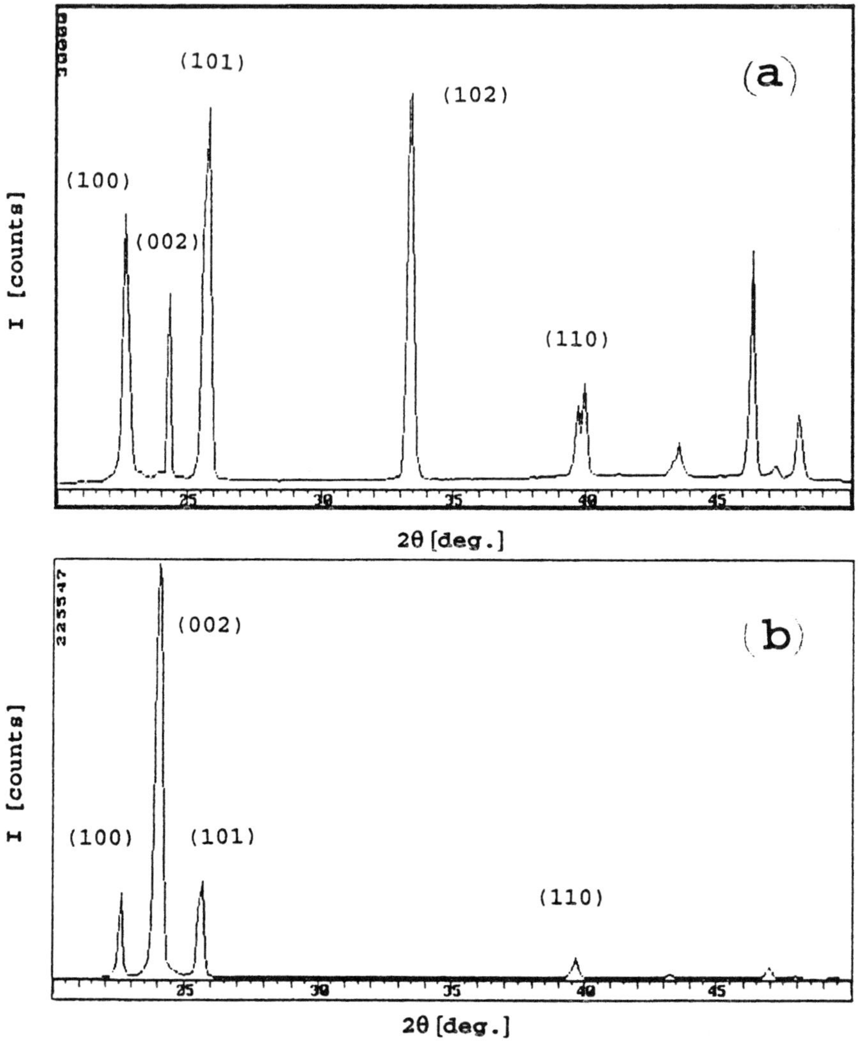

FIG. 7.6. X-ray powder diffractograms (CuK α radiation) of (a) ice frozen from pure water and (b) ice nucleated under a monolayer of $C_{31}H_{63}OH$ at 70% surface coverage.

slow cooling of the water subphase showed an oriented superposition of the six $\{1,0,0\}$ reflections of ice and the four $\{1,1\}$ reflections and two $\{0,2\}$ of the monolayer.[35]

Further demonstration of the epitaxial growth of ice under a monolayer of $C_{31}H_{63}OH$ and insight on the dynamics of ice nucleation under the monolayer was obtained by monitoring by GID the structure of the monolayer on cooling the water subphase from 5°C to freezing and the appearance of hexagonal ice.[28] Cooling the subphase from 5 to 2°C yielded changes in the Bragg rod intensity profiles, indicating a change in tilt angle of the amphiphilic chain from an average angle of 8 deg to one of 3.5 deg. When freezing occurred, the {1, 0, 0} Bragg peak from ice and the {1, 1} Bragg rod from the monolayer could be observed, demonstrating again that the ice crystals are oriented with their (001) face almost parallel to the surface (Fig. 7.7). From the width of the (100) ice diffraction peak, a lateral coherence length of 25 Å of the ice crystals was estimated. Such a low coherence length is indicative of multiple nucleation sites separated by about 50 Å. Consequently, the upper threshold value for the diameter of the critical nucleus of ice under the monolayer can be estimated to be maximally 25 Å, which would correspond to an ice-like hemisphere containing about 100 molecules. On melting the underlying ice, both the {1, 1} and {0, 2} peaks of the alcohol monolayer were again observed, indicating that the 2-D crystalline integrity was by and large maintained through the process of ice nucleation, growth, and melting.

7.3.2. The Effect of Chain Length (Coherence Length) on Ice Nucleation

Among the aliphatic alcohols, we observed an increase in ice nucleation temperature with increase in chain length. It would seem reasonable that in the uncompressed state, monolayer films of long amphiphiles $n > 22$, with more methylene-methylene interactions, are more highly ordered and more crystalline than those obtained from the short-chain alcohols, which tend to be more fluid like.[34] Indeed a correlation has been made between the degree of 2-D crystallinity and molecular interaction energy for a variety of different amphiphilic molecules on water.[36] Such a correlation can also be made for the alcohols according to GID and IR measurements. The GID measurements on the series $C_nH_{2n+1}OH$, for $n = 16, 20, 23, 30$, and 31 indicate that the extent of lateral order (i.e., crystalline coherence length) of these spontaneously formed 2-D crystalline clusters decreases with shorter chain length (Table 7.1). Moreover, the percentage amount of crystalline material is also somewhat dependent on chain length. According to an analysis of the GID data, it appears that the monolayers with $n = 23, 30$, and 31 contain about the same amount of crystalline material, but monolayers with $n = 16$ and 20 contain less crystalline material.[32]

From IR measurements of long-chain hydrocarbons it has been demonstrated that the frequencies of the antisymmetric and symmetric CH_2 stretching vibrations are conformation-sensitive and may be empirically correlated with order (i.e., the *trans*-gauche character) of the hydrocarbon

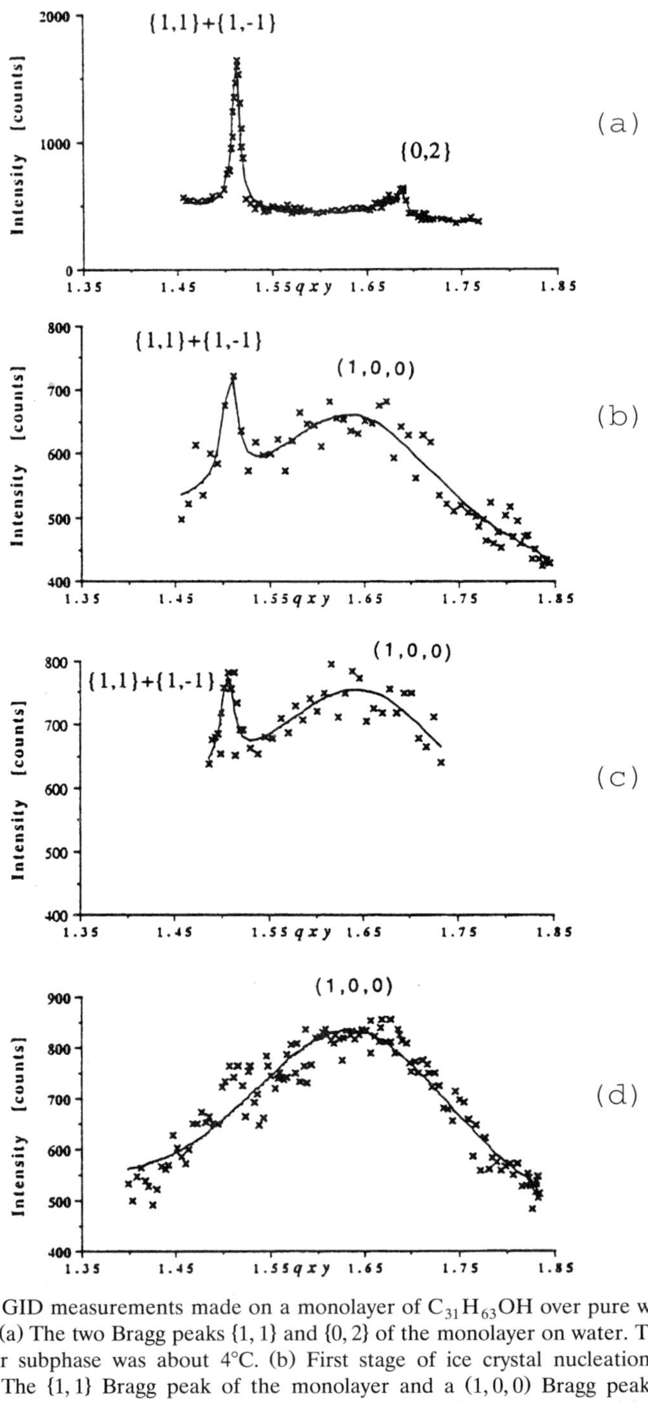

FIG. 7.7. GID measurements made on a monolayer of $C_{31}H_{63}OH$ over pure water cooled to freezing. (a) The two Bragg peaks {1, 1} and {0, 2} of the monolayer on water. Temperature of the water subphase was about 4°C. (b) First stage of ice crystal nucleation under the monolayer. The {1, 1} Bragg peak of the monolayer and a (1, 0, 0) Bragg peak of ice are visible. Temperature of the subphase was slightly below 0°C. (c) Same as part (b) after a time interval of about 15 min. (d) Same as part (c) but after a further 15 min. Only the (1, 0, 0) Bragg peak of ice is visible.

chain.[37] External reflectance IR spectra of the aliphatic alcohol monolayers at an area per molecule of 20 Å^2 have been measured at the air/water interface at 20°C. In this series, with increase in chain length, there is a shift of 1.0 to 1.5 cm^{-1} to lower wavenumbers and a decrease of 2 to 4 cm^{-1} in bandwidth, in both the antisymmetric and symmetric stretching bands, indicating that the chains become more ordered (Fig. 7.8).

By and large, we may come to the following general conclusion: The longer the chain length of $C_nH_{2n+1}OH$ ($16 < n < 31$), the greater the amount of crystalline material, the longer the extent of lateral coherence length, and the smaller the chain tilt angle t from the vertical axis. Thus it would seem that the more ordered the crystalline domains, the more efficient the monolayer for ice nucleation. However, we should note that the greater the tilt angle t of the molecular chain (i.e., $t \approx 9°$ for $n = 31$ and $t \approx 19$ deg for $n = 16$), the worse the lattice match to ice; thus, the degree of crystallinity and lattice match go hand in hand.

7.3.3. The Effect of Chain Parity on Ice Nucleation

Aliphatic alcohols $C_nH_{2n+1}OH$ that belong to the group with n = even nucleate ice at lower temperatures than those from the n = odd series, which suggests a systematic structural difference between the films of the two series of alcohols. The hydrocarbon chains in these two series of alcohol monolayers pack in similar packing arrangements as exemplified by the GID data of the $C_{30}H_{61}OH$ and $C_{31}H_{63}OH$ monolayers. However, there is an ambiguity that cannot be resolved from the GID data in fixing the (x, y) positions of the oxygen atoms. From the results of the ice nucleation experiments, we may conclude that the oxygen atoms of $C_{30}H_{61}OH$ and $C_{31}H_{63}OH$ are differently positioned. Consequently, the packing of the hydrocarbon chains of $C_{30}H_{61}OH$ and $C_{31}H_{63}OH$ are deduced to be alike from the terminal CH_3 groups down until atoms C(1) and C(2) for the $C_{30}H_{61}OH$ and $C_{31}H_{63}OH$ molecules, respectively. We have arbitrarily chosen the arrangement shown in Fig. 7.4(a) for $C_{31}H_{63}OH$. It follows that $C_{30}H_{61}OH$ will exhibit the "different" arrangement shown in Fig. 7.4(b). The interatomic arrangements of the OH groups of the two monolayers $C_{30}H_{61}OH$ and $C_{31}H_{63}OH$ are almost the same [Figs. 7.4(c) and (d)], but the two crystal structures cannot be superimposed in terms of the arrangements of their headgroup moieties CH_2CH_2OH. This suggests that headgroup orientation may strongly influence nucleation efficiency. We may tentatively conclude that a molecular rearrangement occurs at the headgroup (CH_2CH_2OH) level of the monolayer on induced ice formation, which must be different for the odd and even series, thus influencing

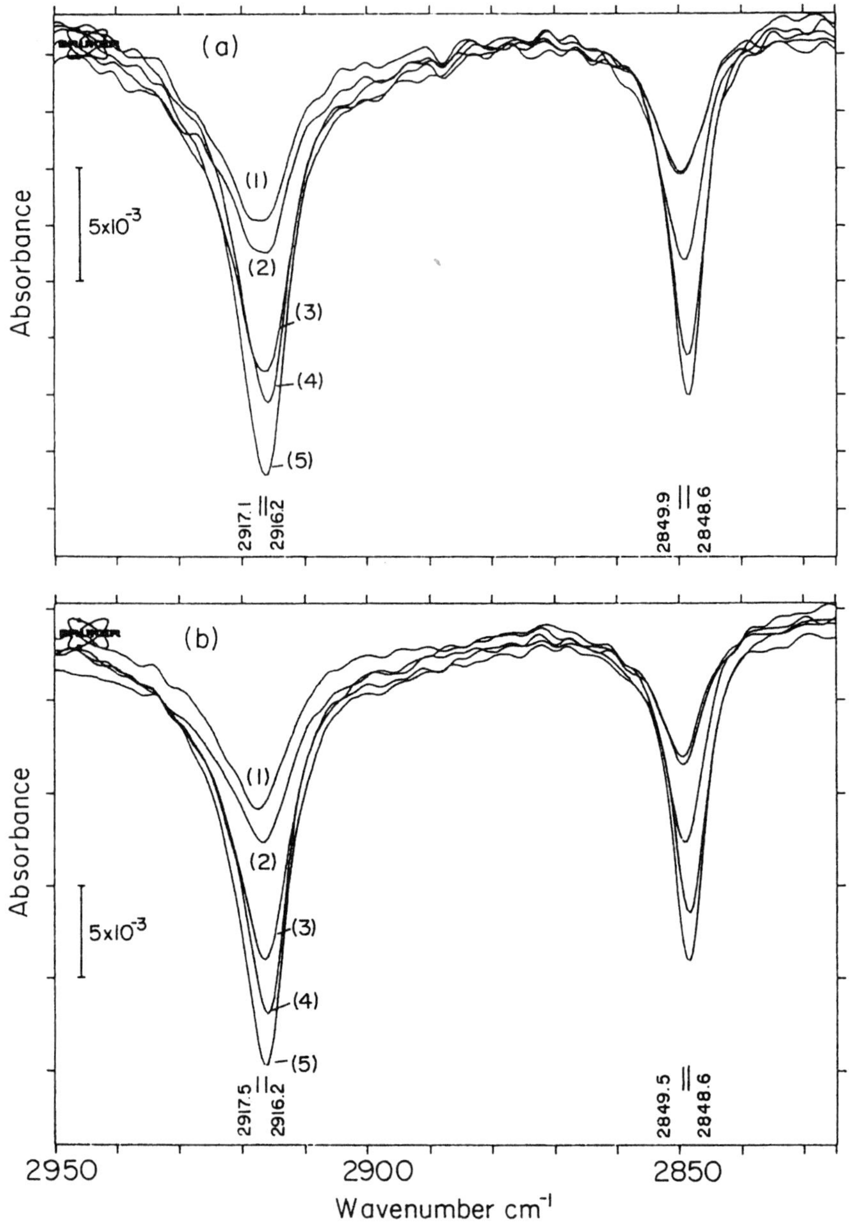

FIG. 7.8. FTIR external reflection spectra of monolayers of aliphatic alcohols $C_nH_{2n+1}OH$ with (1) $n = 16$, (2) $n = 19$, (3) $n = 22$, (4) $n = 24$, and (5) $n = 27$ on water, showing the hydrocarbon stretching region for (a) parallel and (b) perpendicular polarizations.

the free energy of nucleation. A reorganization mechanism has also been invoked in the induced crystallization of *p*-hydroxybenzoic acid with the assistance of *p*-alkoxybenzoic acid monolayers.[26]

7.3.4. Nucleation of Deuterium Oxide Under Monolayers of Aliphatic Alcohols

The aliphatic alcohols $C_nH_{2n+1}OH$ ($n = 16$ to 31) induced freezing of deuterium oxide and yielded results analogous to the experiments carried out on water, but with freezing temperatures proportionally shifted relative to its melting point (mp = +3.8°C). The freezing point curve for the n = odd series increased with chain length, reaching 1°C for $n = 31$, and in the n = even series a plateau is reached at −4°C for $n = 22$ (Fig. 7.9). These results are consistent with the calculation[38] that in the liquid state at low temperatures, the average cluster size in D_2O is larger than in H_2O. If we assume that the alcohol monolayer acts in a cooperative manner with the subphase in the stabilization of ice-like clusters at the interface, then the same monolayer should induce nucleation of D_2O at a higher temperature than H_2O.

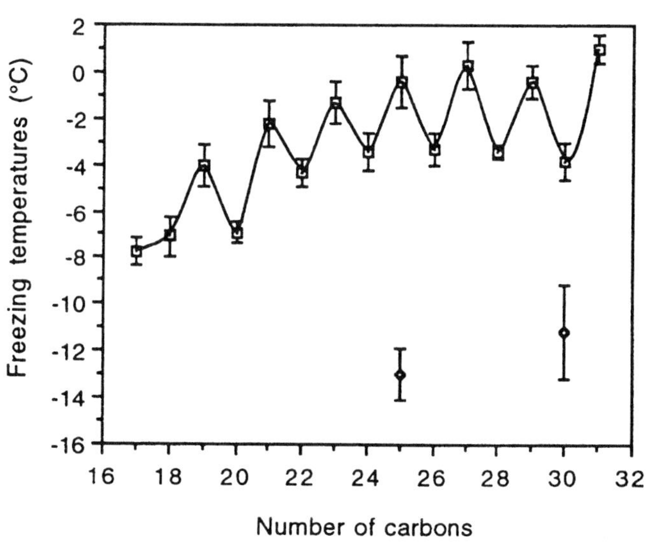

Fig. 7.9. Freezing temperatures of supercooled drops of deuterium oxide covered by monolayers of aliphatic alcohols $C_nH_{2n+1}OH$ (□) and carboxylic acids $C_{n-1}H_{2n-1}CO_2H$ (◇) versus the number of carbon atoms in the chain.

7.4. Nucleation of Ice Under Mixed Monolayers of Aliphatic Alcohols

Strong support for the observation that ice nucleation by the monolayers of $C_nH_{2n+1}OH$ results from specific structural interactions between the monolayer and the nucleating crystal was derived from experiments carried out with mixed monolayers.

7.4.1. Nucleation of Ice Under Mixed Monolayers of Two Miscible Components

Monolayers formed from two miscible hydrocarbon alcohols $C_nH_{2n+1}OH$ and $C_mH_{2m+1}OH$, to $n - m$ ranging from 1 to 4, showed typical solid solution behavior. Threshold freezing temperatures of water drops covered by mixed monolayers of aliphatic alcohols with $n = 29$ and $n = 31$ in various ratios were lower than the freezing points obtained by each of the pure monolayers (Fig. 7.10). There is an asymmetry in the curve of the mixed monolayers. Adding up to 60% of alcohol $C_{29}H_{59}OH$ to monolayer $C_{31}H_{63}OH$ gradually reduced the freezing temperature from

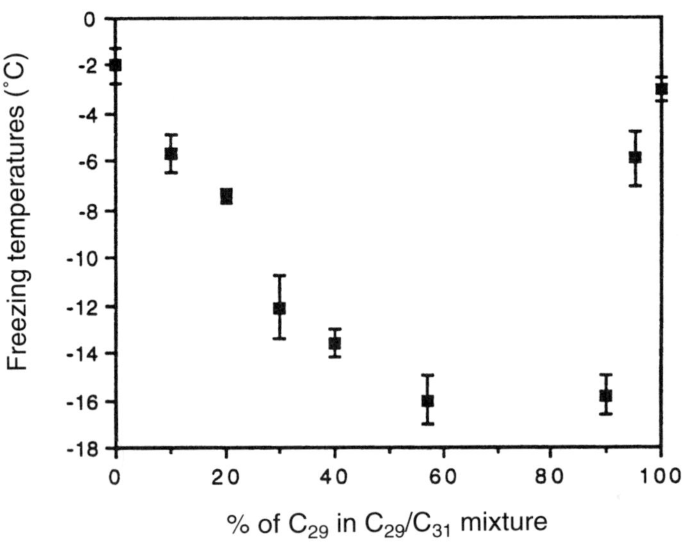

FIG. 7.10. Freezing temperatures of water drops covered by mixed monolayers of alcohols $C_nH_{2n+1}OH$ with $n = 29$ and $n = 31$ in various ratios versus the percentage of $C_{29}H_{59}OH$ in the mixture.

-1 to -16°C. On the other hand, only 10% of $C_{31}H_{63}OH$, was required to reduce the freezing temperature induced by monolayer $C_{29}H_{59}OH$ from -1.5 to -16°C in the mixture. In a different experiment, with two alcohols from the n = even series, $n = 28$ and $n = 30$, each inducing nucleation at the same temperature (-7.5°C), were mixed in various ratios, nucleation occurred at much lower temperatures (Fig. 7.11). Similar behavior was observed with mixed monolayers created by mixing two alcohols differing by only one methylene group ($n = 30$ with $n = 31$), as shown in Fig. 7.12. This behavior suggests that monolayers of two miscible alcohols create surfaces of hydroxyl groups whose structural fit, as observed for the pure phases, had been perturbed. This irregularity perhaps involves a surface roughness at the monolayer/water interface due to the difference in their chain length. GID experiments carried out with two miscible alcohols $C_{27}H_{55}OH$ and $C_{31}H_{63}OH$ have shown that these mixed monolayers in various proportions form 2-D crystalline solid solutions.[33] We may therefore infer that the contaminant molecules are randomly arranged in the monolayer. In such a system the effective coherence length of the OH headgroups of the monolayer must be reduced because the OH groups of the contaminant and the host molecule are not arranged in the

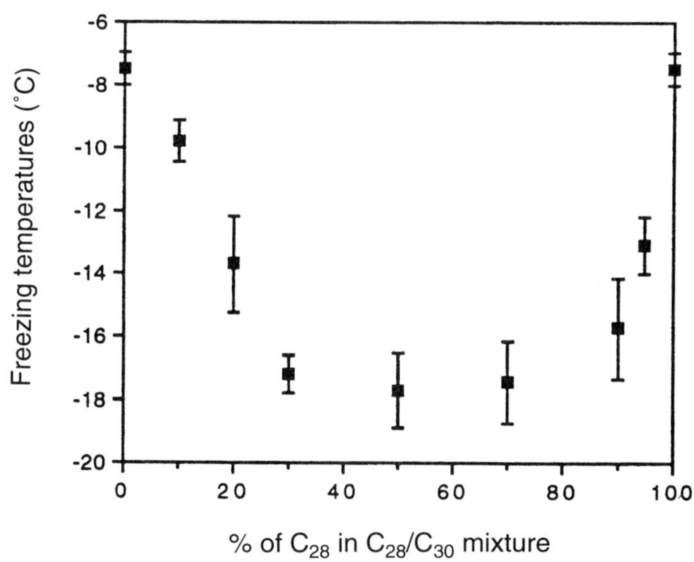

FIG. 7.11. Freezing temperatures of water drops covered by mixed monolayers of alcohols $C_nH_{2n+1}OH$ with $n = 28$ and $n = 30$ in various ratios versus the percentage of $C_{28}H_{57}OH$ in the mixture.

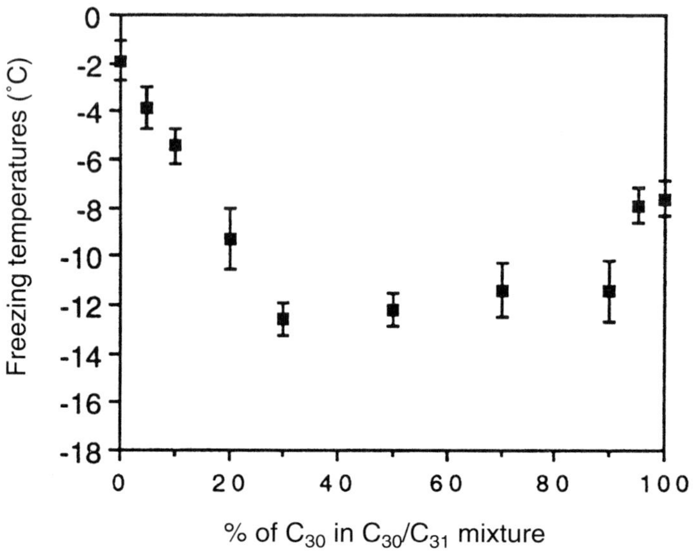

FIG. 7.12. Freezing temperatures of water drops covered by mixed monolayers of alcohols $C_nH_{2n+1}OH$ with $n = 30$ and $n = 31$ in various ratios versus the percentage of $C_{30}H_{61}OH$ in the mixture.

same manner. When a monolayer of the aliphatic alcohol $C_{30}H_{61}OH$ was mixed with the corresponding carboxylic acid $C_{29}H_{59}CO_2H$, which does not nucleate ice, a gradual decrease in freezing temperature was observed (Fig. 7.13).

7.4.2. Nucleation of Ice Under Mixed Monolayers of Two Immiscible Components

In contrast to the preceding binary systems, when the film is created from an aliphatic alcohol and a perfluorinated one, the two components segregate into immiscible crystalline domains. Up to 50% of $C_{10}F_{21}CH_2CH_2OH$ could be mixed in the monolayer of $C_{20}H_{41}OH$ without perturbing its ice nucleation activity as shown in Fig. 7.14. These experiments further demonstrate the effect of the crystalline coherence length on nucleation efficiency. One possible explanation for this observation is that the gradual increase in the percentage of perfluorododecanol above 50% causes a decrease in the crystalline domain size of the aliphatic alcohol without affecting the packing arrangement, which is reflected in a gradual decrease in nucleation efficiency.

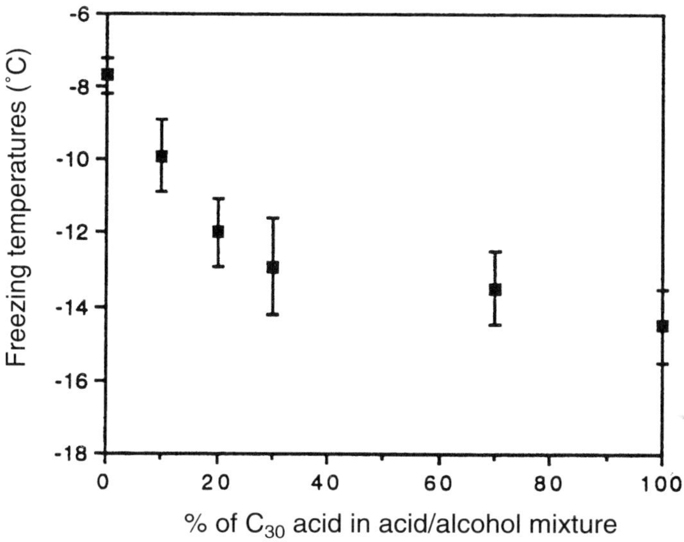

FIG. 7.13. Freezing temperatures of water drops covered by mixed monolayers of the aliphatic alcohol $C_{30}H_{61}OH$ and the corresponding carboxylic acid $C_{29}H_{59}CO_2H$ in various ratios versus the percentage of the acid in the mixture.

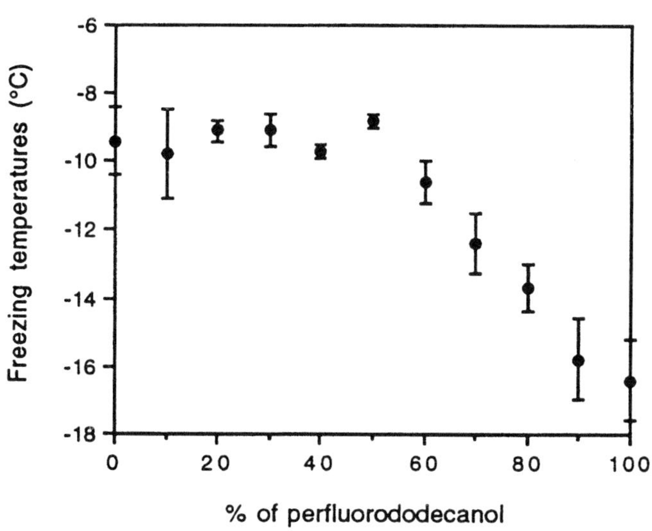

FIG. 7.14. Freezing temperatures of supercooled water drops covered by mixed monolayers of $C_{20}H_{41}OH$ with $C_{10}F_{21}CH_2CH_2OH$ in various ratios versus the percentage of the perfluorinated alcohol in the mixture.

7.5. Nucleation of Ice Under Alcohol Monolayers Bearing Amide or Ester Groups in the Hydrocarbon Chain

The differences in ice nucleation ability of the aliphatic alcohols of the odd and even *n* series are clearly due to structural differences involving their headgroup moieties. This raises the following question: What is the role played by the orientation of the OH group in the ice nucleation process? To probe this point regarding the preferred orientation of the alcohol O–H for best binding to an (001) layer of ice, we studied the ice nucleation activity of a series of amphiphiles bearing functional groups, such as amides and esters, along the aliphatic chain. Previous GID studies carried out on α-amino acid amphiphiles bearing an amide group in the hydrocarbon chain, had shown that the packing arrangement is strongly influenced by the hydrogen bonds between the amide groups interlinking the chains.[20] Thus, in aliphatic alcohols bearing amides or esters along the chain, the interactions between the functional groups should fix the packing arrangement so as to influence the orientation of the OH headgroup at the interface. For such amphiphiles, the areas per molecule, as obtained from $\Pi - A$ isotherms, are in the range of 22 to 24 $Å^2$. The ice nucleation results show no dependence of freezing temperature on length and parity of the total number of atoms in the chain (see Table 7.2). Nevertheless, we can divide the amphiphiles into two groups that nucleate ice at temperatures differing by about 4°C. The structural difference between these two groups is the parity of the hydrocarbon fragment connecting the functional group to the hydroxyl headgroup.

The intermolecular distance between hydrogen-bonded amide groups is ≈ 5 Å, which matches the length of the *a* axis in the crystallites of the $C_nH_{2n+1}OH$ monolayers (see Table 7.1). Thus, in light of the ice nucleation results obtained by the amide alcohols shown in Table 7.2, it is reasonable to assume that in the case of alcohols bearing amide groups, the N–H$\cdots$O hydrogen bonds influence the packing arrangement in a way that determines the orientation of the hydroxyl groups at the water surface. Thus, location of the hydroxyl groups placed an odd or even number of carbons from the amide bond may result in two different hydroxyl orientations relative to the surface. Indeed, a lengthening of the hydrocarbon fragment connecting the amide function to the methyl group by 10 carbons (*11* and *15*) did not have any effect on freezing temperature, whereas one added methylene in the hydrocarbon fragment connecting the amide with the hydroxyl headgroup (*11* and *12*) resulted in an increase of 4°C in the induced freezing temperature.

TABLE 7.2

FREEZING POINTS OF SUPERCOOLED WATER DROPS COVERED BY MONOLAYERS OF ALCOHOLS BEARING AN ESTER OR AN AMIDE GROUP

	Compound	Freezing Points
1	$CH_3(CH_2)_{10}CO_2(CH_2)_{10}OH$	-4.2 ± 0.6
2	$CH_3(CH_2)_{13}CO_2(CH_2)_9OH$	-10 ± 0.8
3	$CH_3(CH_2)_{13}CO_2(CH_2)_{10}OH$	-4.8 ± 0.7
4	$CH_3(CH_2)_{18}CO_2(CH_2)_8OH$	-5.5 ± 0.5
5	$CH_3(CH_2)_{18}CO_2(CH_2)_9OH$	-9.0 ± 0.8
6	$CH_3(CH_2)_{18}CO_2(CH_2)_{10}OH$	-4.0 ± 0.7
7	$CH_3(CH_2)_{18}CO_2(CH_2)_{12}OH$	-4.3 ± 0.9
8	$CH_3(CH_2)_{18}CO_2(CH_2)_{13}OH$	-8.0 ± 0.4
9	$CH_3(CH_2)_{20}CO_2(CH_2)_9OH$	-8.8 ± 0.8
10	$CH_3(CH_2)_{20}CO_2(CH_2)_{10}OH$	-3.0 ± 0.7
11	$CH_3(CH_2)_{10}CONH(CH_2)_{11}OH$	-7.5 ± 0.3
12	$CH_3(CH_2)_{10}CONH(CH_2)_{12}OH$	-3.3 ± 0.5
13	$CH_3(CH_2)_{14}CONH(CH_2)_{11}OH$	-7.4 ± 0.3
14	$CH_3(CH_2)_{14}CONH(CH_2)_{12}OH$	-3.3 ± 0.4
15	$CH_3(CH_2)_{20}CONH(CH_2)_{11}OH$	-7.3 ± 0.8
16	$CH_3(CH_2)_{20}CONH(CH_2)_{12}OH$	-3.8 ± 0.8

It is reasonable to assume in monolayers of mixed amide alcohols ($C_lH_{2l+1}CONHC_mH_{2m}OH + C_kH_{2k+1}CONHC_nH_{2n}OH$) that the hydrogen bonds between the amide groups will fix the orientation of the hydroxyl group depending on the value of n or m. Further support for this idea came from experiments carried out on mixed films of alcohols bearing amide groups at different positions along the hydrocarbon chain. When amide *12* was mixed in various ratios with *11*, which is shorter by one methylene group in the hydrocarbon fragment between the amide and the alcohol, the freezing temperature decreased gradually, reaching a minimum for a 1 : 1 mixture (Fig. 7.15). In contrast, when *12* was mixed with *14*, which is longer by four methylene groups in the hydrocarbon fragment between the amide function and the methyl group, there was almost no change in the freezing temperatures relative to the pure compounds (Fig. 7.16). Furthermore, in contrast to the behavior of mixed aliphatic alcohols, when *12* was mixed with an aliphatic alcohol of the same length ($C_{25}H_{51}OH$), there was no effect on freezing temperatures relative to the pure monolayers (Fig. 7.17). This behavior probably indicates that the two components segregate into immiscible domains, in contrast to mixed aliphatic alcohols and mixed amide-alcohols.

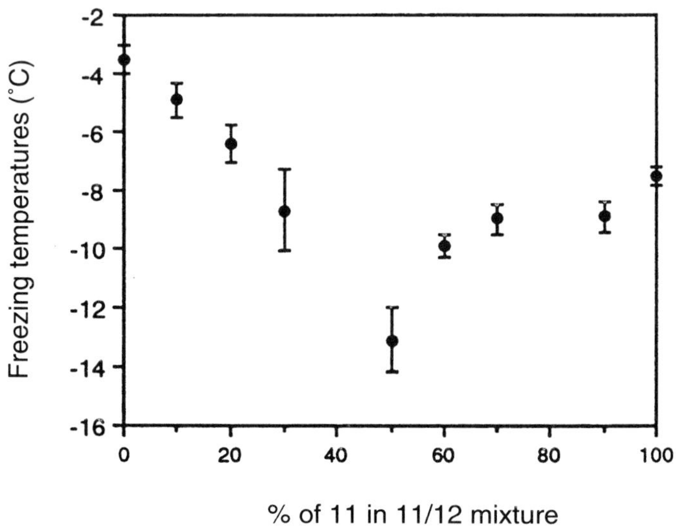

FIG. 7.15. Freezing temperatures of supercooled water drops covered by mixed monolayers of amide-alcohols *11* and *12* in various ratios versus the percentage of *11* in the mixture.

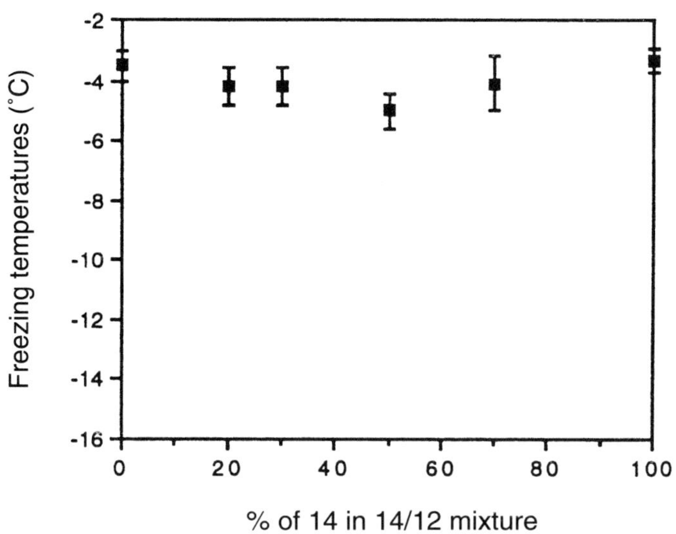

FIG. 7.16. Freezing temperatures of supercooled water drops covered by mixed monolayers of amide-alcohols *12* and *14* in various ratios versus the percentage of *14* in the mixed monolayer.

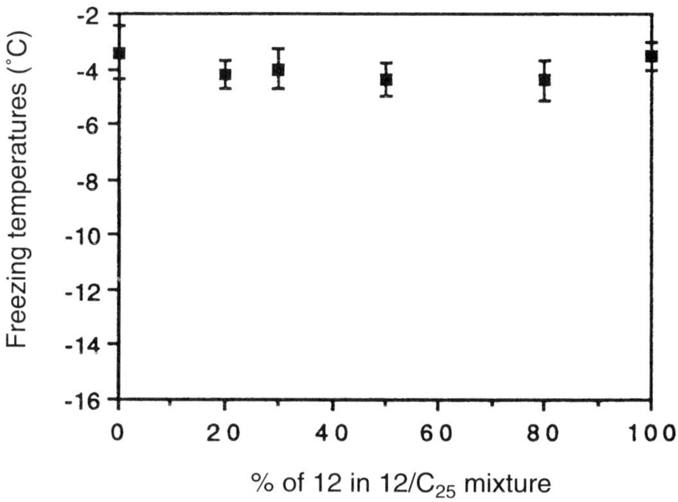

FIG. 7.17. Freezing temperatures of supercooled water drops covered by mixed monolayers of amide-alcohol *12* and the aliphatic alcohol $C_{25}H_{51}OH$ in various ratios versus the percentage of *12* in the mixture.

In a manner akin to the amide-alcohols, when ester groups were introduced along the hydrocarbon chains of the alcohols $C_nH_{2n+1}CO_2C_mH_{2m}OH$, the freezing point was found to depend only on the parity of the hydrocarbon fragment C_mH_{2m}, while the total chain length and parity had no influence on nucleation (Table 7.2).

The behavior observed for mixed ester-alcohols

$$C_nH_{2n+1}CO_2C_mH_{2m}OH + C_kH_{2k+1}CO_2C_lH_{2l}OH,$$

was different from that obtained with mixed amide-alcohols. When *10* was mixed in various ratios with *9*, which is shorter by one methylene group in the fragment connecting the ester to the hydroxyl headgroup, the freezing temperature decreases gradually, reaching a minimum at a 6 : 4 ratio (Fig. 7.18). In comparison, mixing of *10* in various ratios with *6*, which is shorter by two methylene groups in the fragment between the ester and the methyl group, resulted in a gradual decrease in the freezing point but to a lesser extent than in the *10* and *9* mixtures (Fig. 7.19). It is reasonable to assume that when two such amphiphiles are mixed, the interactions between the ester groups of neighboring molecules do not interlink the chains to the same extent as in the case of the amides. In contrast to the mixed amide-alcohol systems, mixed ester-alcohol monolayers differing in the fragment between the ester and the methyl had an effect on the freezing

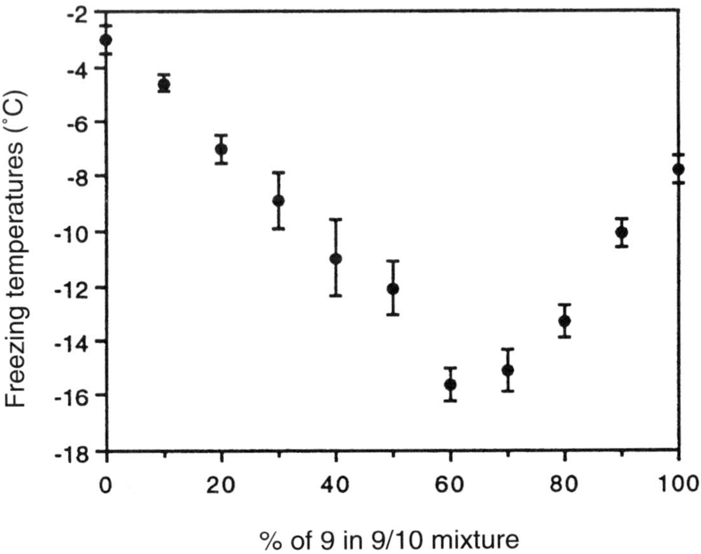

FIG. 7.18. Freezing temperatures of supercooled water drops covered by mixed monolayers of ester-alcohols *9* and *10* in various ratios versus the percentage of *9* in the mixture.

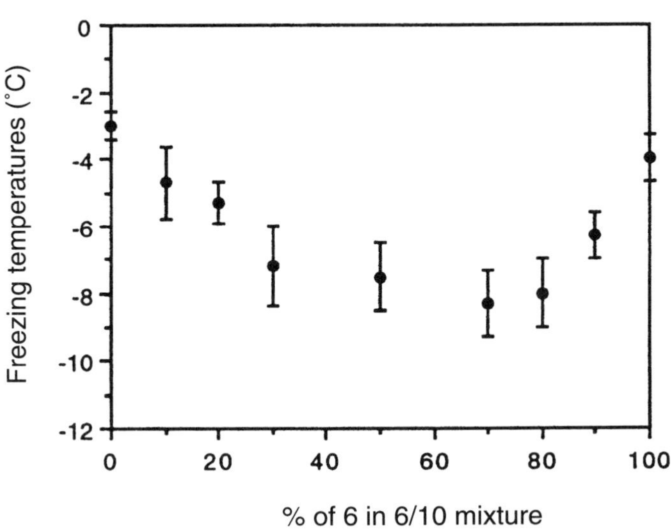

FIG. 7.19. Freezing temperatures of supercooled water drops covered by mixed monolayers of ester-alcohols *6* and *10* in various ratios versus the percentage of *6* in the mixture.

point, suggesting that the ester groups of neighboring molecules of different lengths are not at the same height due to a lack of hydrogen bonding as in the amides. This behavior is more like that of the aliphatic alcohols.

7.5.1. The Effect of Headgroup Orientation on Ice Nucleation

The experiments carried out with amide- and ester-alcohol monolayers demonstrate the effect of headgroup orientation on ice nucleation. That the orientation of the alcohol C–O–H moiety with respect to the water surface plays a definitive role in determining the ice nucleation temperature is supported by the analysis of GID measurements performed on the uncompressed monolayers[39] of $C_{19}H_{39}CO_2C_mH_{2m}OH$ ($m = 9, 10$) at 5°C. These monolayers showed spontaneous formation of crystalline self-aggregation with a high degree of lateral order. The GID patterns of these two monolayers are very similar, indicating similar packing arrangements. They crystallize in a rectangular unit cell with the dimensions listed in Table 7.1. The unit cell contains two molecules related by glide symmetry in the plane group *p*11*g*. The molecules are tilted by an angle of 29 deg from the vertical towards the *a* direction. The area per molecule projected along the molecular axis A_p is 18.6 Å^2, which is almost as low as in densest lamellar crystals.[34] The hydrocarbon chains are arranged in the orthogonal $O_\perp$ motif.[34] From the GID data it was not possible to distinguish between the two orientations shown in Scheme 7.1, which differ by a rotation of 180 deg about the molecular axis. Lattice energy calculations resolved this ambiguity showing a preference for the molecular orientation shown in Fig. 7.20. The result indicates that the energy difference arises primarily from interactions involving the ester moiety. The preferred molecular orientation is similar to that adopted by methyl stearate in its 3-D crystal structure.[39] Thus in the monolayer structure of $C_{19}H_{39}CO_2C_9H_{18}OH$, the C–OH group of the polar headgroup is almost perpendicular to the water

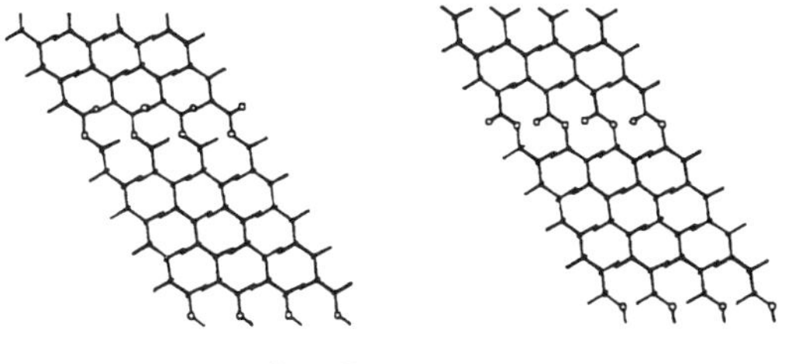

Scheme 7.1.

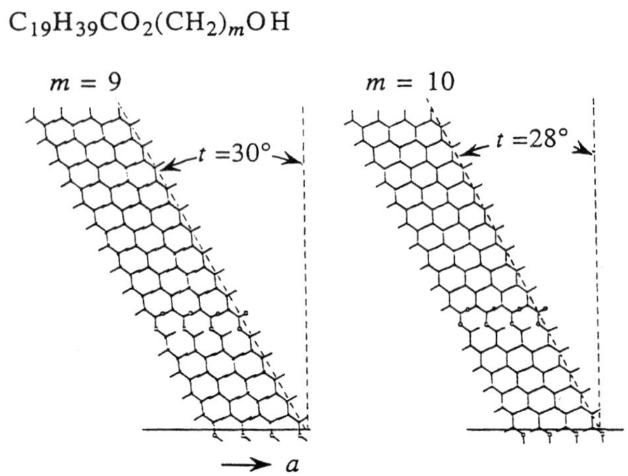

FIG. 7.20. Two-dimensional structures of monolayers of ester alcohols $C_{19}H_{39}CO_2C_mH_{2m}OH$, ($m$ = 9, 10) viewed along the b axis.

surface so that the O–H bond and the two lone-pair lobes of the hydroxyl oxygen atom are equally well exposed to water. In the monolayer structure of the $C_{19}H_{39}CO_2C_{10}H_{20}OH$ amphiphile, the C–OH bond makes an angle of about 20 deg to the water, leaving two possibilities: Either the O–H bond points into the water, or one oxygen lone-pair lobe and the OH group are exposed to water. We may correlate the difference in ice nucleating ability with the difference in orientation of the C–OH group with respect to the water surface. It is necessary to note here that the lattice match between these two alcohol structures to that of the *ab* net of ice is poor. Still to be resolved whether this result suggests that the degree of lattice match between the monolayer and ice need not be large or that there is a gradual change in the (001) layer arrangement of the H_2O molecules in going layer by layer from the alcohol monolayer to the "normal" ice structure. There is also an associated question still to be resolved: Is there a change of tilt of the amphiphilic molecules just prior to ice nucleation?

7.6. Induced Nucleation of Ice by Amphiphilic Alcohols Aggregated at the Oil /Water Interface

Amphiphilic molecules tend to accumulate in the form of monolayers at the interface between water and an immiscible liquid. We have exploited

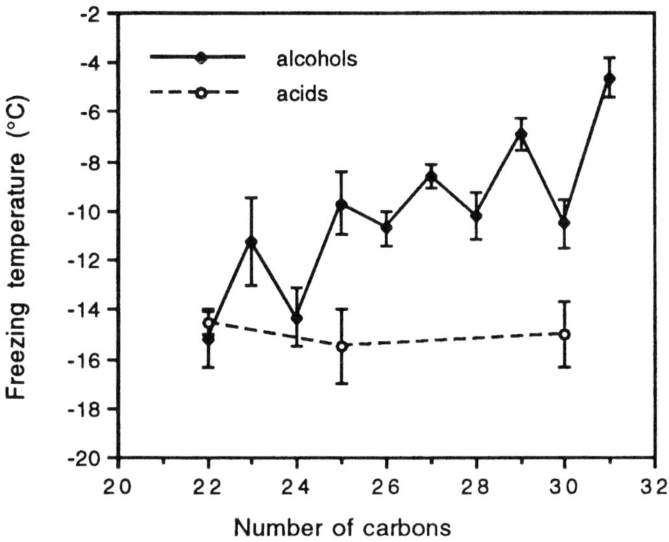

FIG. 7.21. Freezing points of supercooled 10-mL water drops in contact with 2.5×10^{-4} M solutions of aliphatic alcohols ($C_nH_{2n+1}OH$) and acids ($C_{n-1}H_{2n-1}CO_2H$) in 1/1 chloroform-hexane. Reprinted with permission from Popovitz-Biro et al. *J. Am. Chem. Soc.* **113,** 8943–8944 Copyright © 1991 American Chemical Society.

the structural specificity for ice nucleation to probe the oil/water interface on a molecular level.[40] Experiments were carried out by measuring the freezing points of water drops placed in organic solutions of long-chain aliphatic alcohols. As a reference, the solutions of the corresponding carboxylic acids or the pure solvents have been used. Analogous to the air/water interface, aliphatic alcohols $C_nH_{2n+1}OH$ ($n > 22$) nucleate ice at higher temperature than the corresponding carboxylic acids (Fig. 7.21). Moreover, the freezing temperatures were observed to depend on alcohol chain length, on parity, and on the chemical nature of the solvent. The largest difference of about 6°C was obtained between the freezing points induced by alcohols with $n = 30$ and $n = 31$. In view of the air/water studies, these results strongly suggest that the amphiphilic alcohols aggregate at the oil/water interface into crystalline clusters. Furthermore, as at the air/water interface, the surfaces (or templates) created by aggregates of the odd-numbered alcohols seem to have a better structural fit with ice than the even-numbered homologues. The fact that an ice nucleation effect at the oil/water interface was observed with alcohols having a chain length of $n > 22$, whereas at the air/water interface it required only $n > 16$, suggests that longer chains with more hydrophobic interactions are required in order to overcome the disordering effect of the organic solvents.

The analogy with the air/water interface also holds in the case of mixed alcohol solutions. Nucleation experiments performed in solutions containing mixtures of alcohols with $n = 30$ and $n = 31$ in various ratios (Fig. 7.22) showed typical "solid solution" behavior. Similar behavior was obtained with solutions of mixed alcohols with chains having $n = 28$ and $n = 30$ in various ratios, although the freezing points induced by the two pure phases were the same (Fig. 7.23). These results suggest that aggregates of the mixed alcohols have hydroxy groups that create surfaces with a larger structural mismatch to ice than each of the pure phases.

In another series of experiments, the effect of the chemical nature of the organic solvent on ice nucleation was studied. Freezing points induced by solutions of alcohols with $n = 30$ and $n = 31$ in various solvents, have been measured (Table 7.3). When alcohols were dissolved in octane or decane the differences in freezing temperature induced by alcohols with $n = 30$ and $n = 31$ almost completely disappeared. This indicates perhaps that the orientation of the hydroxyl group with respect to the water surface for $n = 30$ and $n = 31$ are the same, which would be in keeping with an "ordered" layer of hydrocarbons above the amphiphilic monolayer. This "ordered" hydrocarbon layer would act as a "buffer" system and so

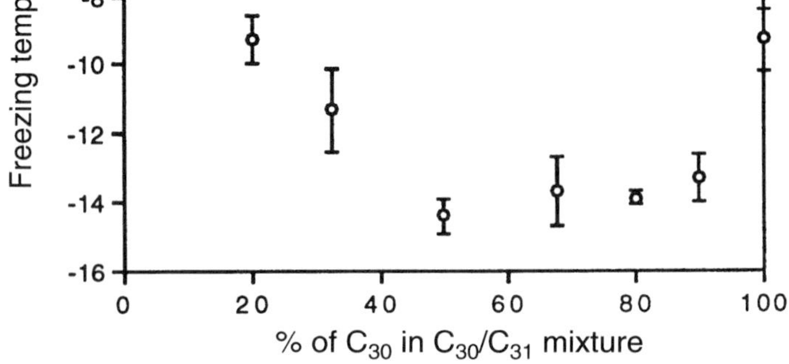

FIG. 7.22. Freezing points of supercooled water drops in contact with 2.5×10^{-4} M solutions of mixtures of C_{30}/C_{31} alcohols in 1/1 chloroform-hexane versus percent of C_{30}.

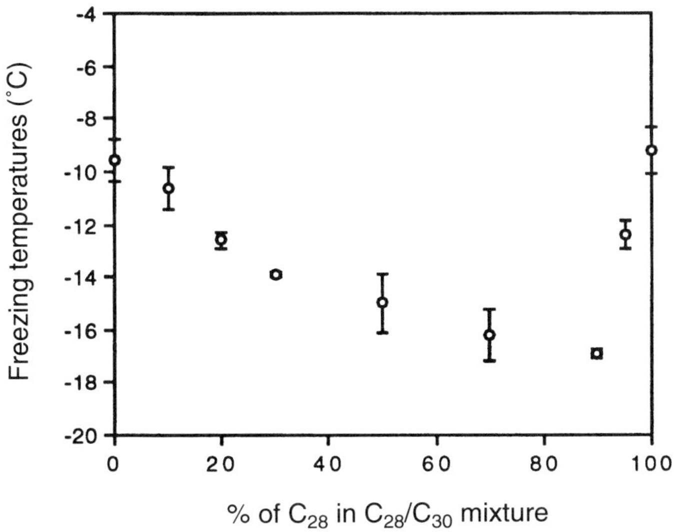

FIG. 7.23. Freezing points of supercooled water drops in contact with 2.5×10^{-4} M solutions of mixtures of C_{28}/C_{30} alcohols in 1/1 chloroform-hexane versus percent of C_{28}.

diminish the original structural differences between the odd and even aggregates. The odd/even effect was observed when the alcohols were dissolved in chloroform or toluene, solvents that are of different chemical nature than the amphiphiles.

The effect of alcohol bulk concentration on the threshold freezing temperatures was also examined, as shown in Fig. 7.24. It is remarkable

TABLE 7.3

FREEZING POINTS OF SUPERCOOLED WATER DROPS IN CONTACT WITH ORGANIC SOLUTIONS OF ALCOHOLS $C_nH_{2n+1}OH$ (n = 30, 31) VERSUS SOLVENTS

Solvent	$n = 30$	$n = 31$	Reference[a]
$CHCl_3$	-9.4 ± 1.0	-4.7 ± 0.6	
$CHCl_3$-hexane	-10.5 ± 1.0	-4.6 ± 0.8	-16.5 ± 2.0
$CHCl_3$-cyclohexane	-8.3 ± 0.2	-4.5 ± 0.8	
Toluene	-9.1 ± 1.0	-3.7 ± 1.0	
Octane	-7.2 ± 0.8	-5.7 ± 1.0	-16.6 ± 1.5
Decane	-6.8 ± 0.8	-6.2 ± 1.4	-15.3 ± 2.0

[a] As references, the freezing points of water drops placed in the pure solvents have been used. Reprinted with permission from Popovitz-Biro et al. *J. Am. Chem. Soc.* **113,** 8943–8944 Copyright © 1991 American Chemical Society.

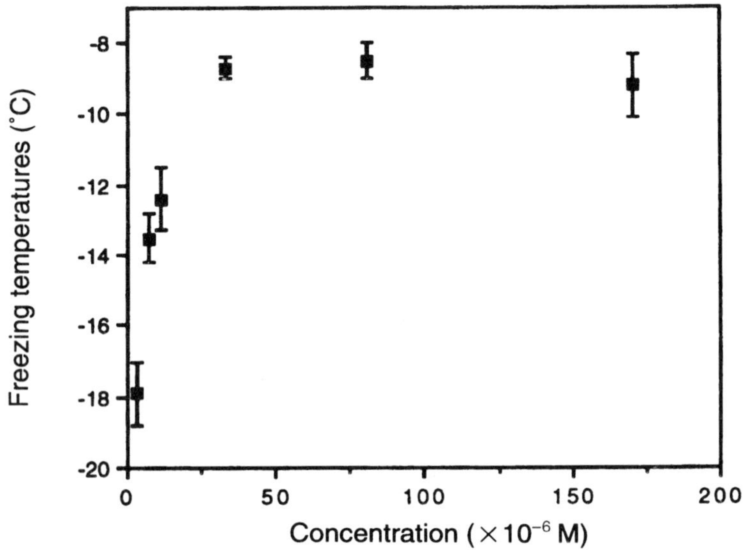

FIG. 7.24. Freezing points of supercooled water drops in contact with solutions of C_{30} alcohol in 1/1 chloroform-hexane versus concentration.

that surface accumulation of C_{30} alcohol at the octane/water interface could not be detected by interfacial tension measurements at bulk concentration of 2.5×10^{-4} M and at 20°C; whereas maximum nucleation efficiencies were reached at alcohol concentrations as low as 5×10^{-5} M. These results suggest very low surface coverage of the alcohols and possibly small domain sizes.

7.7. Induced Nucleation of Ice on Surfaces of Polar Crystals of α-Amino Acids

In the preceding sections we discussed the effect of a (partial) lattice and structural fit between the hydroxyl OH surface of an alcohol monolayer and an (001) layer of hexagonal ice on the freezing temperature of water covered by such a monolayer. Reports in the literature on the induced ice nucleation imply that there are mechanisms other than an epitaxial fit by which auxiliaries might induce ice nucleation. For example, it was reported that the onset temperature of ice nucleation by crystalline powders of chiral-resolved hydrophobic α-amino acids such as valine, leucine, and isoleucine was higher than the temperature induced by the corresponding racemic counterpart.[4,5,41] These results were surprising because the chiral and racemic crystal forms of these amino acids resemble each other in molecular packing and crystal morphology. Moreover, there is no structural and lattice match between these amino acids and ice, thus, nucleation by epitaxy is hardly possible. Therefore, a comparative study of ice nucleation on isostructural faces of single crystals of chiral and racemic α-amino acids was carried out.[19]

7.7.1. Hydrophobic α-Amino Acids

The hydrophobic α-amino acids pack in hydrogen-bonded layers.[42–51] In the chiral crystals, molecules within the layer are related by translation and interlinked by N–H···O hydrogen bonds. In the corresponding racemic crystals, the molecules appear in similar layer motifs as exemplified by L-Val and D,L-Val (Fig. 7.25). The molecular layers are interlinked by N–H ··· O hydrogen bonds to form bilayers. These bilayers are generated in the chiral crystals by twofold symmetry and in the racemic crystals by centers of inversion. The chiral crystals are polar by virtue of the twofold axis parallel to the plane of the bilayer and the racemic crystals are centrosymmetric and therefore nonpolar.

7.7.2. Ice Nucleation on Hydrophobic α-Amino Acid Crystals

All hydrophobic α-amino acids crystallize from aqueous solutions as (001) plates whose two crystal faces are parallel to the plane of the bilayer. According to water contact angles (Table 7.4), these faces are hydrophobic, thus exposing their hydrocarbon residues. Water vapor was condensed at 0°C on these faces of the chiral and racemic crystals, which were simultaneously cooled until freezing occurred. The polar chiral α-amino crystals induce ice nucleation at temperatures higher by 3 to 5°C than the corresponding centrosymmetric racemates (Table 7.4). In a different type of experiment, the substrate crystals were cooled to -15°C before water vapor deposition. Ice crystals with hexagonal morphology and with the *c* axis parallel to the face of the substrate were found to emerge from cracks in the α-amino acid crystals (Fig. 7.26).

Role of Crystal Polarity of α-Amino Acids for Induced Ice Nucleation

In view of the preceding results, a model was proposed to explain ice nucleation on a molecular level. The structure of the plate faces of the polar and nonpolar α-amino acid crystals are depicted schematically in Fig. 7.26. The two faces are essentially the same in structure, exposing hydrocarbon groups. Both faces contain cracks that expose the hydrophobic as well as the hydrophilic groups. In a polar crystal the structures of the two

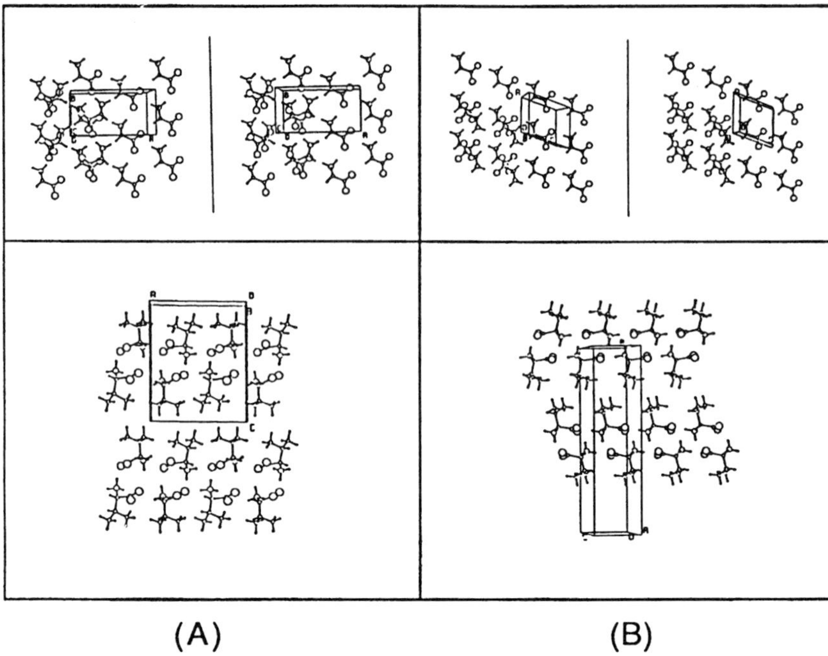

FIG. 7.25. Comparison of packing arrangements of (A) L-Val and (B) D,L-Val. The upper figures show stereoscopic views of the hydrogen-bonded layers, which omit the isopropyl groups. Note the similarity in structures of L- and D,L-Val. In L-Val the two layers are related by pseudorotation symmetry about the polar *b* axis (which is the vertical direction); in D,L-Val the two layers are related by a center of inversion. The lower figures show side views of neighboring hydrogen-bonded bilayers. In (A) L-Val, the neighboring bilayers are related by translation. In (B) D,L-Val, the neighboring bilayers are related by twofold screw symmetry. From Gavish et al., *Science*, **250,** 973–975, 1990, © AAAS.

opposite surfaces within the crack, perpendicular to the polar axis, are radically different. Moreover, were such a surface to carry a net charge, the two opposite faces would carry charges of opposite sign. In contrast, the opposite faces within a crack of a centrosymmetric crystal have similar structures with no net charge on each of these two faces. Therefore, we anticipate that an effective net charge on the opposite faces of cracks along the polar axis can produce an electric field strong enough for the alignment of water molecules into proton-ordered ice-like clusters, which would be polar along its hexagonal *c* axis. According to this model it was predicted that for Ala and Tyr, whose racemates have a polar axis and whose chiral crystal counterparts do not have polar axes, the racemates were better ice nucleators (Table 7.4). This model is in agreement with calculations of interaction energy between polar and nonpolar hexagonal

TABLE 7.4

FREEZING POINTS (FP) AND CONTACT ANGLES (α) OF WATER ON α-AMINO ACID CRYSTALS (α-AMINO OCTANOIC ACID, AOA)

	Chiral-Resolved Crystals			Racemic Crystals		
Amino Acid	FP (°C)	α(deg)	SG	FP (°C)	α(deg)	SG
A:[a]						
Val	-5.6 ± 0.6	80	$P2_1$	-9.9 ± 0.8	80	$P2_1/c$
Leu	-5.5 ± 0.5	89	$P2_1$	-8.1 ± 0.5	89	$P\bar{1}$
Ile	-5.1 ± 0.5	91	$P2_1$	-9.8 ± 0.6	89	$P\bar{1}$
Met	-3.7 ± 0.3	80	$P2_1$	-7.2 ± 0.1	79	$P2_1/a$
Norleu	-4.1 ± 0.3	88	$C2$	-6.7 ± 0.2	88	$P2_1/a$
tert-Leu	-5.8 ± 0.6	75	$P1$	-8.6 ± 0.5	73	$P2_1/c$
AOA	-4.1 ± 0.2	89	$C2$	-7.3 ± 0.3	89	[b]
B:						
Tyr	-6.6 ± 0.3		$P2_12_12_1$	-1.1 ± 0.2		$Pna2_1$
Ala	-7.5 ± 0.6		$P2_12_12_1$	-2.6 ± 0.3		$Pna2_1$

[a] The space groups (SG) of the amino acids listed in (A) exhibit a polar axis for the chiral resolved (L) compounds and nonpolar axes for the racemic (D,L) compounds.

[b] The x-ray diffraction photographs of racemic AOA showed pronounced diffuse scattering, indicative of molecular disorder, so that the space group could not be unambiguously assigned. Indications were very strong that the crystal structure is monoclinic, centrosymmetric $P2_1/c$.

ice-like clusters contained within the crevice of both the L and the D,L forms (Table 7.5).[52] The interaction energy involving D,L-Ala and a polar ice-like cluster was lower than the corresponding value for D,L-Ala and a nonpolar ice cluster or L-Ala crystal with either polar or nonpolar ice cluster. Thus, it is the favorable electrostatic energy of a polar ice-like crystallite in an electric field that assists the stabilization of the ice cluster. This molecular approach appears to be consistent with a macroscopic model proposed by Wilen[52]: The higher freezing point of ice in the polar crystals is because the dielectric constant of ice is greater than that of water and thus an electric field lowers the free energy of an ice nucleus. The connection between this macroscopic model and the molecular model outlined earlier is that with an assumed polar ice nucleus, one has in effect increased the dielectric constant of the intercalated ice medium.

7.8. Conclusion and Outlook

The phase transition of supercooled water into ice is a complex process and appears to occur by a variety of routes. Here we have presented two

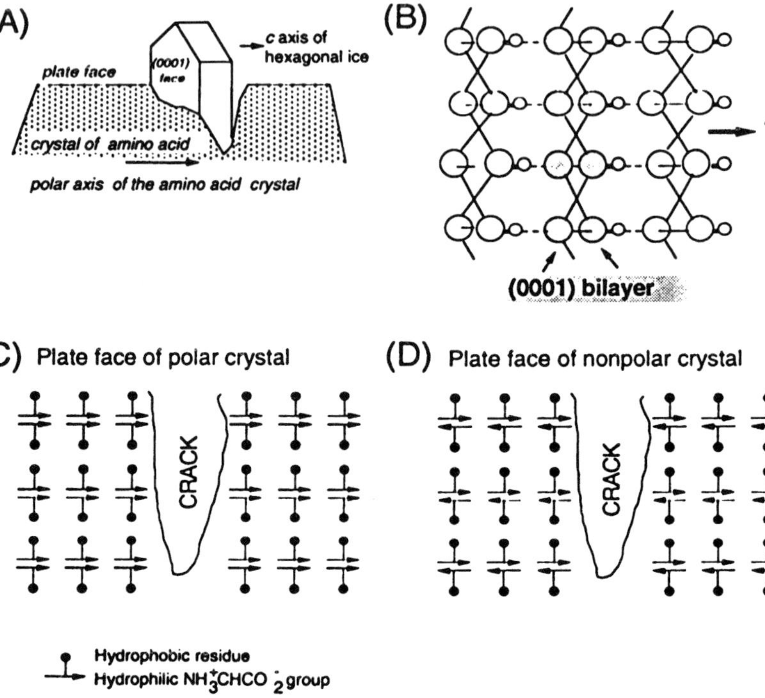

FIG. 7.26. (A) Schematic view of hexagonal ice crystals (exaggerated for clarity) emerging from the plate face of an α-amino acid crystal. (B) Molecular packing of a hexagonal ice form in which the O–H···O hydrogen bonds along the unique axis are assumed to be proton ordered and so the structure is polar along this direction. The three remaining O–H···O bonds per molecule within each (0001) bilayer are assumed to be proton disordered. (C) Schematic view of a polar α-amino acid crystal composed of hydrogen-bonded bilayers. The polar axis is along the horizontal direction. The opposite faces within the crack expose different groups. (D) Schematic view of the centrosymmetric α-amino acid crystal. The opposite faces within the crack are equivalent. From Gavish et al., *Science*, **250**, 973–975, 1990, © AAAS.

different approaches. In one we have used the self-organization of amphiphilic alcohols, which form 2-D crystalline domains at the air/water interface. The alcohol hydroxyl groups that make contact with the top water layer help organize and stabilize the bound water molecules into ice-like clusters. The nucleation temperature depends on many factors including the lattice and structural match between the monolayer and hexagonal ice and the orientation of the O–H bond and lone pairs toward the water surface.

TABLE 7.5

CALCULATED INTERACTION ENERGY E BETWEEN THE D,L AND L FORMS OF ALANINE WITH $n = 2$ OR 3 BILAYERS OF ICE INTERCALATED WITHIN A CREVICE OF ALANINE[a]

n	$E_{D,L}$	E_L	ΔE
		Polar Ice	
2	−4.0	−2.7	−1.3
3	−3.3	−1.9	−1.4
		Nonpolar Ice	
2	−1.2	−1.3	0.1
3	−1.1	−1.0	−0.1

[a] Polar and nonpolar ice models were used. Energies are in kcal/mol.

In the second approach, it has been shown that polar ice-like clusters inside the crevices of polar crystals are stabilized by virtue of an electric field. It is possible that some systems may take advantage of both epitaxy and electric field to induce freezing into ice. For example, it has been assumed that crystals of AgI, which appear in a hexagonal lattice with cell dimensions very similar to hexagonal ice, induce ice nucleation by epitaxy. However, since silver iodide is a polar crystal, strong electric fields in crevices along the polar axis may be a contributing factor for induced ice nucleation. A second system of interest is the frost bacteria *Pseudomonas syringae*[9] which induces freezing of water at −4°C and can cause serious damage to crops. Here the question is open as to the mode of action of the active site of the ice-nucleating protein in the bacteria. A model has been suggested in which the protein constitutes a hexagonal entity through which ice nucleation can occur by some kind of epitaxy.[54] On the other hand, one may envisage that α-helices, which incorporate large dipole moments and possess charged groups such as NH_3^+ and CO_2^- at the ends, may generate electric fields strong enough for ice-like clusters to be stabilized. In conclusion, clarification of the different pathways of ice nucleation by using model systems may help unravel the mode of operation of complex biological systems or the silver iodide phases containing Na or K additives, which are of practical importance for rain precipitation.

Acknowledgments

This project was supported by the Minerva Foundation, Munich, Germany.

References and Notes

1. Bryant, G. W., Hallett, J., and Mason, B. J. (1959). *J. Phys. Chem. Solids* **12,** 189.
2. Davis, B. L., Johnson, L. R., and Moeng, F. J. (1975). *J. Appl. Meteor.* **14,** 891.
3. Langer, G., Rosinski, J., and Bernsen, S. (1963). *J. Atmos. Sci.* **20,** 557.
4. Parungo, F. P., and Lodge, J. P. (1967). *J. Atmos. Sci.* **24,** 274.
5. Barthakur, N., and Maybank, D. J. (1963). *Nature* **200,** 866.
6. Fukuta, N., and Mason, B. J. (1963). *J. Phys. Chem. Solids* **24,** 715.
7. Levin, Z., Yankofsky, S. A., Pardes, D., and Magal, N. (1987). *J. Climate. Appl. Meteor.* **26,** 1188.
8. Green, R. L., and Warren, G. J. (1985). *Nature* **317,** 645.
9. Caple, G., Sands, D. C., Layton, R. G., Zucker, W. V., and Sneider, J. R. (1986). *J. Theor. Biol.* **119,** 37.
10. Vonnegut, B. (1947). *J. Appl. Phys.* **18,** 593.
11. Corrin, M. L., and Nelson, J. A. (1968). *J. Phys. Chem.* **72,** 643.
12. Schnell, R. C., and Vali, C. (1972). *Nature* **263,** 163.
13. Lindow, S. E., Arny, D. C., and Upper, C. D. (1978). *Phytopathology* **68,** 523.
14. Maki, L. R., Galyon, E. L., Chien, M. C., and Colwell, D. R. (1974). *Appl. Microbiol.* **28,** 456.
15. Thangaraj, K., Palanisamy, M., Gobinathan, R., and Ramasamy, P. (1986). *J. Mater. Sci. Lett.* **35,** 326.
16. Pruppacher, H. R., and Pflaum, J. C. (1975). *J. Colloid Interface Sci.* **52,** 543.
17. Shichiri, T., and Nagata, T. (1981). *J. Crystal Growth* **54,** 207.
18. Gavish, M., Popovitz-Biro, R., Lahav, M., and Leiserowitz, L. (1990). *Science* **250,** 973.
19. Gavish, M., Wang, J.-L., Eisenstein, M., Lahav, M., and Leiserowitz, L. (1992). *Science* **256,** 815.
20. Landau, E. M., Grayer Wolf, S., Levanon, M., Leiserowitz, L., Lahav, M., and Sagiv, J. (1989). *J. Am. Chem. Soc.* **111,** 1436.
21. Weissbuch, I. (1993). *J. Phys. Chem.* **97,** 12848.
22. Zhao, X. K., Yang, J., McCormick, L. D., and Fendler, J. H. (1992). *J. Phys. Chem.* **96,** 9933.
23. Landau, E. M., Popovitz-Biro, R., Levanon, M., Leiserowitz, L., Lahav, M., and Sagiv, J. (1986). *Mol. Cryst. Liq. Cryst.* **134,** 323.
24. Mann, S., Heywood, B. R., Rajam, S., and Birchall, J. D. (1988). *Nature* **334,** 692.
25. Heywood, B. R., and Mann, S. (1992). *J. Am. Chem. Soc.* **114,** 4682.
26. Weissbuch, I., Berkovic, G., Leiserowitz, L., and Lahav, M. (1990). *J. Am. Chem. Soc.* **111,** 5874.
27. Jacquemain, D., Leveiller, F., Weinbach, S., Lahav, M., Leiserowitz, L., Kjaer, K., and Als-Nielsen, J. (1991). *J. Am. Chem. Soc.* **113,** 7684–7691.
28. Majewski, J., Popovitz-Biro, R., Kjaer, K., Als-Nielsen, J., Lahav, M., and Leiserowitz, L. (1994). *J. Phys. Chem.* **98,** 4087.
29. Pauling, L. (1935). *J. Am. Chem. Soc.* **57,** 2680.
30. Peterson, W., and Levy, H. A. (1957). *Acta Cryst.* **10,** 70.
31. Jacquemain, D., Grayer Wolf, S., Leveiller, F., Lahav, M., Leiserowitz, L., Deutsch, M., Kjaer, K., and Als-Nielsen, J. (1990). *J. Am. Chem. Soc.* **112,** 7724.
32. Popovitz-Biro, R., Wang, J. L., Majewski, J., Shavit, E., Leiserowitz, L., and Lahav, M. (1994). *J. Am. Chem. Soc.*, **116,** 1179.
33. Wang, J. L., Levellier, F., Jacquemain, D., Kjaer, K., Als-Nielsen, J., Lahav, M., and Leiserowitz, L. (1994). *J. Am. Chem. Soc.*, **116,** 1192.

34. Small, D. M. (1986). "The physical chemistry of lipids." In: *Handbook of Lipid Research*. Plenum Press, New York.
35. Majewski, J. Margulis, L., Jacquemain, D., Leveiller, F., Böhm, C., Arad, T., Talmon, Y., Lahav, M., and Leiserowitz, L. (1993). *Science*, **261**, 899.
36. Jacquemain, D., Grayer-Wolf, S., Leveiller, F., Frolow, F., Eisenstein, M., Lahav, M., and Leiserowitz, L (1992). *J. Am. Chem. Soc.* **114**, 9983.
37. Dluhy, R. A., and Cornell, D. G. (1990). *Fourier Transform Infrared Spectroscopy in Colloid and Interface Science. ACS Symposium Series 447* (D. R. Scheuing, ed.). American Chemical Society, Washington, DC, p. 192.
38. Kavanau, J. L. (1964). *Water and Solute-Water Interactions*. Holden-Day, San Francisco.
39. Aleby, S., and von Sydow, E. (1960). *Acta Cryst*. **13**, 487.
40. Popovitz-Biro, R., Lahav, M., and Leiserowitz, L. (1991). *J. Am. Chem. Soc.* **113**, 8943.
41. Power, B. A., and Power, R. F. (1962). *Nature* **194**, 1170.
42. Benedetti, C., Pedone, A., and Sirigu, A. (1973). *Acta Crystallogr. Sec. B* **29**, 730.
43. Mathieson, A. M. (1952). *Acta Crystallogr*. **5**, 332.
44. Mathieson, A. M. (1953). *Acta Crystallogr*. **6**, 399.
45. Coll, M., Solans, X., Font-Altaba, M., and Subirana, J. A. (1986). *Acta Crystallogr. Sec. C* **42**, 599.
46. di Blasio, B., Pedone, C., and Sirugu, A. (1975). *Acta Crystallogr. Sec. B* **31**, 601.
47. Mallikarjunan, M., and Thyagaraja Rao, S. (1969). *Acta Crystallogr. Sec. B* **25**, 296.
48. Torii, K., and Iitaka, Y. (1970). *Acta Crystallogr. Sec. B* **26**, 1317.
49. Torii, K., Iitaka, Y. (1971). *Acta Crystallogr. Sec. B* **27**, 2237.
50. Torii, K., and Iitaka, Y. (1973). *Acta Crystallogr. Sec. B* **29**, 2799.
51. Weissbuch, I., Frolow, F., Addadi, L., Lahav, M., and Leiserowitz, L. (1990). *J. Am. Chem. Soc.* **112**, 7718.
52. Wilen, L. (1993). *Science* **259**, 1469.
53. Lahav, M., Eisenstein, M., and Leiserowitz, L. (1993). *Science* **259**, 1469.
54. Warren, G. J., Corrotto, L., Wolber, P. (1986). *Nucl. Acids. Res.* **14**, 8047.

Ion-Selective Monolayer Membranes Based on Self-Assembling Tetradentate Ligand Monolayers on Gold Electrodes: Nature of the Ionic Selectivity

SUZI STEINBERG, YITZHAK TOR, ABRAHAM SHANZER, and ISRAEL RUBINSTEIN

The Weizmann Institute of Science, Department of Materials & Interfaces and Department of Organic Chemistry
Rehovot 76100, Israel

8.1. Introduction

Currently interest is focused on the study of organized monomolecular films on solid supports, because such systems provide a unique opportunity to construct custom-designed microstructures of molecular dimensions. Specifically, self-assembling monolayers comprising sulfur-containing molecules prepared on gold substrates have been studied by numerous groups.[1-20]

The majority of these studies focused on the structural properties of such systems. It should be emphasized, however, that a major objective in this field is the preparation of organized, solid-supported monomolecular systems, designed to perform specific tasks. For self-assembling monolayers, this aspect has so far received only limited attention. Notable examples are self-assembled skeletonized monolayers for recognition of molecular shapes,[21,22] the use of self-assembling octadecyl monolayers for the construction of ultramicroelectrodes,[2] the construction of size-selective cyclodextrin monolayers on gold,[6a] structural modification of electrochemically grown polyaniline by a self-assembled monomolecular underlayer,[23] charge discrimination in ionic binding to certain monolayers,[10a] and specific recognition and binding of streptavidin by biotin-functionalized monolayers.[16]

ISBN 0-12-523485-6

In recent years we have been engaged in an effort to construct and study ion-selective self-assembling monolayers, i.e., monolayer systems that would recognize a selected ion in the presence of other ions in solution. Our approach to this problem has been to combine solid-supported monolayer techniques with structural and functional elements of natural bilayer membranes. Several basic elements are involved here: organization, choice of the monolayer components, choice of the support, and a strategy for detection. The required organization is achieved by the use of the self-assembly technique (see Ref. 1 and references therein). As for monolayer components, we have adopted the fundamental principle frequently found in natural membranes, namely, "receptor molecules" embedded in a blocking monolayer matrix. The solid support of choice is gold, because it provides a dual function: effective anchoring of the monolayer-forming molecules, while also serving as an electrode. The underlying gold electrode provides the electrochemical signal required to probe the selective binding of ions, in the form of a voltammetric current.

Based on these principles, we studied the construction of artificial ion-selective monolayer membranes on gold electrodes.[24–26] The assemblies were prepared by spontaneous adsorption of a "receptor" molecule [e.g., 2,2′-thiobisethyl acetoacetate (TBEA) (Fig. 8.1)] and a surface sealing component. We used as sealing components either a self-assembled long-chain amphiphile (*n*-octadecyl mercaptan, OM) or an ultrathin polymeric component (electropolymerized 1-naphthol). Such monolayer membranes on gold electrodes were shown to produce a selective electrochemical

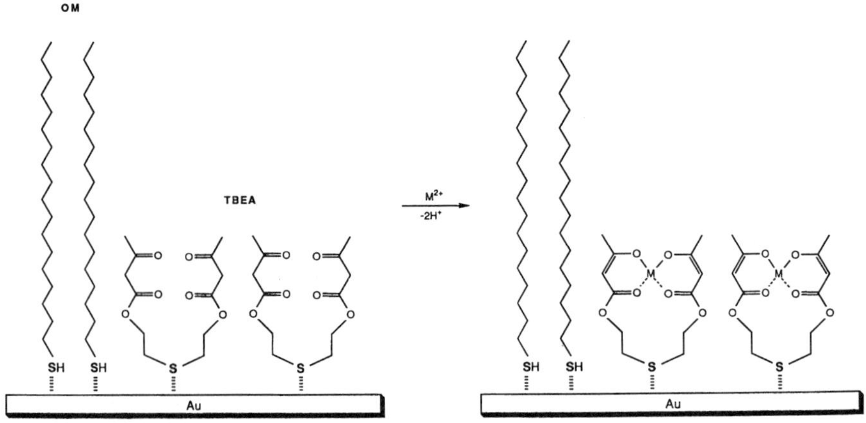

FIG. 8.1. Schematic representation of the binding of a divalent cation (e.g., Cu^{2+}) to Au/(TBEA + OM). The ion binds to TBEA in the enol form, generating a neutral complex.

response for Cu^{2+} ions by forming a 1 : 1 complex with TBEA on the surface (as shown schematically in Fig. 8.1 for a divalent ion, M^{2+}). No response was observed for Fe^{3+} ions, which do not form a complex with TBEA in the monolayer. This provided a strong indication that only those ions capable of forming stable four-coordinated complexes with β-diketonates may bind to a TBEA monolayer, producing a selective electrochemical response. It has been emphasized that this kind of selectivity can only be achieved with monolayer membranes that effectively fulfill both the selectivity and blocking requirements.

The major part of this chapter deals with two key issues that are of fundamental interest. The first concerns the mechanism of the selective response, and the question of whether the observed electrochemical response is due to selective binding to TBEA molecules or rather to the presence of pinholes. Experimental results presented later, including ionic competition experiments and a study of the behavior of monolayers with intentionally added pinholes, provide strong evidence for a mechanism based on coordination to TBEA in the monolayer, as originally suggested by us.[24]

The second issue concerns the nature of ion binding to molecules arranged in monolayer assemblies. It is shown that the molecular assembly imparts enhanced selectivity to the binders when compared with the situation in bulk solution. Moreover, complexation to the monolayer is shown to be governed by special selectivity rules, arising from geometric and size restrictions induced by the monolayer organization.

Several factors that affect the stability and performance of the Au/(TBEA + sealing component) system are also explored, including the gold substrate and the nature of the sealing component. In this context we describe results on possible "healing" of defective monolayers.

8.2. Experiments

8.2.1. Materials

Bicyclohexyl (BCH; Aldrich, 99%), n-hexadecane (HD; Aldrich, 99%), an Isopar G (Hunt Chemical Corp.) were purified by passing through a column of basic alumina (ICN Biomedicals). Chloroform (Bio-Lab, AR) was similarly treated and then stabilized with 1% absolute ethanol (Bio-Lab, AR). The n-octadecyl mercaptan (OM; Aldrich, 98%) was recrystallized from absolute ethanol, and n-octadecyl trichlorosilane (OTS; Aldrich, 95%) was used as received and stored under nitrogen. Silane-containing solutions were always freshly prepared and used within 3 h after preparation. H_2SO_4, $HClO_4$, $CuCl_2$, $FeCl_3$, $Cu(ClO_4)_2$, $ZnCl_2$, and Na_2SO_4 were reagent grade. All aqueous solutions were prepared with triply distilled water.

8.2.2. Synthetic Procedures for bis-β-ketoester Ligands

The TBEA[24] and 3,3′-thiobispropyl acetoacetate (TBPA)[24] were prepared by 4-dimethylaminopyridine (DMAP) catalyzed reaction of diketene with 2,2′-thiobisethanol and 2,2′-thiobispropanol, respectively (Scheme 8.1) as follows: To a chilled solution of dialcohol (10 mmol) in tetrahydrofuran (THF, 15 mL), under nitrogen atmosphere, was added DMAP (0.16 mmol), followed by diketene (22 mmol). Stirring was continued for 2 h at 0°C and then for 12 h at room temperature. The THF was evaporated, and the orange residue was subjected to two chromatographic purifications on silica gel 60, using the flash chromatography technique. The first chromatographic purification (25% to 50% ethyl acetate/*n*-hexane) afforded a yellow liquid, which was then divided into smaller portions, and further purified by a second column (0% to 0.5% MeOH/CH_2Cl_2). The analytically pure products (50% to 70% yield) were obtained as colorless to slightly yellow viscous liquids.

Note the following spectral and analytical data:

TBEA: IR (neat) $\nu_{c=o}$ 1744, 1718 cm^{-1}; ^1H NMR (270 MHz, CD_3CN) δ 2.19 (*s*, 6H, $COCH_3$), 2.79 (*t*, 4H, $J = 6.6$ Hz, SCH_2), 3.49 (*s*, 4H, $COCH_2CO$), 4.23 (*t*, 4H, $J = 6.6$ Hz, OCH_2).

Elem. anal. calcd. for $C_{12}H_{18}O_6S$: C, 49.64; H, 6.25; S, 11.04. Found: C, 49.65; H 6.33; S, 11.11.

TBPA: IR (neat) $\nu_{c=o}$ 1742, 1716 cm^{-1}; ^1H NMR (270 MHz, CD_3CN) δ 1.87 (*m*, 4H, $CH_2\underline{CH_2}CH_2$, $J = 6.3$, 7.2 Hz), 2.18 (*s*, 6H, $COCH_3$), 2.57 (*t*, 4H, $J = 7.2$ Hz, SCH_2) 3.48 (*s*, 4H, $COCH_2CO$), 4.17 (*t*, 4H, $J = 6.3$ Hz, OCH_2).

Elem. anal. calcd. for $C_{14}H_{22}O_6S$: C, 52.81; H, 6.96; S, 10.07. Found: C, 53.02; H, 7.06; S, 9.74.

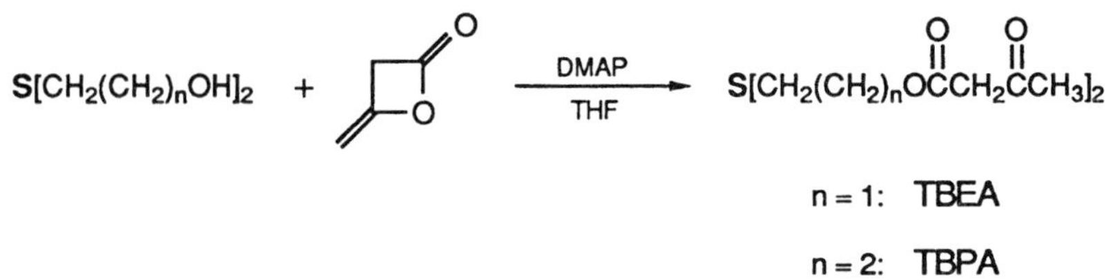

Scheme 8.1.

8.2.3. Gold Electrodes

Glass microscope cover slides (Deckglass), chemically cleaned (with acetone, then dried under Ar stream) were subjected to 2 min of Ar plasma at ~ 0.15 mbar and 2.5 mA (Edwards S150B Sputter Coater). Without breaking the vacuum, ~ 1000 Å of gold were sputter deposited at 10 mA and the same pressure. Unless otherwise indicated, the slides were then subjected to a temperature ramp of 10°C/min up to 400°C, left for 10 min at 400°C, and allowed to cool to room temperature. The gold substrates were cleaned immediately prior to use by O_2 plasma (same conditions as earlier).

8.2.4. Monolayer Preparation

TBEA and TBPA were adsorbed by immersion for 3.5 h in a solution of 3.3×10^{-2} M TBEA (or TBPA) in BCH : $CHCl_3$ 4 : 1 (v/v). OM was adsorbed by immersion for 3 h in a solution of 2.0×10^{-2} M OM in either BCH or Isopar G. Mixed monolayers, TBEA + OM, were prepared by adsorption from a solution of 2.0×10^{-2} M TBEA + 2.0×10^{-2} M OM in BCH : $CHCl_3$ 4 : 1 (v/v). OTS was adsorbed by immersion for 60 to 80 min in a solution of 2.0×10^{-3} M OTS in either BCH or Isopar G; this adsorption was occasionally followed by additional short adsorptions (10 to 15 min each) to improve the barrier efficiency (see later). The slides were thoroughly rinsed with chloroform after each adsorption step.

8.2.5. Contact Angles

Contact angles were determined by the sessile drop technique using a Rame-Hart Model 100 goniometer at ambient temperature and humidity. Typical accuracy was ± 2 deg. The liquid drop was formed at the tip of a syringe needle extended with a fine PTFE tubing.

8.2.6. Electrochemical Measurements

The electrochemical instrumentation included a potentiostat (Model 303-C) and an electrochemical programmer, both from the Department of Chemistry, Technion, Haifa, and a Houston Instruments Model 100 X-Y recorder. The gold substrates were mounted as working electrodes in a conventional three-electrode cell. The exposed area was 0.63 cm^2, except in Fig. 8.2 (1.0 cm^2). The counter electrode was a Pt wire. A KCl-saturated calomel electrode (SCE) was used in perchloric acid solutions; a mer-

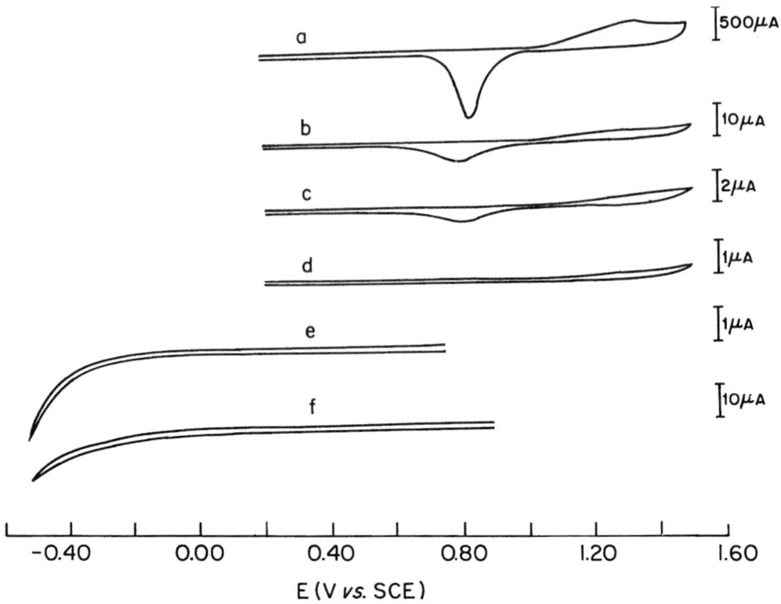

FIG. 8.2. Cyclic voltammograms in 0.10 M H_2SO_4. Electrode area for (a)–(d) is 1.0 cm^2 and for (e) and (f) 0.63 cm^2. Scan rate: 0.10 V/s. (a) Bare Au electrode; (b) after 2 h OM adsorption; (c) after additional 1 h OM adsorption; (d) after additional 0.5 h OM adsorption; (e) the electrode of part (d) in 1.0 mM Cu^{2+}; and (f) the electrode of part (d) in 3.0 mM Fe^{3+}.

curous sulfate reference electrode (+0.400 V versus SCE) was used in sulfuric acid solutions. For the sake of clarity, all potentials are reported versus a SCE. First scan voltammograms are always shown in the figures; subsequent scans are practically identical.

8.3. Results and Discussion

8.3.1. THE GOLD SURFACE

It is expected that the nature of the gold substrate, i.e., its cleanliness, roughness, and crystallographic orientation, would affect the monolayer adsorption and properties. In the present study, sputtered gold films were annealed for 10 min in an oven at 400°C. This treatment decreases the surface roughness substantially, as recently shown by transmission electron microscopy.[27] This is also evidenced from the amount of charge required to reduce a gold oxide layer in a cyclic voltammetric experiment in acid solution[1]; comparison of the voltammogram for annealed gold [Fig. 8.2(a)] with those for nonthermally treated gold[1,2] shows a decrease in the gold surface area by a factor of 3 to 4, indicating a considerable smoothening of the surface.

The gold thermal treatment has a pronounced effect on the barrier properties of the blocking monolayer component. As noted in previous reports,[1,2] adsorption of octadecyl monolayers onto sputter-deposited, nonthermally treated gold substrates always left microscopic pinholes in the resultant monolayers. As shown in Figs. 8.2(a) through (d), the situation is quite different with annealed gold. A 2-h adsorption of OM provides a surface coverage of 98.1% [Fig. 8.2(b)]; subsequent OM adsorption for 1 h increases the coverage to 99.8% [Fig. 8.2(c)]; an additional 0.5 h adsorption provides an OM monolayer for which no water penetration can be detected electrochemically. Such OM monolayers on gold are also totally impermeable to ions such as Cu^{2+} or Fe^{3+}, as shown in Fig. 8.2(e) and (f).

The gold annealing has a dual effect on the structure and performance of the monolayer membranes. First, the superior barrier properties of blocking monolayers, which constitute the blocking component of the ion-selective monolayer membranes, are essential for achieving the high selectivity. Second, the thermal treatment substantially improves the organization of monolayer membranes, as evidenced by their wettability properties. An example of the effectiveness of the thermal treatment in improving monolayer membrane structural properties is given in Table 8.1, where the contact angles for an Au/(TBEA + OM) electrode are shown, with and without gold annealing. The marked improvements in the contact angles for annealed gold appear to be the result of improved surface cleanliness and smoothness.[28]

TABLE 8.1

EFFECT OF GOLD THERMAL TREATMENT ON THE CONTACT ANGLES OF AU / (TBEA + OM). [ADVANCING (a) AND RECEDING (r) ANGLES ARE INDICATED.]

Electrode Treatment	Contact Angles		
	H_2O	BCH	HD
Before annealing	(a) 120	42	36
	(r) 90	20	12
After annealing	(a) 105	60	57
	(r) 104	59	56

8.3.2. Variation of the Blocking Component

We have previously shown that monolayer membranes of the TBEA + OM type display the desired high ionic selectivity.[24] However, the stability of the systems on long-term operation is not entirely satisfactory. Thus, partial removal of monolayer constituents, and in particular parts of the blocking component (OM), occurs on continuous voltage cycling. A similar instability occurs (with no cycling) in the presence of certain species in solution. For example, Ag^{+} ions (and to a lesser extent, Cu^{2+} ions), which appear to penetrate at monolayer imperfections, bind to thiol groups on removal of the molecules from the gold surface.

To improve the stability of the monolayer membranes, we replaced the monomeric blocking component OM with the polymerizable octadecyl silane derivative, OTS, or a mixture of OTS and OM.[1] TBEA + OM monolayers on gold are generally prepared by co-adsorption from a solution 1 : 1 in both components[24]; however, OTS is anticipated to react with the enol form of TBEA present at equilibrium in the solution[24]; therefore, a different procedure has been employed for OTS, i.e., a sequential adsorption protocol. Monolayer membranes with various blocking elements were studied, i.e., Au/(TBEA + OTS), prepared by TBEA adsorption followed by OTS adsorption; Au/(TBEA + OM + OTS), prepared by simultaneous adsorption of TBEA and OM, followed by OTS adsorption; and Au/(TBEA + OTS + OM), prepared by sequential adsorption of the three components, respectively.

Note that these procedures are based on the experimental observation that OTS does not react with TBEA in the monolayer. This may be explained by the silane headgroup of OTS being sterically banned from approaching the TBEA oxygens in the cavity defined by the monolayer packing (see later).

All three types of OTS-containing monolayer membranes on gold display the same electrochemical response as the original Au/(TBEA + OM) electrodes,[24] as exemplified in Fig. 8.3 for an Au/(TBEA + OTS) electrode in Cu^{2+} solution. All the characteristic features that distinguish Cu^{2+} reduction at a monolayer membrane [Fig. 8.3(c)] from its reduction at a bare Au electrode [Fig. 8.3(a)] are similar to the Au/(TBEA + OM) case[24]; i.e., (1) complete absence of Cu underpotential deposition (UPD) peaks around 0.25 V and (2) a sizable overpotential for Cu deposition and a marked voltammetric loop upon reversing the potential scan direction, indicating an energy barrier for Cu^{2+} reduction, which is diminished on Cu deposition in the monolayer.

It is clear from the relatively large amount of charge passed during the reduction [Figs. 8.3(b) and (c)] that Cu reduction in the monolayer is

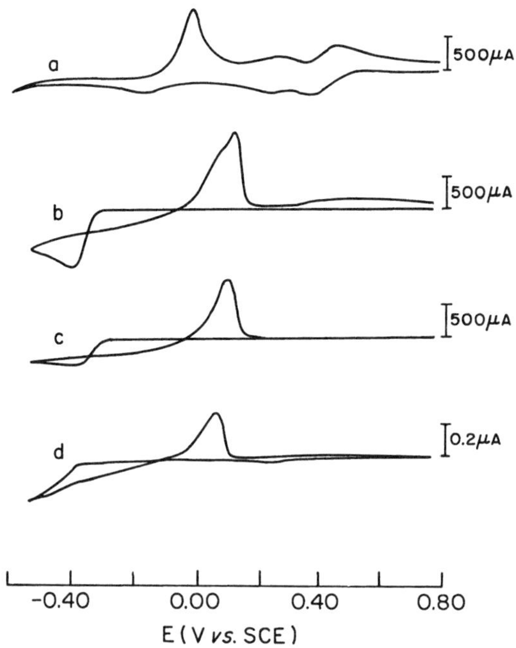

FIG. 8.3. Cyclic voltammograms in 0.10 M H_2SO_4 (electrode area: 0.63 cm^2. Scan rate: 0.10 V/s) for (a) bare Au in 1.0 mM Cu^{2+} + 3.0 mM Fe^{3+} (the peaks around 0.45 V correspond to the $Fe^{3+/2+}$ couple); (b) Au/(TBEA + OTS) (1-h OTS adsorption) in 1.0 mM Cu^{2+}; (c) after additional 0.5-h OTS adsorption, in 1.0 mM Cu^{2+}; (d) the electrode of part (c), in 1.0×10^{-6} M Cu^{2+} + 3.0×10^{-3} M Fe^{2+}.

followed by bulk Cu deposition, presumably on top of the organic monolayer. The oxidation charge is usually smaller than the reduction charge, indicating some loss of Cu during the deposition process. The electrode can, however, be cycled continuously for prolonged periods of time with no apparent change in the shape of the voltammogram.

Figure 8.3(d) presents a rather dramatic demonstration of the ionic discrimination of an Au/(TBEA + OTS) electrode between Cu^{2+} and Fe^{2+} ions (the preferred coordination geometry of Fe^{2+} is octahedral,[29] same as that of Fe^{3+}). As shown in Fig. 8.3(d), the response of the electrode to trace amounts of Cu^{2+} ions (1.0×10^{-6} M) is totally unaffected by the presence of a large concentration of Fe^{2+} ions (3.0×10^{-3} M). Thus, the Au/(TBEA + OTS) electrode appears perfectly inert toward Fe^{2+} even at the exceedingly high current sensitivity used to detect the Cu^{2+} ions. This pronounced selectivity can be used for trace analysis of binding ions, as described elsewhere.[26]

As noted earlier, a major goal in using OTS as a blocking component in monolayer membranes was to improve system stability. Indeed, the long-term stability of OTS-containing monolayer membranes is substantially improved compared to Au/(TBEA + OM) electrodes. Of the various combinations of blocking components used, the highest stability is obtained with Au/(TBEA + OTS) electrodes. For example, an electrode of the kind shown in Fig. 8.3(c) was cycled continuously for 12 h without any sign of deterioration of the electrochemical behavior. Electrodes of the Au/(TBEA + OTS + OM) and Au/(TBEA + OM + OTS) types are somewhat less durable than Au/(TBEA + OTS), but are still considerably more stable than Au/(TBEA + OM), e.g., in experiments requiring prolonged exposure to Ag^+ or Cu^{2+} ions.

It was expected that the new adsorption procedures involving OTS would also influence the wetting properties of the monolayers, since they would effect different packing modes of the various monolayer constituents. We have previously reported unusually high contact angles for Au/(TBEA + OM), typically 108, 59, and 57 deg for water, BCH, and HD, respectively, with very small hysteresis.[24] We have suggested that these high values (particularly for the organic solvents) reflect a particular spatial distribution of the monolayer components.[30]

As seen from Table 8.2, unusual wetting properties are only observed in cases that involve TBEA + OM adsorption, while those that involve TBEA + OTS (as well as monolayers of OM or OTS alone[1]) show "normal" contact angles. This difference appears to reflect a specific distribution of the TBEA component in TBEA + OM monolayers,

TABLE 8.2

CONTACT ANGLES FOR VARIOUS MONOLAYER MEMBRANES ON GOLD ELECTRODES. [ADVANCING (a) AND RECEDING (r) ANGLES ARE INDICATED.]

Monolayer on Gold	Contact Angles		
	H_2O	BCH	HD
TBEA + OTS	(a) 109	57	46
	(r) 107	54	44
TBEA + OTS + OM	(a) 109	58	49
	(r) 108	56	48
TBEA + OM	(a) 109	60	57
	(r) 108	59	55
TBEA + OM + OTS	(a) 109	60	56
	(r) 108	59	55

attributed to the simultaneous adsorption. The detailed structure of these monolayers is still unknown.

8.3.3. Monolayer "Healing"

A common procedure for preparing self-assembling monolayers on solid supports comprises a single adsorption step for a certain period of time. This, however, is not necessarily the best procedure, because the resultant monolayers do not, in many cases, possess the desired structural properties. Factors such as surface roughness, grain boundaries, and impurities are likely to result in monolayer defects and various kinds of imperfections. Such defects are evident in electrochemical measurement involving penetration to the underlying electrode, as seen in Fig. 8.2(b) for water penetration through pinholes in an OM monolayer on gold (as gold oxide formation/removal peaks), and in Fig. 8.3(b) for Cu^{2+} penetration through pinholes in a TBEA + OTS monolayer on gold (as the ill-defined Cu UPD stripping current).

A simple and generally effective procedure for improving the barrier properties of such monolayers is repeated adsorption steps. As shown in Figs. 8.2(c) and (d) and 8.3(c), subsequent adsorption steps may result in marked improvement in the blocking characteristics of the monolayers, i.e., complete disappearance of the peaks due to pinhole currents.

It can be reasonably assumed that the mechanism responsible for the improvement in monolayer structural properties upon repeated adsorption steps is a certain kind of a dissolution-readsorption process, i.e., removal of molecules from defective or poorly organized sites and readsorption in a more ordered and compact form. This process can be rendered more efficient by the use of certain agents, which, if allowed to penetrate and interact with the sulfur headgroups, remove molecules from defective sites at the surface. A logical choice would be Ag^+ ions, known to bind strongly to sulfur-containing compounds, and thus anticipated to serve this purpose quite effectively.

This expectation can be realized, as exemplified by the "healing" process shown in Fig. 8.4. The TBEA + OM monolayer in Fig. 8.4(a) is slightly defective, as is evident from the traces of Cu UPD current at the positive branch of the voltammetric baseline. As shown in Fig. 8.4(b), exposure of the electrode to an Ag^+ solution results in an increase in the Cu UPD current and a marked decrease (~ 0.10 V) in the overpotential for Cu^{2+} reduction,[12] indicating the removal of small portions of the monolayer. A short subsequent adsorption of OM alone produces superior electrochemical response, shown in Fig. 8.4(c), with no detectable Cu UPD and a high overpotential for Cu^{2+} reduction, ~ 0.03 V more negative than the

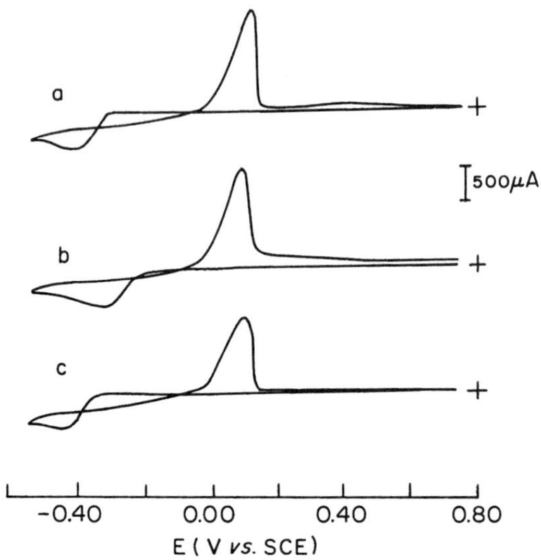

FIG. 8.4. Cyclic voltammograms in 0.10 M H_2SO_4 + 1.0 mM Cu^{2+} (electrode area: 0.63 cm^2; scan rate: 0.10 V/s) for (a) Au/(TBEA + OM), prepared by 3.2-h adsorption; (b) after 15 min in 0.10 M H_2SO_4 + 1.0 mM Ag^+; and (c) after additional 20-min OM adsorption.

reduction peak in Fig. 8.4(a). Moreover, the electrode of Fig. 8.4(c) is much less sensitive to prolonged exposure to Ag^+ solutions than the same electrode prior to the "healing" process [Fig. 8.4(a)].

Note that the same effect as that induced by exposure to Ag^+ ions, though considerably slower, is observed on exposure of thiol monolayers to Cu^{2+} ions. This accounts for the slow deterioration in the electrochemical behavior of Au/(TBEA + OM) electrodes observed upon operation in Cu^{2+}-containing solutions, and can be rather easily corrected by additional OM adsorption. Also, the similarity in the magnitude of the Cu^{2+} reduction currents before and after "healing" (with either Ag^+ or Cu^{2+}) suggests that little, if any, TBEA is removed from the surface during this process.

8.3.4. Ionic Competition at Monolayer Membranes

Experiments involving the electrochemical response of TBEA-containing monolayer membrane electrodes in the presence of competing ions were designed to provide additional insight into the selective binding model. Zn^{2+} and Cd^{2+} ions are expected to compete with Cu^{2+} for monolayer binding sites (i.e., TBEA moieties). This competition would depend on the relative concentrations of the competing ions and their binding affinity to TBEA in the monolayer.

An example of such ionic competition is shown in Figs. 8.5(a), (c), and (d). The peaks for 1.0 mM Cu^{2+} are suppressed in the presence of Zn^{2+} ions in the solution; and at a high enough Zn^{2+} concentration (8.0 mM) the Cu^{2+} peaks completely disappear. The same effect is observed in similar experiments involving competition of Cd^{2+} and Cu^{2+} ions, i.e., total suppression of the Cu^{2+} peaks in the presence of, for example, 9.0 mM Cd^{2+}.[31] Conversely, the presence of Fe^{3+} [Fig. 8.5(b)] and Ce^{3+} [Fig. 8.5(e)] has little effect on the intensity of the Cu^{2+} peaks at Au/(TBEA + OTS).[32] (The presence of any of these above mentioned ions at any concentration has no effect on Cu deposition at a bare gold electrode.)

The replacement of Cu^{2+}-loaded monolayer binding sites with Zn^{2+} or Cd^{2+} ions is fully reversible. For example, monolayer-bound Zn^{2+} ions can be removed by immersing the electrode in an EDTA solution for 40 min. The Cu^{2+} peaks are fully recovered following this treatment. Alternatively, Cu^{2+} replacement by Zn^{2+} or Cd^{2+} can be reversed on immersion of the electrodes in a 1.0 mM Cu^{2+} solution. The recovery of the Cu^{2+} peaks is immediate in the case of bound Cd^{2+}, and takes considerably

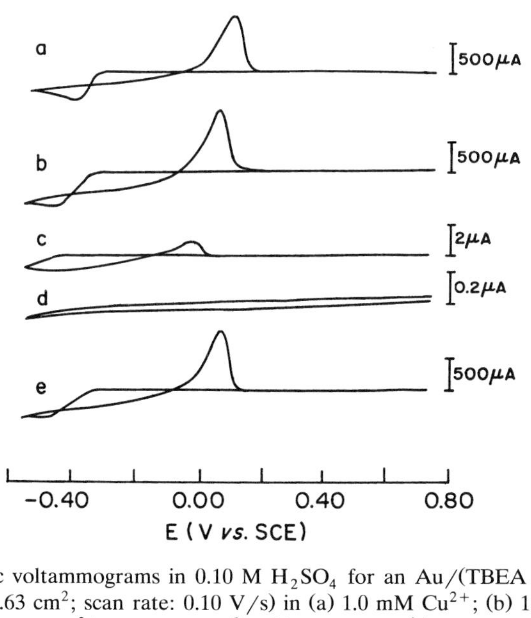

FIG. 8.5. Cyclic voltammograms in 0.10 M H_2SO_4 for an Au/(TBEA + OTS) electrode (electrode area: 0.63 cm^2; scan rate: 0.10 V/s) in (a) 1.0 mM Cu^{2+}; (b) 1.0 mM Cu^{2+} + 3.0 mM Fe^{3+}; (c) 1.0 mM Cu^{2+} + 6.0 mM Zn^{2+}; (d) 1.0 mM Cu^{2+} + 8.0 mM Zn^{2+}; and (e) 1.0 mM Cu^{2+} + 6.0 mM Ce^{3+}.

longer (~ 25 min) for bound Zn^{2+} (Fig. 8.6). This apparent difference in the affinity of Zn^{2+} and Cd^{2+} toward TBEA in a monolayer membrane is attributed to the smaller ionic radius of Zn^{2+} and its larger binding constant to β-diketonate ligands (see Table 8.3), and is discussed further later in this chapter.

The ionic competition results provide strong support for the selective binding model. The systematic competition observed clearly indicates that the binding of selected divalent ions to the ligand TBEA is responsible for the ionic discrimination, whereas it is practically impossible to envision ionic competition at monolayer pinholes. Indeed, as shown section 8.3.7, Cu^{2+} reduction at monolayer pinholes is not affected by the presence of Zn^{2+}, i.e., no ionic competition is observed at pinholes.

8.3.5. Complexation in a Two-Dimensional Molecular Configuration

As we have noted,[24] the discrimination in the binding of Cu^{2+} and Fe^{3+} is a geometric effect, deriving from the two-dimensional (2-D) arrangement of the ligand molecules, TBEA, in the monolayer.[40] Thus, TBEA in solution forms a stable complex with Fe^{3+}, whereas a TBEA monolayer membrane would not accommodate the ions. While oligomeric TBEA-Fe^{3+} complexes (3 : 2 or higher aggregates), which satisfy the coordination and

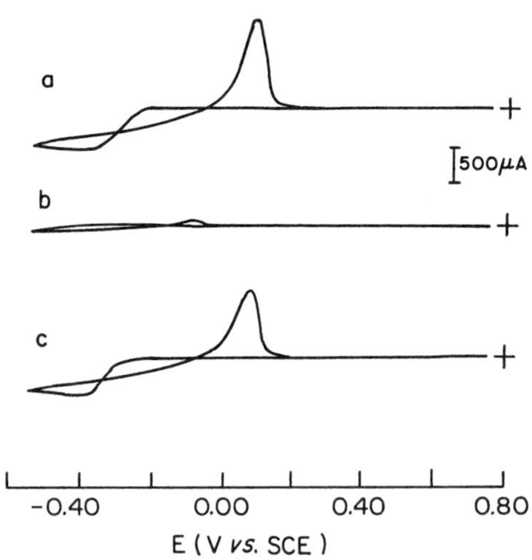

Fig. 8.6. Cyclic voltammograms in 0.10 M H_2SO_4 for an Au/(TBEA + OTS + OM) electrode (electrode area: 0.63 cm^2; scan rate: 0.10 V/s). (a) In 1.0 mM Cu^{2+}; (b) in 1.0 mM Cu^{2+} + 3.5 mM Zn^{2+}; and (c) the electrode of part (b) after 25 min in 1.0 mM Cu^{2+}.

TABLE 8.3

BINDING OF VARIOUS IONS TO TBEA MONOLAYER MEMBRANES

Ion	Effective Ionic Radius[a] (Å)	Coordination Number	Log K[b]	Binding to TBEA Monolayers
ZN^{2+}	0.60	4	4.9 (1) 3.9 (2)	+ + +
Cu^{2+}	0.57	4	8.2 (1) 6.7 (2)	+ +
Cd^{2+}	0.78	4	3.8 (1) 2.7 (2)	+ +
Pb^{2+}	0.98	4	6.4 (1 + 2)	+
VO^{2+}		4	8.6 (1) 7.0 (2)	–
Fe^{3+}	0.645	6	9.8 (1) 8.7 (2) 7.0 (3)	–
Fe^{2+}	0.61 (l-spin) 0.78 (h-spin)	6	5.1 (1) 3.6 (2)	–
Ce^{3+}	1.01	> 6	5.3 (1) 4.0 (2) 3.4 (3)	–

[a] From Ref. 33.

[b] Binding constants to acetylacetone (K) are average values from Refs. 34 through 39. The number n in parentheses represents the n'th binding constant.

geometric requirements of the ion, may be formed in solution, this is prohibited in the 2-D monolayer configuration, which only allows for the formation of 1 : 1 complexes.

To further investigate this point, two other electroactive cations were studied, Pb^{2+} and VO^{2+}. Both ions form complexes with β-diketonate ligands in solution with quite large binding constants (see Table 8.3). On the other hand, the binding of either ion to TBEA in the monolayer membrane is expected to be sterically hindered: Pb^{2+} is a large ion (its ionic radius is almost twice that of Cu^{2+}; see Table 8.3). This should inhibit Pb^{2+} penetration into the binding cavities of TBEA, the size of which is determined by the monolayer packing. As for VO^{2+}, its preferred coordination geometry is a tetragonal pyramid,[29] i.e., the vanadium would be situated at the apex of a pyramid defined by the four TBEA binding oxygens, with the V=O oxygen extended away perpendicular to the plane defined by the binding oxygens (i.e., parallel to the gold surface, in a compact TBEA monolayer). While such a configuration is perfectly reasonable in solution, it would be sterically prohibited in an organized array of closely spaced TBEA molecules.

The results for Pb^{2+} are shown in Fig. 8.7(a) and (b). (The electrolyte has been changed to $HClO_4$ due to the insolubility of Pb^{2+} in sulfate solutions.) The shape of Pb^{2+} voltammetry at an Au/(TBEA + OM + OTS) electrode [Fig. 8.7(b)] is strikingly similar to that of Cu^{2+} [Fig. 8.7(c)], essentially in every respect: the complete absence of Pb UPD peaks, a large overpotential for Pb^{2+} reduction, and a voltammetric loop. [The potential is not scanned to higher (negative) overpotentials due to the onset of water decomposition in the Pb^{2+} case.] This result emphasizes the generality of the electrochemical response of TBEA monolayer membranes to bound ions.

The data in Fig. 8.7, showing deposition and dissolution of bulk Pb and Cu, do not, however, provide an indication as to the relative amount of bound Pb^{2+} compared to Cu^{2+}. To obtain this information, a series of electrochemical stripping experiments was conducted. This includes dipping of an electrode (coated with an adsorbed TBEA monolayer membrane) in a solution containing a certain ion, for example, Cu^{2+} or Pb^{2+},

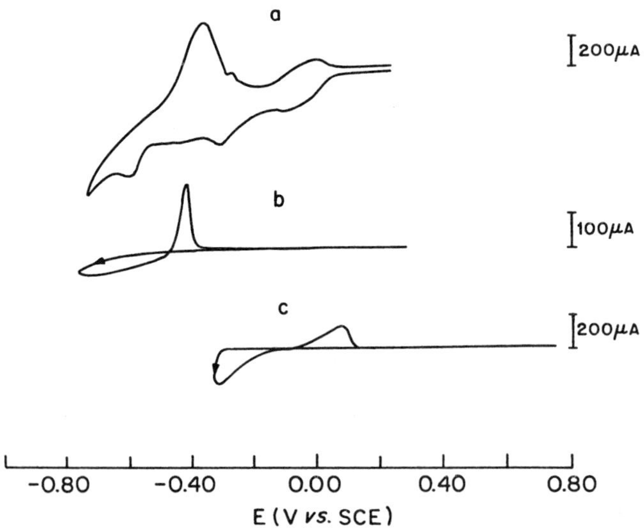

FIG. 8.7. Cyclic voltammograms (electrode area: 0.63 cm^2; scan rate: 0.10 V/s) for (a) bare Au in 0.10 M $HClO_4$ + 1.0 mM Pb^{2+}; (b) Au/(TBEA + OM + OTS) in 0.10 M $HClO_4$ + 1.0 mM Pb^{2+}; and (c) Au/(TBEA + OM + OTS) in 0.10 M H_2SO_4 + 1.0 mM Cu^{2+}.

then transferring the electrode to a background electrolyte solution, where a cyclic voltammogram is recorded, showing the peaks for ions immobilized in the monolayer. A detailed description of the procedure and the results is given elsewhere[25]; suffice it to note here that in such stripping experiments, the magnitude of the currents obtained for Pb^{2+} is considerably smaller relative to Cu^{2+} under similar conditions, indicating a much smaller number of bound Pb^{2+} ions.[25] The reason for this apparent difference in binding affinities to TBEA in the monolayer appears to be the difference in ionic radius; i.e., by "pushing" itself into a (small) binding site, the (large) Pb^{2+} ion blocks adjacent sites for binding of more ions.

Figure 8.8 presents the behavior of the monolayer system in the presence of VO^{2+} ions. Evidently, an Au/(TBEA + OM + OTS) electrode, which exhibits the usual Cu^{2+} reduction at the monolayer membrane [Fig. 8.8(c)] is totally blocking for VO^{2+} ions [Figs. 8.8(a) and (b)]. In light of the earlier discussion regarding VO^{2+} coordination geometry, it is clear that in this case the severe steric limitations imposed by the 2-D, densely packed monolayer indeed prevent any binding of VO^{2+} ions to TBEA monolayer membranes.

8.3.6. Ionic Selectivity

Table 8.3 summarizes our results on the binding of various ions to TBEA monolayer membranes. It also contains pertinent literature data for

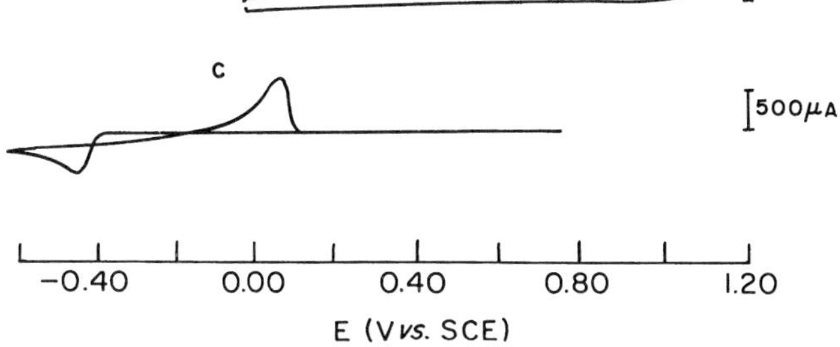

FIG. 8.8. Cyclic voltammograms in 0.10 M H_2SO_4 (electrode area: 0.63 cm^2; scan rate: 0.10 V/s) for (a) bare Au in 1.0 mM VO^{2+}; (b) Au/(TBEA + OM + OTS) in 1.0 mM VO^{2+}; and (c) the electrode of part (b) in 1.0 mM Cu^{2+}.

these ions, i.e., ionic radius at the indicated oxidation states and binding constants with β-diketonate ligands in solution. The last column (the present work) is given on a relative basis, from direct voltammetric experiments or ionic competition experiments, as detailed earlier.

Inspection of the data in Table 8.3 reveals that the scale of binding efficiencies of metal ions to TBEA monolayer membranes is distinctly different from that observed in solution. The series of decreasing binding efficiency to the monolayer is $Zn^{2+} > Cu^{2+} \sim Cd^{2+} > Pb^{2+} \gg VO^{2+}, Fe^{3+}$; whereas in solution the binding constants are in the following order: $Fe^{3+} > VO^{2+} > Cu^{2+} > Zn^{2+} > Cd^{2+} \sim Pb^{2+}$. This difference is attributed to the stereochemical constrains imposed by the monolayer matrix, in terms of cavity geometry and cavity size. The role of the geometry is illustrated by the lack of binding to iron. Fe^{3+} is of a similar size to Zn^{2+} and Cu^{2+} and has a higher binding affinity to diketonates in solution. The lack of Fe^{3+} binding to TBEA monolayers derives, most likely, from its requirement of an octahedral, rather than tetragonal, binding geometry.

The effect of cavity geometry is further emphasized in the higher binding affinity to the monolayer of Zn^{2+} compared with Cu^{2+}, although Cu^{2+} is the stronger binding ion in solution. Both metal ions bind tetragonally to acetoacetonates, but with different geometries: Cu^{2+} adopts a square planar geometry; Zn^{2+} assumes a distorted tetrahedral arrangement.

Size discrimination, as discussed, may explain the higher binding affinity to the monolayer of Cd^{2+} compared with Pb^{2+}, because the binding constants and binding geometries of the two metal ions in solution are similar.

8.3.7. Effect of Intentional Monolayer Pinholes

Fletcher[41] suggested that the selective electrochemical response observed with Au/(TBEA + blocking component) is due to penetration of ions at monolayer pinholes. As discussed in our response to Fletcher's comments,[42] this possibility is unlikely. Diffusion-controlled faradic currents are substantially enhanced at randomly distributed pinholes[2]; a situation in which the bulk Cu peaks are prominent, while Cu UPD and $Fe^{3+/2+}$ peaks are totally absent requires highly unusual size and spatial distribution of pinholes. It seems highly improbable that two different ligands (TBEA and TBPA[24]), a variety of blocking elements (OM, OTS, OTS + OM, OM + OTS, naphthol polymer[24]), and different adsorption procedures would lead to monolayers with the same special kind of pinhole distribution.

To further study the electrochemical behavior of Cu^{2+} and Fe^{3+} at randomly distributed monolayer pinholes, blocking monolayers of pure OM were prepared, in which a controlled amount of pinholes was deliberately introduced. This was achieved by immersion of a perfectly blocked Au/OM electrode for 40 min in a solution of 1.0 mM Cu^{2+} in 0.10 M H_2SO_4. The result is a slight removal of OM from the surface, to leave an Au/OM electrode of about 99.9% surface coverage (determined by the gold oxide method, as in Fig. 8.2). As shown in Fig. 8.9, at this very low pinhole content, the $Fe^{3+/2+}$ peaks cannot be observed, while Cu peaks still persist, apparently due to Cu deposition in the pinholes. Thus, a kind of "apparent ionic selectivity" may be obtained with very carefully controlled monolayer pinholes.

However, some fundamental differences exist between the "pinhole selectivity" and that observed with TBEA monolayer membranes. First, the faradic currents for Cu deposition/dissolution in the former case are merely 4% of the intensity observed with TBEA-containing monolayers. [Compare Fig. 8.9(a) with Fig. 8.3(c) or 8.4(c); the pinhole current intensity for Cu^{2+} may be increased by increasing the number of intentional pinholes, which, however, immediately results in the appearance of $Fe^{3+/2+}$ peaks and loss of the "apparent selectivity."] Moreover, the shape of the voltammogram in Fig. 8.9(a) is typical of normal Cu deposition at bare gold electrodes, i.e., prominent Cu UPD features and no unusual overpotential for bulk Cu deposition.

This exercise demonstrates that the unique features associated with Cu^{2+} reduction at TBEA-containing monolayer membranes cannot be

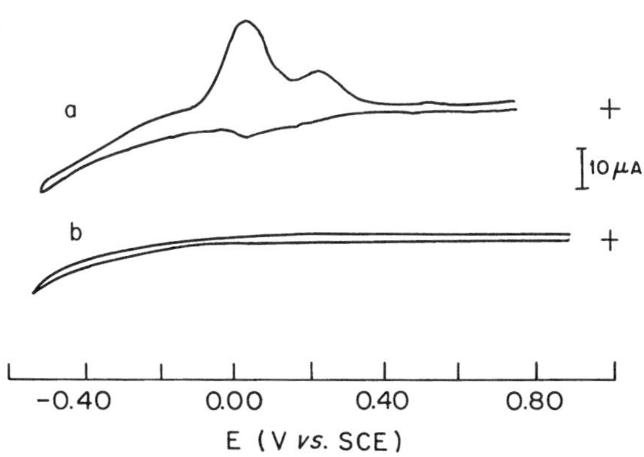

FIG. 8.9. Cyclic voltammograms in 0.10 M H_2SO_4 for an Au/OM electrode, 99.9% coverage (see text) (electrode area: 0.63 cm^2; scan rate: 0.10 V/s) in (a) 1.0 mM Cu^{2+} and (b) 3.0 mM Fe^{3+}.

observed in the absence of the active component TBEA. Also, as noted earlier, Cu deposition at intentional monolayer pinholes is unaffected by Zn^{2+}, i.e., essentially the same voltammograms as in Fig. 8.9(a) are obtained in the presence of Zn^{2+} ions in solution. These results, therefore, strongly support the selective binding mechanism.

8.4. Conclusions

We have shown that monolayer membranes comprising the ligand TBEA and an inert blocking component (OM, OTS, or a combination thereof) on gold electrodes provide a unique example of organized monomolecular systems designed to perform a specific function, i.e., recognition and selective electrochemical response for certain ions. The performance of these systems derives from their ability to fulfill two functions simultaneously: (1) selectively binding certain ions while (2) denying other ions access to the electrode. Thus, highly selective response for certain divalent ions (e.g., Cu^{2+} or Pb^{2+}) is observed in the presence of large concentrations of other ions (e.g., Fe^{2+} or Fe^{3+}).

The mechanism responsible for the ionic selectivity was studied using two complementary types of experiments: ionic competition (i.e., the response to a certain ion in the presence of another ion in solution) and electrochemical behavior at intentionally induced monolayer pinholes. The pronounced ionic competition observed with TBEA-based monolayers on the one hand, and the lack of any competition at monolayer pinholes on the other hand, provide strong support for a selectivity mechanism based on the binding of selected ions to TBEA molecules in the monolayer. It was also shown that the use of a polymerizable blocking component (OTS) substantially improves the lifetime of the monolayer membranes. Moreover, signs of deterioration of the performance can be easily reversed by applying "monolayer healing" procedures. Such monolayer systems may thus be useful as sensing elements for trace amounts of certain ions in the presence of large concentrations of nonbinding ions.[26]

A major conclusion of the present work concerns the selectivity considerations. As can be expected, the requirements for binding of an ion to TBEA in a monolayer membrane are quite different from the case of binding in solution. We have shown that the 2-D arrangement of closely packed TBEA molecules in the monolayer membrane defines the coordination geometry and the cavity dimensions for ionic binding. The latter appears to be an interesting example of a cooperativity effect: Inadequate matching of the ionic size of a bound ion to the cavity dimensions (either smaller or larger) introduces a local structural disturbance in the monolayer. This disturbance is transferred to neighboring molecules by virtue of the monolayer packing, giving rise to a lower effective binding constant.

Such considerations, playing a prominent role in the case of TBEA monolayer membranes, will have to be taken into account in the design of future monolayer systems based on binding, penetration, and specific interactions.

Some aspects of TBEA monolayer membrane systems still remain unclear, for example, the observation that $Fe(CN)_6^{4-}$ ions produce a sizable electrochemical signal at Au/(TBEA + blocking component) electrodes, which appear completely blocking toward nonbinding cations. This emphasizes the fact that penetration mechanisms in self-assembling monolayers are still not entirely understood.

Acknowledgments

I. Rubinstein acknowledges support for this work from the European Community, Grant No. 2655 (through the Israel National Council for Research and Development).

References and Notes

1. Sabatani, E., Rubinstein, I., Maoz, R., and Sagiv, J. (1987). *J. Electroanal. Chem.* **219,** 365.
2. Sabatani, E., and Rubinstein, I. (1987). *J. Phys. Chem.* **91,** 6663.
3. (a) Bain, C. D., Troughton, E. B., Tao, Y.-T., Evall, J., Whitesides, G. M., and Nuzzo, R. G. (1989). *J. Am. Chem. Soc.* **111,** 321. (b) Bain, C. D., Evall, J., and Whitesides, G. M. (1989). *J. Am. Chem. Soc.* **111,** 7155. (c) Bain, C. D., and Whitesides, G. M. (1989). *J. Am. Chem. Soc.* **111,** 7164. (d) Laibinis, P. E., Fox, M. A., Folkers, J. P., and Whitesides, G. M. (1991). *Langmuir* **7,** 3167. (e) Hickman, J. J., Laibinis, P. E., Auerbach, D. I., Zou, C., Gardner, T. J., Whitesides, G. M., and Wrighton, M. S. (1992). *Langmuir* **8,** 357. (f) Laibinis, P. E., and Whitesides, G. M. (1992). *J. Am. Chem. Soc.* **114,** 1990.
4. (a) Nuzzo, R. G., Fusco, F. A., and Allara, D. L. (1987). *J. Am. Chem. Soc.* **109,** 2358. (b) Porter, M. D., Bright, T. B., Allara, D. L., and Chidsey, C. E. D. (1987). *J. Am. Chem. Soc.* **109,** 3359. (c) Nuzzo, R. G., Dubois, L. H., and Allara, D. L. (1990). *J. Am. Chem. Soc.* **112,** 558.
5. (a) Chidsey, C. E. D., Bertozzi, C. R., Putvinski, T. M., and Mujsce, A. M. (1990). *J. Am. Chem. Soc.* **112,** 4301. (b) Chidsey, C. E. D., and Loiacono, D. N. (1990). *Langmuir* **6,** 682. (c) Chidsey, C. E. D. (1991). *Science* **251,** 919. (d) Camillone III, N., Chidsey, C. E. D., Liu, G.-Y., Putvinski, T. M., and Scoles, G. (1991). *J. Chem. Phys.* **94,** 8493.
6. (a) Porter, M. D. (1989). *Chem. Eng. News* May 1, 32. (b) Stole, S. M., and Porter, M. D. (1990) *Langmuir* **6,** 1199. (c) Widrig, C. A., Chung C., and Porter, M. D. (1991). *J. Electroanal. Chem.* **310,** 335. (d) Widrig, C. A., Alves, C. A., and Porter, M. D. (1991). *J. Am. Chem. Soc.* **113,** 2805. (e) Alves, C. A., Smith, E. L., and Porter, M. D. (1992) *J. Am. Chem. Soc.* **114,** 1222.

7. (a) Finklea, H. O., Avery, S., and Lynch, M. (1987). *Langmuir* **3,** 409. (b) Bunding Lee, K. A., Mowry, R., McLennan, G., and Finklea, H. O. (1988) *J. Electroanal. Chem.* **246,** 217. (c) Finklea, H. O., Snider, D. A., and Fedyk, J. (1990). *Langmuir* **6,** 371. (d) Finklea, H. O., and Hanshew, D. D. (1992). *J. Am. Chem. Soc.* **114,** 3173.
8. (a) Diem, T., Czajka, B., Weber, B., and Regen, S. L. (1986). *J. Am. Chem. Soc.* **108,** 6094. (b) Fabianowski, W., Coyle, L. C., Weber, B. A., Granata, R. D., Castner, D. G., Sadownik, A., and Regen, S. L. (1989). *Langmuir* **5,** 35.
9. (a) Ulman, A., Eilers, J. E., and Tillman, N. (1989). *Langmuir* **5,** 1147. (b) Evans, S. D., and Ulman, A. (1990). *Chem. Phys. Lett.* **170,** 462. (c) Evans, S. D., Sharma, R., and Ulman, A. (1991). *Langmuir* **7,** 156. (d) Evans, S. D., Goppert-Berarducci, K. E., Urankar, E., Gerenser, L. J., and Ulman, A. (1991). *Langmuir* **7,** 2700.
10. (a) Sun, L., Johnson, B., Wade, T., and Crooks, R. M. (1990). *J. Phys. Chem.* **94,** 8869. (b) Sun, L., and Crooks, R. M. (1991). *J. Electrochem. Soc.* **138,** L23. (c) Thomas, R. C., Sun, L., and Crooks, R. M. (1991). *Langmuir* **7,** 620.
11. (a) Tarlov, M. J., and Bowden, E. F. (1991). *J. Am. Chem. Soc.* **113,** 1847. (b) Tarlov, M. J. (1992). *Langmuir* **8,** 80.
12. (a) De Long, H. C., and Buttry, D. A. (1990). *Langmuir* **6,** 1319. (b) De Long, H. C., Donohue, J. J., and Buttry, D. A. (1991). *Langmuir* **7,** 2196.
13. (a) Rowe, G. K., and Creager, S. E. (1991). *Langmuir* **7,** 2307. (b) Creager, S. E., Hockett, L. A., and Rowe, G. K. (1992). *Langmuir* **8,** 854.
14. Miller, C., Cuendet, P., and Gratzel, M. (1991). *J. Phys. Chem.* **95,** 877.
15. Samant, M. G., Brown, C. A., and Gordon II, J. G. (1991). *Langmuir* **7,** 437.
16. Haussling, L., Ringsdorf, H., Schmitt, F.-J., and Knoll, W. (1991). *Langmuir* **7,** 1837.
17. Kwan, W. S. V., Atanasoska, L., and Miller, L. L. (1991). *Langmuir* **7,** 1419.
18. Li, Y., Huang, J., McIver, R. T., and Hemminger, J. C. (1992). *J. Am. Chem. Soc.* **114,** 2428.
19. Collard, D. M., and Fox, M. A. (1991). *Langmuir* **7,** 1192.
20. Uosaki, K., Sato, Y., and Kita, H. (1991). *Langmuir* **7,** 1510.
21. Sagiv, J. (1979). *Isr. J. Chem.* **18,** 346.
22. Kim, J.-H., Cotton, T. M., and Uphaus, R. A. (1988). *Thin Solid Films* **160,** 389.
23. Rubinstein, I., Rishpon, J., Sabatani, E., Redondo, A., and Gottesfeld, S. (1990). *J. Am. Chem. Soc.* **112,** 6135.
24. Rubinstein, I., Steinberg, S., Tor, Y., Shanzer, A., and Sagiv, J. (1988). *Nature* **332,** 426.
25. Steinberg, S., Tor, Y., Sabatani, E., and Rubinstein, I. (1991). *J. Am. Chem. Soc.* **113,** 5176.
26. Steinberg, S., and Rubinstein, I. (1992). *Langmuir* **8,** 1183.
27. Golan, Y., Margulis, L., and Rubinstein, I. (1992). *Surf. Sci.* **264,** 312.
28. A different annealing procedure, which also produces smooth gold surfaces, is described in Ref. 25.
29. Cotton, F. A., and Wilkinson, G. (1988). *Advanced Inorganic Chemistry*, 5th ed. Wiley, New York.
30. Bain *et al.*[3b] argued that these contact angles indicate that the monolayers consist primarily of OM. Direct electrochemical determination of the amount of Cu^{2+} bound to TBEA + OM + OTS monolayers on gold[25] substantiates our original claim[24] that the fraction of TBEA in the mixed monolayer is approximately one-half, indicating that the unusual contact angles measured for Au/(TBEA + OM) are indeed due to a unique molecular distribution.
31. It is evident from Figs. 8.5(a), (c), and (d) that the ionic binding (and competition) is not governed by simple equilibrium rules. This is probably the result of structural modification of the monolayer upon ionic binding.

32. The small increase in the reduction current in Fig. 8.5(b) compared with 8.5(a) indicates some reduction of Fe^{3+} mediated by deposited Cu.
33. Shannon, R. D. (1976). *Acta Cryst.* **A32,** 751.
34. Izatt, R. M., Haas, Jr., C. G., Block, B. P., and Fernelius, W. C. (1954). *J. Phys. Chem.* **58,** 1133.
35. Izatt, R. M., Fernelius, W. C., and Block, B. P. (1955). *J. Phys. Chem.* **59,** 80.
36. Izatt, R. M., Fernelius, W. C., Haas, Jr., C. G., and Block, B. P. (1955). *J. Phys. Chem.* **59,** 170.
37. Izatt, R. M., Fernelius, W. C., and Block, B. P. (1955). *J. Phys. Chem.* **59,** 235.
38. *Stability Constants of Metal-Ion Complexes*. Special Publication No. 25, Supplement No. 1, Part 2 (A. E. Martell, compiler). The Chemical Society, London, 1971, pp. 365–368.
39. Christensen, J. J., and Izatt, R. M. (1983). In: *Handbook of Metal Ligand Heats*. Marcel Dekker, New York, pp. 396–405.
40. The selectivity in the ionic binding to TBEA monolayers has been further confirmed by electrochemical stripping experiments as well as by ellipsometric and wettability measurements. These results are described elsewhere.[25]
41. Fletcher, S. (1989). *Nature* **337,** 216.
42. Rubinstein, I., Steinberg, S., Tor, Y., Shanzer, A., and Sagiv, J. (1989). *Nature* **337,** 217, 514.

Specific Recognition at Functionalized Interfaces: Direct Force Measurements of Biomolecular Interactions

DEBORAH LECKBAND

State University of New York at Buffalo, Department of Chemical Engineering, Buffalo, New York 14260

JACOB ISRAELACHVILI

University of California at Santa Barbara, Department of Chemical and Nuclear Engineering and Materials Department, Santa Barbara, California 93106

9.1. Introduction

The physical forces between biological molecules are often highly specific, in contrast to nonspecific interactions such as metallic, van der Waals, electrostatic (ionic), and hydrophobic forces.[1] The extraordinary specificity of these interactions has been exploited in the development of biosensors and in the affinity purification of biologicals. Most engineering processes and materials rely on nonspecific interactions, primarily due to the instability of some biological molecules under the harsh conditions typically used in industrial processes. Additionally, the cost of these biological materials is high.

Currently, the ability to synthesize more robust biomimetic recognition molecules is still in an embryonic stage when compared with the specificity and diversity of biological systems. The development of more versatile and stable biomimetics, however, requires a fundamental understanding of the forces involved in specific recognition as well as their range of influence. We describe here experimental methods for directly measuring biospecific interactions and provide a unique picture of the magnitudes and ranges of

ISBN 0-12-523485-6

the forces influencing the interaction. We also discuss a number of additional parameters such as the supporting bilayer fluidity that influences the specific interactions between surfaces. Finally, we consider some current applications as well as propose new uses for specific recognition molecules in materials engineering.

9.1.1. Direct Force-Measuring Techniques

We used a surface forces apparatus (SFA) to measure the forces and adhesion of various lipid-protein membranes supported on molecularly smooth solid surfaces (of mica). Details of the SFA technique are described elsewhere.[1,2] Briefly, two thin sheets of atomically smooth mica are glued to the curved surfaces of two glass disks. The disks are then mounted in the apparatus facing each other. The separation D between the two surfaces is measured to 1 Å resolution by viewing a colored fringe pattern produced when white light passes through them. These fringes also give the shapes of the two surfaces and any local deformations, which can be continually monitored in real time at the angstrom level during an interaction. The two surfaces can be made to approach or separate from each other with control to $\sim$ 1 Å by means of a variety of mechanical and piezoelectric distance controls. Normalized forces are determined with a resolution of 0.1 mN/m from the measured deflection of a spring.

9.1.2. Preparation of Surfaces for Force Measurements

The two surfaces were prepared as previously described.[3,4] The receptor used in this work is the protein streptavidin (MW 60,000) which, for brevity, will be referred to as *avidin*.[5] The corresponding ligand is biotin, which has been covalently immobilized to a lipid molecule via a six-carbon spacer (Fig. 9.1). The mica surfaces are coated with supported lipid bilayers to which the receptors and ligands are bound (Fig. 9.1). At 33°C, the biotin surface exposes a fluid dilauryl-phosphatidyl ethanolamine (DLPE) monolayer containing dipalmitoyl-phosphatidyl ethanolamine (DPPE) conjugated to biotin ligand molecules at surface coverages of 0.5 or 5 mol%. The avidin surface exposes 45-Å-thick avidin receptor proteins, each of which has four biotin binding sites—two on either side of the molecule.[6–8] Avidin binds to the surface-exposed biotin via two binding sites, leaving the two unoccupied sites exposed to the medium (Fig. 9.1).

The modified avidin and biotin surfaces are subsequently installed in the apparatus. In the case of biotin-avidin interactions, the final configuration is as shown in Fig. 9.1. In other cases, experiments were conducted between two identical biotin or two avidin surfaces. The distances reported

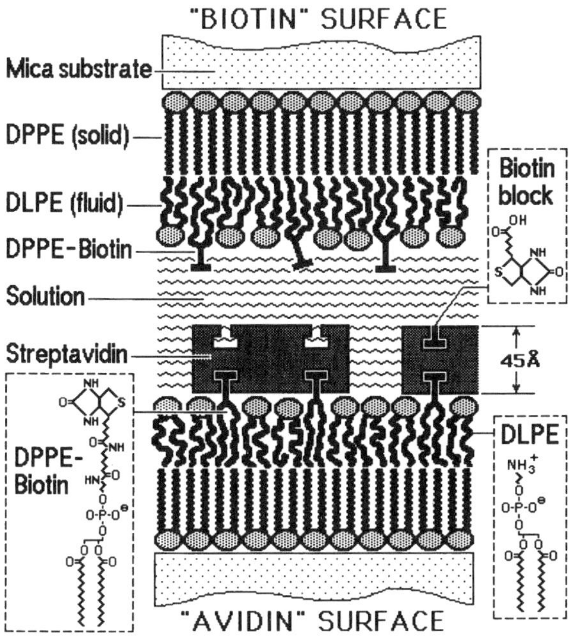

FIG. 9.1. Illustration of the avidin and biotin surfaces used in the experiments.

throughout this work refer to the separation between the surfaces of the outer lipid monolayers.[3,4] Thus, biotin-avidin contact occurs at $D = 45$ Å, the thickness of a single protein monolayer.[6]

9.2. Results

The interactions between two avidin or between two biotin surfaces do not exhibit any unusual or specific interactions. In both cases, the force laws were determined by nonspecific electrostatic, attractive van der Waals, and oscillatory steric forces. The adhesion between the latter surfaces was always weak, consistent with van der Waals forces. In contrast, an avidin and a biotin surface exhibited a very strong, short-range binding attraction at least an order of magnitude larger than expected for a van der Waals force alone. In contrast to specific recognition interactions between molecules in solution, as we now describe, the characteristics of the intersurface interactions are determined not only by the specific avidin-biotin interactions, but also by specific molecular rearrangements on the surfaces.

9.2.1. Specific Recognition

A great deal is known concerning the contact interactions conferring specificity in biological recognition. In both biological recognition and in biomimetic systems, specificity is a consequence of precise geometric and chemical complementarity between the binding site (receptor) and the target molecule (ligand). Historically, this has been referred to as the "lock-and-key" binding mechanism. However, in addition to the geometric complementarity, the surface properties of receptor and ligand also strongly influence the overall interaction potential, altering bimolecular association rates and modulating the affinity constants. As described here, both the specific and nonspecific forces play key roles in the interactions, but their ranges of influence cover different distance regimes.

Figure 9.2 shows the force law for the interaction of the avidin (receptor) and biotin (ligand) surfaces. The force profile is determined by the nonspecific forces at long range, the short-range specific interactions, as well as by the structural properties of the supporting bilayer. At pH 7.4, both the receptor and ligand surfaces are negatively charged since the isoelectric point of streptavidin is pH 6. As a consequence, a nonspecific repulsive double-layer force dominates at separations of $D > 55$ Å, and there is no apparent attractive force between the two surfaces. However, at $D < 55$ Å, the short-range specific attractive forces dominate all repulsive forces, causing the surfaces to jump toward each other into strong adhesive contact.

It is well known from crystallographic studies that short-range van der Waals and, particularly, hydrogen-bonding interactions are the primary sources of the extremely tight avidin-biotin bond.[7,8] Thus, the short-range component of the strong intersurface attractive force is attributed to the known specific avidin-biotin contact interactions. These data clearly demonstrate that the avidin-biotin interaction is influenced by both specific and nonspecific interactions, but the nonspecific double-layer forces determine the long-range interactions, whereas the specific recognition interactions dominate at short ranges.

9.2.2. Steric Interactions and Surface Mobility

Because the receptor and ligand molecules on opposing surfaces are typically not in registry when the surfaces are brought together, an additional short-range steric repulsive force is often observed [Fig. 9.2(a), inset]. This short-range repulsion is attributed to the steric effects produced by the protruding biotin groups as they interact with the avidin surface before encountering the receptor binding site. The steric force observed in Fig. 9.2 occurs at a separation distance $D = 63$ Å, consistent

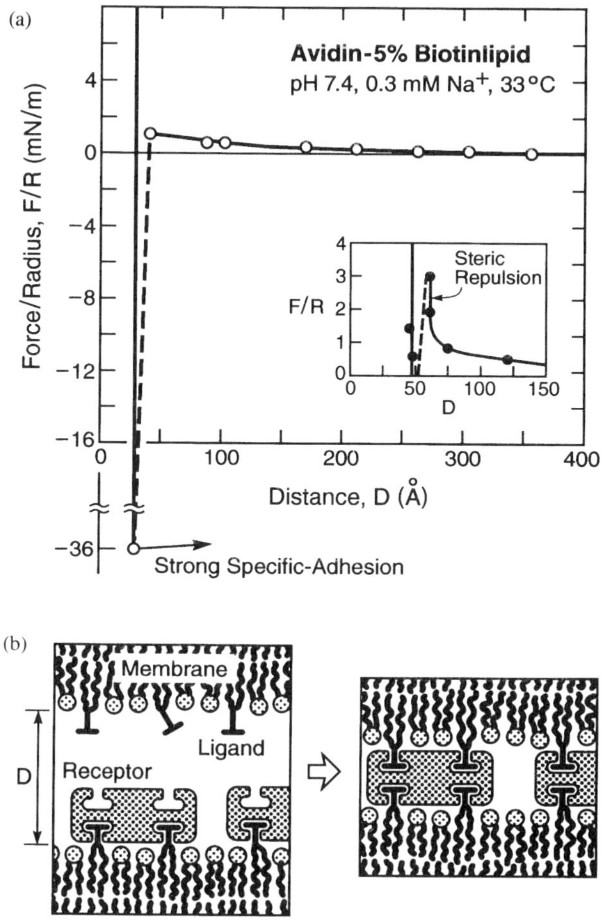

FIG. 9.2. (a) Force versus distance curve for the interaction of the avidin and biotin at a pH 7.4 and 33°C. (b) Illustration of the relative molecular orientations during force measurements.

with the biotin dimensions.[9] This type of thermal or soft steric repulsion is in contrast to the stronger and sharper steric repulsion that occurs once the molecules of the two surfaces are fully in contact with each other (compare the hard, steep repulsion at D = 45 Å). Since the supporting surface must be mobile or fluid in order for these molecular rearrangements to occur, this behavior depends critically on the properties of the supporting bilayer.

As a consequence of the time dependence of the reorientations (Fig. 9.2 and inset), avidin-biotin interactions exhibit two types of force curves depending on the rate of approach of the two surfaces. When the surfaces are rapidly brought into contact, the molecules have insufficient time to achieve their equilibrium, aligned configuration, and a steric repulsive force is measured. However, if the surfaces are brought into contact slowly, such that the opposing molecules mutually align before contact, the steric repulsion is not observed and the true equilibrium force law is measured (Fig. 9.2). The time scale of these molecular rearrangements can be directly determined from the time required for the steric force to collapse. This represents the time required for the ligand to "find" the binding site. In the avidin-biotin case, this occurred after ~ 15 s. One can easily see that the time scale of the rearrangements will depend strongly on the diffusivity in the outer monolayer (i.e., on the membrane fluidity).

9.2.3. Effects of Temperature and Membrane Fluidity

The effect of receptor-ligand mobility on the dynamics of molecular rearrangements and membrane-membrane adhesion was studied by measuring how the short-range adhesion force changed when the temperature was raised from lower than to greater than the chain melting temperature (T_c ~ 30°C) of the lipids in the supporting bilayers. Figure 9.3 illustrates the adhesion measured above and below the chain melting temperature.

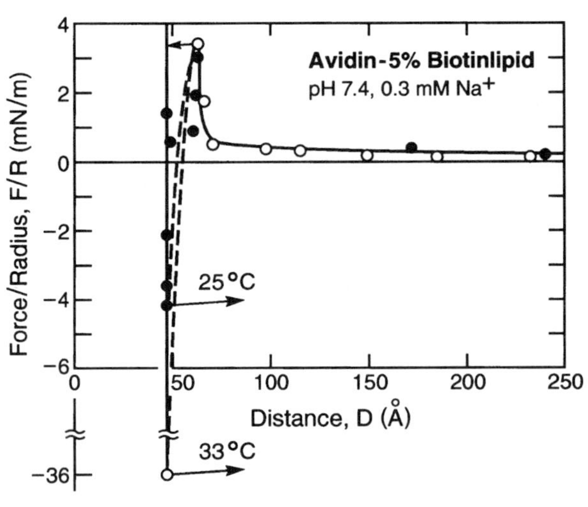

Fig. 9.3. The dependence of intermembrane adhesive strength on the membrane fluidity at 33°C (open circles), above the fluid/solid transition temperature of 30°C, or at 25°C (solid circles).

There were two related consequences of the restricted mobility with *frozen* membranes (at $T < T_c$). The soft steric short-range repulsion did not collapse with time, exactly as expected if the membrane components could no longer freely diffuse in the plane of the membranes. This provides further evidence that the soft steric repulsion is due to the nonequilibrium interactions of protruding surface groups. Furthermore, the surfaces do not "jump in" from $D = 55$ Å, but must be forced into contact. A second, related effect of reduced surface mobility was a dramatic decrease in the intermembrane adhesion. At $T > T_c$, the maximum adhesion of -36 mN/m was measured. In contrast, at 25°C ($T < T_c$), the adhesion was reduced nearly sevenfold to -4.7 mN/m. Both effects can be attributed to the reduced diffusion and rates of molecular rearrangement in the less fluid bilayers, which resulted in reduced and less rapid intermembrane bond formation.

These results clearly illustrate that specific interactions between *surfaces* depend strongly on the bilayer fluidity as well as on the specific interactions. This is in sharp contrast to receptor-ligand associations in solution where the bimolecular associations are single collision events; namely, binding occurs without the requirement of concerted molecular rearrangements.

This fluidity dependence of the specific receptor-mediated intersurface adhesion implies that such adhesion between materials will not immediately achieve the maximum value, but requires a slower "annealing" period during which the binders rearrange and lock. Additionally, biosensor detection of bivalent ligands will be a two-step reaction. The ligand first binds the surface, and receptor diffusion to the second ligand site results in complete, tight binding. In cases of weak initial (or single-site) binding, the ligand may detach before the second in-plane binding event occurs. The detection limits and response time of the sensor will in general depend on both the single-site affinities and the in-plane diffusivity.

9.2.4. Steric Inhibition

Receptor binding to surface-bound ligands requires that the ligand have sufficient flexibility and extensivity normal to the interface to penetrate the receptor binding site to the same extent that occurs in solution. For example, biotin is completely enclosed by the avidin binding site, and the terminal carboxylate barely protrudes from the avidin surface.[8] Crystal structures also show that the binding sites lie in a 15-Å-deep recess on the protein surface; consequently, hard steric repulsion between the outer streptavidin surface and flat "ligand surface" will occur 15 Å from the active site cleft. Immobilized biotin with a tether shorter than 15 Å will fail to penetrate fully the active site and achieve tight streptavidin binding. This is demonstrated by the finding that avidin binds tightly to surface-attached biotin with a six-carbon spacer, only weakly with shorter spacers, and not at all to directly coupled biotin.[10]

Antibodies exhibit similar behavior when binding to functionalized lipid monolayers; namely, the introduction of a long spacer between the hapten and bilayer surface is required to achieve strong binding.[11,12] This principle also applies to the crosslinking of two different receptor molecules. The latter instance is relevant to the formation of multitiered protein assemblies where the total thickness and the identity of each tightly bound layer is strictly controlled.[13] As an example, antifluorescein antibodies do not crosslink avidin molecules via a bifunctional biotin-fluorescein crosslinker unless the spacer between the biotin and fluorescein is sufficiently long to accommodate the steric requirements of both the avidin and antibody binding sites.[14] Furthermore, bis-biotin crosslinking agents with spacers of less than 12 carbons (6 carbons/avidin) between the two terminal biotin moieties does not crosslink avidin.[15] Coupling proteins to surfaces or to each other is, in principle, identical to crosslinking two surfaces, in that the protein surface roughness near the binding site must be considered when designing the coupling moiety. This applies to protein crosslinking as well as to the functionalization of monolayers for interfacial recognition.

9.2.5. Effect of Ligand and Receptor Surface Concentration

The intermembrane binding strength was also measured as a function of the biotin (ligand) surface concentration, both with fluid and frozen bilayers. Binding strengths were measured at two biotin-lipid surface concentrations of 5% and 0.5% (mole/mole) which differ by an order of magnitude. At 5% biotin lipid (1200 $Å^2$/biotin), there is a 2.3-fold excess of biotin headgroups relative to the avidin binding site density (2500 $Å^2$/site). Assuming a 1 : 1 binding correlation between the number of avidin sites and biotin, a tenfold decrease in the biotin lipid coverage is expected to decrease the intermembrane adhesion by 77%. At 25°C the tenfold reduction in the biotin coverage resulted in an 82 $\pm$ 10% reduction in the membrane adhesion. Thus, at 25°C the adhesion appears to vary linearly with the ligand surface density. In contrast, at 33°C, the adhesion decreased by only 60 $\pm$ 10%, slightly less than encompassed by the limits of error. These results are easily explained by nonideality effects within the monolayer. On the frozen membranes at 25°C, the biotin headgroups are physically isolated and headgroup interactions (i.e., in-plane attraction or excluded area repulsion) that give rise to nonideal behavior are unimportant; consequently, on frozen membranes, the surface biotin activity should vary linearly with the biotin coverage, as observed experimentally. In contrast, at 33°C the biotin mobility facilitates molecular interactions, and the nonideal interactions play a more significant role.

Nonlinear dependences of receptor binding on the ligand surface density have been reported with regard to antibody binding to functionalized monolayers. Both in-plane ligand attractive interactions and the excluded area (repulsive interactions) of the antibody (receptor) have been invoked to explain the deviations from ideality for different systems.[11,16] Consequently, particularly at high ligand densities, both the in-plane ligand interactions and the excluded area of the ligand and/or receptor could have pronounced effects on the receptor binding and should be considered when examining the binding or adhesion as a function of the ligand density.

Since the molecules in the contact region are in equilibrium with semi-infinite reservoirs of ligand and receptor on the same surfaces, intermembrane bond formation will induce diffusion of free receptors and ligand into the contact region, in accordance with LeChatelier's principle. In principle, this diffusion would continue until the equilibrium distribution between bound and free species was attained. The establishment of this distribution is, however, restricted since the resulting high affinity intermembrane bonds confine the "locked" molecules to the contact zone. Diffusion into the area will ultimately be opposed by repulsive excluded area interactions. Consequently, the equilibrium distribution of bound and free streptavidin on the surfaces and the resulting equilibrium adhesion cannot be achieved.

9.2.6. Effect of Protein Structure on Molecular Recognition

It has been proposed that long-range attractive forces, in addition to the short-range contact interactions, are important in recognition.[17-19] The functional significance of these forces lies in their reduction of the overall interaction potentials and the resulting increase in the association kinetics, and such long-range attractive forces have been implicated in bimolecular trajectory steering and in the enhancement of association rates.[19,20] In contrast to the short-range contact interactions, the long-range forces are intimately linked to the properties of the external protein surface. In particular, the electrostatic potential field is determined by the topological map of charges over the protein exterior. Asymmetric charge distributions may give rise to focusing electrostatic potential fields which steer ligands into binding pockets.[17,18,21] At intermediate range, the hydrodynamic forces associated with the surface topology may also steer ligand trajectories.[22]

The charged amino acids on the streptavidin surface have different pKs, and the effects of amino acid titration (charge perturbations) on the avidin–biotin forces were investigated in measurements at pH 6 and at pH 7.2.[23] Only histidine (His) has a pK at 6.5, whereas the other amino acids have either pKs <5 or pKs >8. Thus, differences between the force–distance curves measured at pH $6 < pK_{His}$ and at pH $7.2 > pK_{His}$ would directly reflect the titration of the solvent-accessible histidines of avidin. The corresponding changes in the avidin electrostatic surface potential can be obtained from data fits to the theoretical double-layer force curves. Moreover, since there are only two exposed histidines, these measurements would demonstrate the force sensitivity to single amino-acid perturbations, and indicate the potential impact of directed mutations on bimolecular interactions.

Figure 9.4 shows the resulting force curves obtained at pH 7.2 (open circles), pH 6.0 (filled circles), and at pH 6.0 and increased ionic strength (open squares). At pH 7.2 (Figs. 9.2 and 9.4), the profile at $D > 100$ Å was determined by a monotonic repulsive force. Conversely, when the pH was reduced to pH 6.0, the long-range repulsive force switched from repulsive to purely attractive. The resulting behavior can be entirely described by the pH-dependent changes in the electrostatic properties of the avidin surface. The data were fit to the superposition of the van der Waals and electrostatic double-layer forces—the DLVO (Derjaguin, Landau, Verwey, and Overbeck) theory—and the avidin electrostatic potentials at the

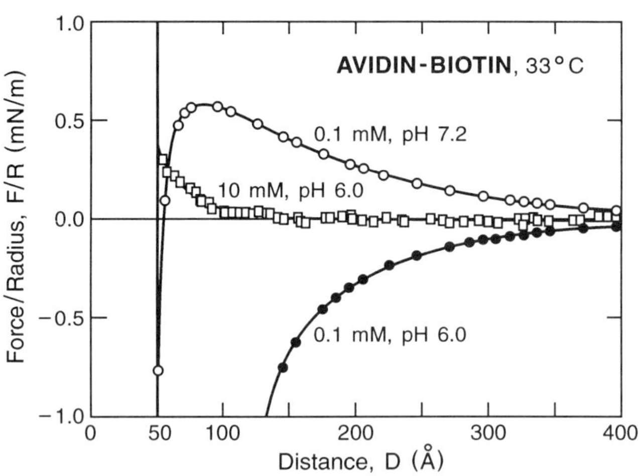

FIG. 9.4. Force–distance curves for avidin–biotin interactions at pH 7.2 (open circles) and at pH 6.0 in 0.1 mM salt (filled circles) and in 10 mM salt (open squares).

different pH values were determined from those fits. At pH 7.2, the fitted electrostatic potentials of the avidin and biotinylated membrane surfaces were −38 mV and −60 mV, respectively. The former value corresponded to −0.8 unit charge per protein.

Thus, the repulsive long-range force was attributed to the electrostatic double-layer repulsion between the negatively charged biotin membrane and the negatively charged avidin binding surface. The screening of the long-range force following an increase in the ionic strength (Fig. 9.4) verified the electrostatic origin of the force. Conversely, at pH 6.0, the measured avidin electrostatic surface potential was +29 mV, or +0.5 unit charge per protein. Thus, the pH-dependent changes were consistent with the titration of the exposed avidin histidines.[23] Moreover, a net change of only 1.3 unit charge per protein dramatically altered the force profile from repulsive to attractive at $D > 100$ Å, and demonstrated that small surface perturbations can significantly impact protein recognition interactions. On the basis of these measurements one would predict that the avidin–biotin binding rate at pH 6 would be accelerated relative to that at pH 7.2.

The resulting kinetic rate enhancement conferred on the receptor-ligand association by these differences in the intermediate- to long-range forces can be estimated from[24]

$$\frac{k_{6.0}}{k_{7.2}} = \exp\frac{\Delta W}{KT} \tag{9.1}$$

where $k_{6.0}$ is the rate constant obtained at pH 6.0 in the presence of the attractive double-layer force, $k_{7.2}$ is the rate constant at pH 7.2, ΔW is the potential difference at the different pH values, $W_{7.2}-W_{6.0}$, K is the Boltzmann constant, T is the absolute temperature. Thus, a reduction in the repulsive potential at pH 6, $W_{6.0}$, relative to $W_{7.2}$ gives a positive value for ΔW, and $k_{6.0}$ will be greater than $k_{7.2}$. Note, however, that the influence of ΔW is scaled by KT, the thermal energy. The potential difference ΔW can be determined directly from the measured force profiles by making use of the relationship

$$\frac{W}{R} = -\int_0^\infty \frac{F(x)}{R}\,\mathrm{d}x. \tag{9.2}$$

For additive forces, the radius R simply scales the interaction energy. Thus, the corresponding *intermolecular* energy between soluble biotin and avidin can be estimated by the multiplication of Eq. (9.2) by r, the approximate geometric average radius of avidin and biotin. Thus, one can estimate changes in rate constants by use of directly measured force profiles together with Eqs. (9.1) and (9.2). For example, from the data shown in Fig. 9.4, we predict that the pH-dependent changes in the force could enhance the bimolecular association rate by up to ~100 fold at the ionic strength of these measurements.[23] It is important to note, however, that bulk kinetic rates reflect the receptor–ligand interaction potentials averaged over the entire receptor surface. Therefore, since these force profiles reflect a single relative orientation of the receptor and ligand, they can only be used to estimate relative changes in bulk kinetic rates. These measurements nevertheless demonstrate that the long-range forces that control the binding kinetics depend intimately on the protein surface properties.

In these measurements, the long-range avidin–biotin force was primarily due to double-layer interactions between the charged avidin surface and the biotin-lipids. Moreover, the force profiles were highly sensitive to the avidin surface properties, and hence, to the avidin surface composition. These results are consistent with previous examples of protein stucture-dependent rate enhancements and bimolecular trajectory steering.[19,25] Such nonlocal influences also extend to recognition at solid surfaces or membranes where, for example, the electrostatic properties of the underlying support can also influence the apparent biological activity of immobilized or membrane-bound receptors.[26] Although the magnitude of the electrostatic double-layer force depends on the electrostatic properties of both the receptor and ligand, its range is determined by the ionic strength of the solution. Consequently, at physiological saline concentrations, the forces will be intermediate in range with a Debye length of 8–10 Å. Under the latter conditions, though shorter range than that described in this chapter, double-layer force will still be influential at intermediate to large separations.

In general, one can see from Eq. (9.2) that the magnitude of the potential relative to KT, not the *measured* range, determines the effective range of the force. Only potentials on the order of KT will have a significant influence on the interaction; consequently, although a force may be detectable up to a few hundred angstroms, it may be insignificant when judged by the KT criterion. To illustrate this point, Fig. 9.5 shows the hypothetical scaled potential as a function of separation for the interaction of the avidin and biotin surfaces. Although the potential is attractive at separations up to ~150 Å, the normalized potential only exceeds unity at $D < 85$ Å. Thus, the attractive force will significantly influence binding only at $D < 100$ Å, not over the entire *measurable* range of ~150 Å. Thus, the scaled potential is the appropriate criterion for determining the length scale over which a force will have a significant influence.

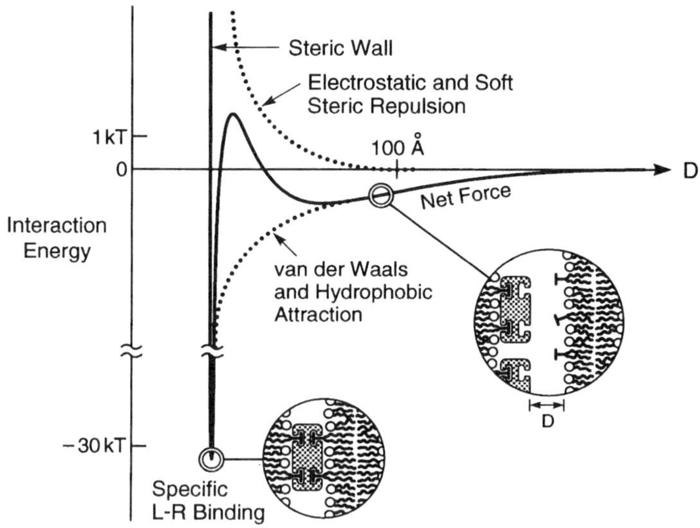

FIG. 9.5. Characteristic force-distance or energy-distance curve for ligand-receptor (L-R) interactions.

9.2.7. Disruption of Strong Intermembrane Bonds

Attempts to determine the avidin-biotin bond strength by measurement of the intermembrane pull-off force result in tearing biotin molecules from the membrane surface.[3,4] Cohesive failure between nonspecifically adhering surfaces is well described by adhesion theories.[27] However, the separation mechanism of two membranes coupled via specific bonds is less clear. Although different membrane receptors exhibit binding affinities varying over several orders of magnitude, attempts to disrupt the receptor-ligand bonds mechanically always resulted in the receptor extraction from the membrane, independent of the magnitude of the bond energy relative to the extraction energy. This behavior is easily explained in terms of the forces involved in the competing fracture mechanisms. The distance over which the energy is expended on surface separation, not the energy alone, determines the pull-off force because $F = -\mathrm{d}E/\mathrm{d}x$. This is illustrated in the following analysis, which has been briefly discussed previously by Bell.[28]

Assume that a force is applied on a time scale much shorter than the bond lifetime. The force required to disrupt the bond is $F_B = -\Delta E/\Delta x_B$, where ΔE is the receptor-ligand bond energy and Δx_B is the bond length.

Assume also that the energy cost of extracting a lipid-bound ligand from the membrane is comparable to the bond energy ΔE. However, this energy is expended over the length of the lipid chain, Δx_L, as the chain is pulled from the membrane. The ligand extraction force is $F_L = -\Delta E/\Delta x_L$. The force required for ligand extraction relative to that of bond cleavage is $F_L/F_B = \Delta x_B/\Delta x_L < 1$. Thus, the lipid-bound ligand will be preferentially pulled from the membrane. This will always be the case as long as the ratio $(\mathrm{d}E_L/\mathrm{d}x_L)/(\mathrm{d}E_B/\mathrm{d}x_B)$ is less than unity. Note that this condition also applies in cases where $E_B < E_L$ as long as the effective bond length is sufficiently short. From this result it is apparent that the adhesion between specifically coupled membranes is more likely to be determined by the lipid anchor than by the specific intermembrane bonds, consistent with experimental observations.

9.3. Discussion

These direct force measurements demonstrate that it is possible to study directly highly specific biomolecular interactions at the molecular level. Furthermore, it is possible to observe the associated molecular rearrangements in real time. The results on the avidin-biotin system show that in addition to the expected strong specific short-range interaction, intermediate- to long-range forces also influence the binding. As expected, long-range van der Waals and electrostatic forces operate between the active receptor and ligand. The long-range electrostatic force between avidin and biotin depends on the avidin structure, and this most likely plays a significant role in steering bimolecular trajectories. At pH 7.2, the short-range attractive force is opposed by the electrostatic repulsive force, which together give rise to an energy barrier at some finite separation. The energy barrier does not affect the final strong adhesion energy, but it would have a marked effect on the bimolecular association rates.

Intermembrane interactions were also shown to be modulated by the fluidity of the supporting membranes and by the receptor and ligand densities. The increased membrane fluidity correlated directly with a stronger intermembrane adhesion, presumably because the increased lateral mobility and greater ease of molecular reorientations result in a greater number of bonds forming per unit time. Nonideal in-plane interactions also influence the dependence of adhesion on ligand density. These results have important consequences for understanding the role of cell surface dynamics in cell attachment processes and the resulting adhesion strength.

9.4. Conclusion and Implications for Future Bioengineering Materials Technology

It is clear that nature has here developed an extremely efficient mechanism whereby noncovalent adhesive junctions having the strength of covalent bonds can be switched on, or mechanically "locked," quickly and with minimal expenditure of energy. While the exact biological function of the avidin-biotin system is unknown, it is generally accepted that these types of bonds can be switched off ("unlocked") equally easily (e.g., by a change in the pH, ionic strength, or by light stimulation). This modulated recognition has been employed in separations technology (affinity chromatography), biosensor design, and diagnostics. In many cases, the affinity can be reversibly modified by alterations in the solution conditions or by the addition of "effector molecules," which modulate the affinity or compete with the bound ligand.[20] Often the tight binding can only be "unlocked" by irreversible denaturation of the receptor and/or ligand, as in the case of avidin or antibodies, thereby providing a somewhat environmentally insensitive adhesive. It is conceivable that such "molecular Velcro" mechanisms can be developed for a wider range of technological applications.

One such possible application is in producing new types of paints or adhesive coatings, where the coated film will adhere strongly to a particular substrate surface via these specific bonds. Moreover, when the film needs to be removed, a simple "unlocking" mechanism could be triggered, for example, by irradiating the surface with light of a particular wavelength so the paint will simply fall off.

This is an example of how controlling interactions at the molecular level could result in a new class of *tunable material* that possesses a range of different properties that can be "tuned" by some external stimulus.

Acknowledgments

We thank the National Science Foundation (Grant No. CTS-9015537), the National Institutes of Health (GM13300), and Boehringer-Mannheim GmbH (Werk Tützing, FRG) for financial support.

References

1. Israelachvili, J. (1991). *Intermolecular and Surface Forces*, 2nd ed. Academic Press, New York.
2. Marra, J., and Israelachvili, J. (1985). *Biochem.* **24,** 4608.

3. Helm, C., Knoll, W., and Israelachvili, J. (1991). *Proc. Natl. Acad. Sci.* **88,** 8169.
4. Leckband, D. E., Schmitt, F. J., Knoll, W., and Israelachvili, J. (1992). *Science* **255,** 1419.
5. Bayer, E., and Wilchek, M. (1990). *Meth. Enzymol.* **184,** 49.
6. Darst, S., Ahlers, M., Meller, P. H., Kukalek, E. W., Blankenburg, R., Ribi, H. O., Ringsdorf, H., and Kornberg, R. D. (1991). *Biophys. J.* **59,** 387.
7. Hendrickson, W. A., Pahler, A., Smith, J., Satow, Y., Merritt, E. A., and Phizackerley, R. P. (1989). *Proc. Natl. Acad. Sci.* **86,** 2190.
8. Weber, P. C., Wendoloski, M. W., Pantoliano, M. W., and Salemme, F. R. (1992). *J. Am. Chem. Soc.* **114,** 3197.
9. Metzler, D. (1977). *Biochemistry*, Academic Press, New York, p. 433.
10. Blankenburg, R., Meller, P., Ringsdorf, H., and Salesse, C. (1989). *Biochemistry* **28,** 8214.
11. Timbs, M. M., Poglitsch, C. L., Pisarchick, M. L., Sumner, M. T., and Thompson, N. L. (1991). *Biochim. Biophys. Acta* **1064,** 219.
12. Cooper, A. D., Balakrishnan, K., and McConnell, H. M. (1981). *J. Biol. Chem.* **256,** 9379.
13. Ahlers, M., Blankenburg, R., Haas, H., Möbius, D., Möhwald, H., Müller, W., Ringsdorf, H., and Siegmund, H.-U. (1991). *Adv. Mater.* **3,** 39.
14. Herron, J., Müller, W., Paudler, M., Riegler, H., Ringsdorf, H., and Suci, P. (1992). *Langmuir* **8,** 1413.
15. Green, N. M., Konieczny, L., Toms, E. J., and Valentine, R. C. (1971). *Biochem. J.* **125,** 781.
16. Tamm, L., and Bartoldus, I. (1988). *Biochemistry* **27,** 7453.
17. Allison, S. A., Ganti, G., and McCammon, J. A. (1985). *Biopolymers* **24,** 1323.
18. Wendolowski, J. J., Matthew, J., Weber, P. C., and Salemme, S. (1987). *Science* **238,** 794.
19. Margoliash, E., and Bosshard, H. R. (1983). *Trends. Biochem. Sci.* **8,** 316.
20. Creighton, T. E. (1989). *Proteins*. W. H. Freeman and Company, New York.
21. Sharp, K. A., and Honig, B. (1990). *Ann. Rev. Biophys. Chem.* **19,** 301–322.
22. Brune, D., and Kim, S. (1994). *Proc. Natl. Acad. Sci.* **91,** 2930–2934.
23. Leckband, D., Schmitt, F.-J., Knoll, W., and Israelachvili, J. (1994). *Biochemistry* **33,** 4611–4624.
24. Berry, R. S., Rice, S. A., and Ross, J. (1980). *Physical Chemistry*. John Wiley & Sons, New York, pp. 1160.
25. Stayton, P., and Sligar, S. (1990). *Biochem.* **29,** 7381.
26. MacRitchie, R. (1978). *Adv. Protein Chem.* **32,** 283.
27. Johnson, K. L., Kendall, K., and Roberts, A. D. (1971). *Proc. R. Soc. London A* **324,** 301.
28. Bell, G. I. (1978). *Science* **200,** 618.

Formation of Recognition Patterns by Langmuir-Blodgett Techniques

TILMAN SCHWINN

Technische Universität München, Physikdepartment E22, 85748 Garching, Germany

SVEN-PETER HEYN

VDI / VDE Technologiezentrum, 14513 Teltow, Germany

MARTIN EGGER

Boehringer Mannheim GmbH, Tutzing, Germany

HERMANN E. GAUB

Technische Universität München, Physikdepartment E22, 85748 Garching, Germany

10.1. Introduction

The very intense research conducted during recent years to develop novel detection schemes has led to a variety of extremely sensitive techniques for the quantitative detection of macromolecular molecules.[1-7] Most of them utilize immunochemical techniques, some of them in combination with surface sensitive techniques. The major problem that all of these very sensitive techniques have to overcome is nonspecific adsorption. Particularly at low analyte concentrations, this contributes significantly to the error in the measurements.[8,9] It is obvious that relative measurements against various controls help—at least partially—to overcome this problem. Although the difference between the signal and control contains the information about the specifically bound amount, the accuracy of the

ISBN 0-12-523485-6

measurement depends on the magnitude of the signal compared to the control.

If the binding experiment is performed at a surface, a control area is desirable next to the area, where the binding is measured. Patterns of binding areas and control areas help to reduce the contributions of instrumental variations to the error.[10] Such a pattern may in principle be of any shape. Regular periodic patterns, however, offer additionally a drastic improvement in the contrast of the measurement and allow an easy Fourier space analysis. A periodic pattern, for example, yields the desired value—the difference between signal and control—directly in the amplitude of the first Fourier component. An analogy to this proposed detection scheme may be seen in the lock-in amplifier, a frequently used instrument for noise reduction in low signal detection. In the lock-in amplifier a noisy dc signal is modulated and then detected at higher frequencies. In the detection scheme discussed here, the noisy signal is spatially modulated and detected at higher spatial frequencies. All the lower wavelength fluctuations and inhomogeneities of the detector do not contribute anymore to the signal. The limitation will be imposed by the lateral resolution of the detection. As in most cases when the detection is done optically, the characteristic dimensions of the pattern should for practical reasons be larger than 10 μm.

Several techniques are available for the formation of regular patterns on solid surfaces.[11] In this chapter, we focus on ideas that make use of the self-organization of patterns at the air/water interface.[12–14] We develop some general ideas on the thermodynamics of such patterns and report on experiments that we have performed to design functional recognition patterns by means of the Langmuir-Blodgett (LB) technique.

With the introduction of microfluorescence techniques[15] for the investigation of surfactant films at the air/water interface, a wealth of new information about these systems and about their lateral organization became available. Among the first systems that were investigated with this new approach was the fluid-solid coexistence of lipids.[12,16–18] A minor fraction of fluorescently labeled lipids, which is enriched in the fluid phase, gives rise to a pronounced contrast between the fluid and the solid phase and allows the investigation of the macroscopic pattern of the two phases. It was soon understood that not only the inherent molecular properties, which were mainly responsible for the symmetries of the solid domains, but also long-range fields play a significant role in the formation of certain patterns.[12,14,19] It turned out that the electrical polarization of the air/water interface is locally modified in different ways by the solid and the fluid phase, which gives rise to a strong dipole net interaction among the solid domains. This predominantly repulsive interaction between the domains results in a superlattice of the domains. Under conditions where a strong line tension between the fluid and the solid phase favors the formation of equally sized domains, the resulting hexagonal superlattice can be of a surprisingly regularity.[12] In other cases a deliberately reduced line tension may lead to the formation of a well-pronounced stripe pattern as superlattice.[12,20]

More recently films in the fluid-fluid coexistence have been investigated both experimentally as well as theoretically.[14,21,22] Here the internal and the external interactions due to the electrical polarization in combination with the line tension between the two phases determine the shape of the domains. One of the systems investigated in this study shows certain phenomenological similarities with a fluid-fluid coexistence and may possibly be described in these terms. However, we would rather discuss our system in terms of a bilayer with vertical coupling.

As the molecular basis for the formation of what we call a *recognition pattern*, we chose two amphiphilic receptors (Fig. 10.1): the well-known streptavidin-biotin lipid complex[23,24] and a hybrid molecule between a phospholipid and the binding site of an antibody, which we have designed and synthesized for these studies.[25] Both molecules share certain thermodynamic properties, although they exhibit a completely different pattern at the air/water interface.

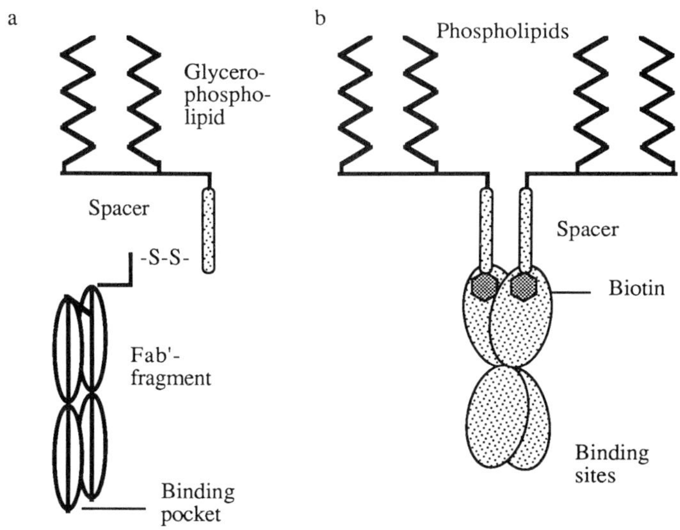

FIG. 10.1. Schematics of the lipid-anchored model receptors: (a) Fab′-lipid and (b) streptavidin-biotin lipid complex.

10.2. Materials and Methods

10.2.1. Streptavidin-Biotin Lipid Complex

The biotinylated lipid DPPE-X-biotin (N-(6-((biotinoyl)amino)-hexanoyl)-dipalmitoyl-L-a-phosphatidylethanolamine) and the fluorescein-labeled streptavidin were purchased from Molecular Probes, Eugene, Oregon. All other chemicals, unless otherwise noted, were from Sigma. All substances were used without further purification. The lipids were dissolved in chloroform or in chloroform-methanol (4 : 1).

A lipid mixture of dipalmitoylphosphatidylcholine (DPPC) and 5% DPPE-X-biotin was spread on a home-built microfluorescence film balance[26] with a 500 mM NaCl (p.a. grade) subphase, pH = 7, $T = 20°C$, using ultrapure Millipore filtered water (Bedford, Massachusetts). Streptavidin was diluted in 100 mM NaCl solution at a concentration of 1 mg/mL from which 10 to 100 mL was added to the electrolytic subphase (15 mL) with a syringe (Hamilton) behind the barrier after compression to the coexistence region. The incubation times varied from 2 h up to 24 h. Then the monolayer was expanded to the gas-fluid coexistence region. After 5 to 20 min of equilibration, the monolayer was compressed again before the microfluorescence images were recorded.

10.2.2. Fab-Lipid Films

The synthesis of the Fab-lipid [Fig. 10.1(a)] and its characterization is given elsewhere in detail.[25] Briefly, recrystallized monoclonal anti-DNP antibodies were Texas-red labeled and cleaved with papain. The hinge of the resulting Fab2′ fragments was reduced and the resulting Fab′-fragments were reacted with an activated spacer lipid in a vesicle suspension. The Fab′-lipid vesicles were purified by affinity chromatography.

Monomolecular films containing Fab′-lipids were made following the procedure given in Ref. 27. Briefly, a suspension of vesicles from DMPC containing a fraction of 0.6% Fab-lipid was allowed to form a monolayer at the air/water interface of a small trough kept at 37°C. With increasing surface pressure, the monolayer was driven via the surface of a strip of humid filter paper to the surface of the trough of a microfluorescence film balance.[26] The humidity of the filter paper was controlled such that attached vesicles were stripped off the monolayer. The filter paper was removed once the pressure on the trough had reached a value of 20 mN/m. The monolayer was then slowly compressed and imaged in parallel by fluorescence microscopy.

10.2.3. Results and Discussion

Starting from a vesicular suspension, films containing Fab′-lipids and DMPC as matrix lipid were allowed to form at the air/water interface.[10] It was confirmed by atomic force microscopy[28,29] and by surface plasmon microscopy[3] that films formed under the given experimental conditions are of monomolecular thickness. It was furthermore confirmed that when the thickness of the water film on the bridge between spreading trough and LB trough is kept small enough, the film is essentially free from attached vesicles. As shown in Fig. 10.2(a), at a pressure of less than 20 mN/m, such a film appears homogeneous when observed by microfluorescence. Thickness measurements by surface plasmon spectroscopy[2,3] revealed that in this phase the average lateral density of the Fab′-lipid is smaller than 1/100 nm.

Upon compression the film spontaneously undergoes a phase transition resulting in a protein-rich phase that appears bright and a phase where the

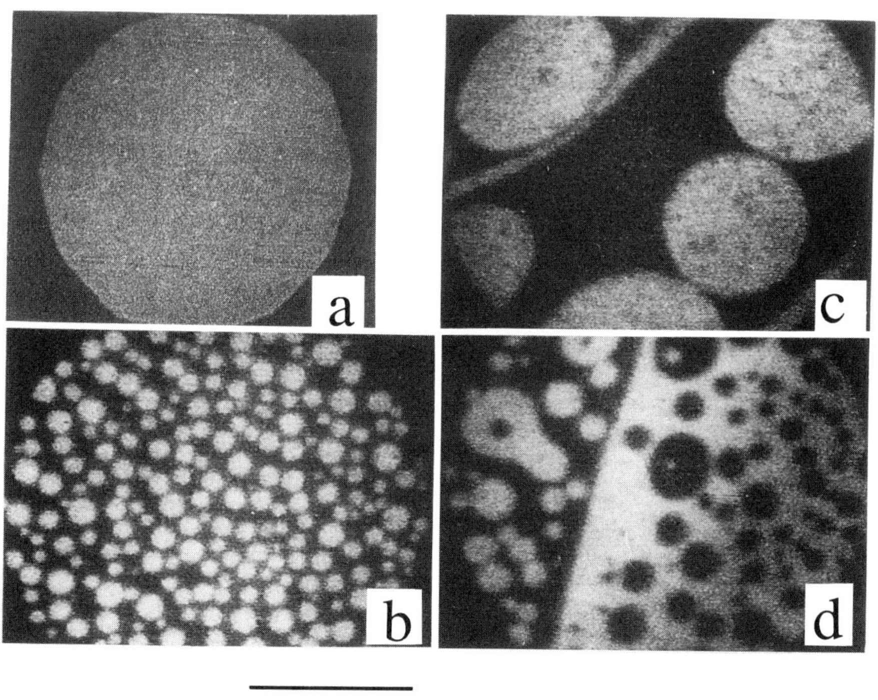

FIG. 10.2. Microfluorescence images of a Fab′-lipid film at the air/water interface: (a) In the homogeneous phase and (b)–(d) after the gas-fluid transition.

Fab′-lipid is depleted, which appears dark [Figs. 10.2(b), (c), and (d)]. No sign of a phase transition can be seen in the pressure area diagram; the pressure monotonically increases with compression. The lateral density of the Fab′-lipid was determined by quantitative microfluorescence and surface plasmon microscopy to be larger than 0.02 nm^2 and smaller than $1/230\ nm^2$ for the bright and the dark phases, respectively.[3] None of the resulting phases is solid. The small domains, which were created right at the phase transition, fuse without further compression and form larger coherent areas of both phases. A detailed investigation of this process provides further information on the nature of the phases. In Fig. 10.3 a sequence of the fusion of two smaller bright domains to a single larger domain is shown. After approximately 30 s, the macroscopic fusion process is completed. Both domains had initially been equally bright but we had deliberately bleached one of the domains. This enabled us to investigate the lateral mixing of bleached and unbleached Fab′-lipids. A plot of the intensity profiles across the interface line between both domains is shown in Fig. 10.4.

These profiles can be quantitatively analyzed by standard diffusion equations. The symmetry of the experiment reduces the problem to a one-dimensional (1-D) equation. The analytical solution, which is basically the complement of the Gaussian error integrals,[30] was fitted numerically to the measured profiles, where the diffusion coefficient D was the only fit parameter. The diffusion coefficient of the Fab′-lipids determined from each intensity profile differed little and turned out to be $D_{\text{Fab}'\text{-lipid}} = 1.5 \pm 0.3\ \mu m^2/s$. This indicates that the distortion of the geometry due to the fusion and the accompanying lateral flow did not affect the profile in the center and justifies our simple 1-D ansatz. The measured diffusion coefficient is only somewhat smaller than that of a lipid molecule in the fluid phase. This indicates strongly that the Fab′-lipids are not aggregated in this phase and move freely in the matrix of the fluid DMPC monolayer. This finding is in agreement with the measured low lateral density of the Fab′-lipids.

The pressure at which the first domains were observed was found to be dependent on the lateral concentration of the Fab-lipids. This is understandable if we assume that the phase transition is triggered by the lateral interaction of the Fab′-fragments, where the matrix lipid plays only the role of a two-dimensional (2-D) solvent. In this model we can treat the Fab′-film independently from the lipid film. The lateral pressure, which is measured with the Wilhelmy system, is then the sum of the pressure in both films. It is to be expected that this total pressure is dominated by the lipid. This would explain our finding that the pressure area isotherm is insensitive to the phase transition in the protein leaflet. Taking into

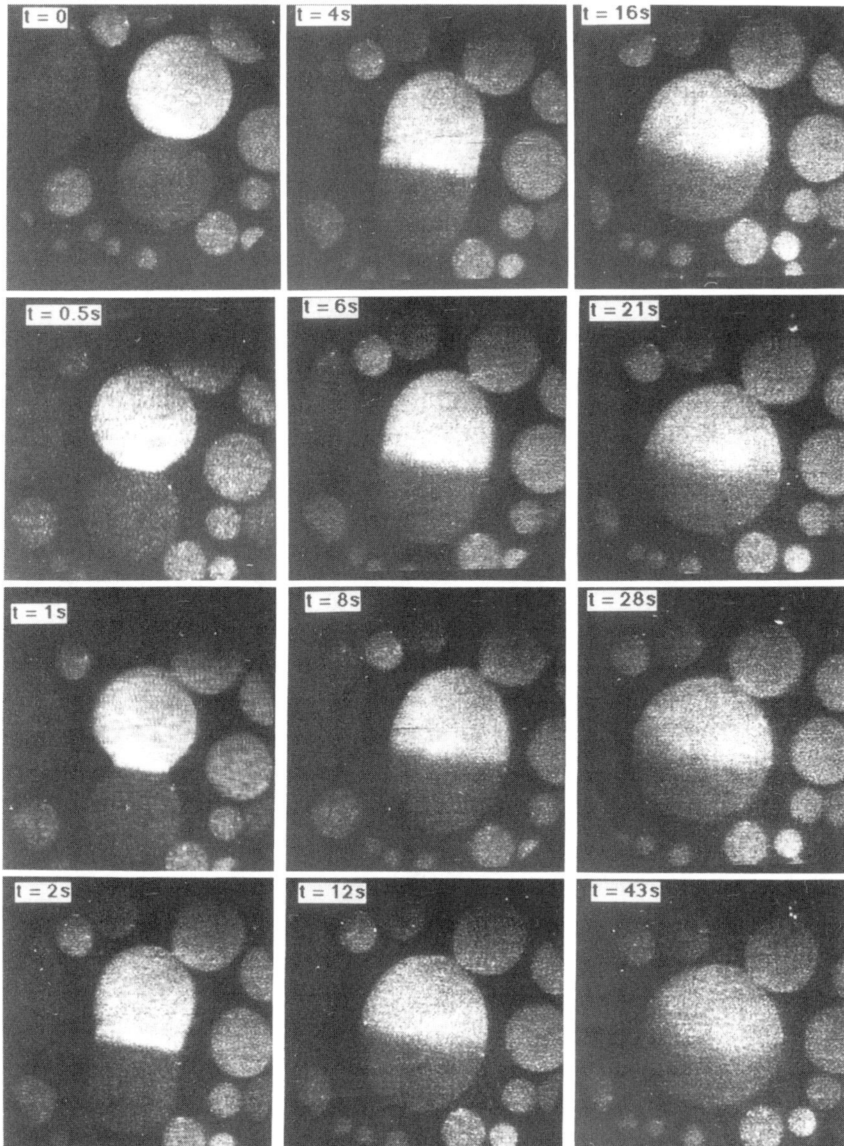

FIG. 10.3. Sequence of the fusion of two fluid Fab′-lipid domains.

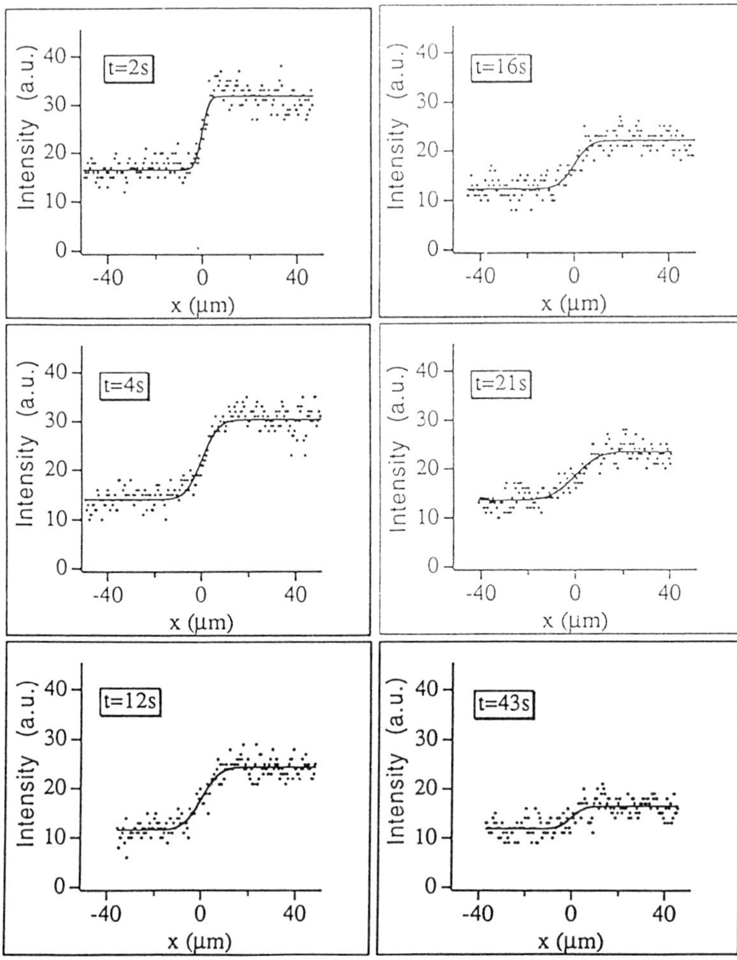

FIG. 10.4. Analysis of the diffusion of the Fab′-lipids from Fig. 10.3.

account the low lateral density of the Fab′-fragments in the homogeneous phase, we have to regard this phase as gas-like. In the simplest case the transition would be a gas-fluid transition. The fact that we never see small seeds of the fluid phase that grow upon compression but that we always observe a sudden breakup of the homogeneous phase indicates that a marked undercooling of the gas occurs. This gas-fluid transition is reversible: Upon decompression, the domains dissolve again gradually.

One way to prove the model would be to increase the lateral density of the Fab-lipids such that the following transition to a solid phase can be induced. Unfortunately, the lateral concentration of the Fab′-lipid is limited due to the way they are synthesized. The excess of matrix lipid from the inner side of the vesicle limits the lateral density to values well below a solid density. We have therefore investigated this aspect of the thermodynamics of lipid-anchored proteins at the air/water interface using a similar system where higher lateral densities can be achieved. Figure 10.1(b) shows the complex of two biotinylated lipids and a streptavidin molecule. This system was very intensely investigated by various groups. The conditions at which a 2-D crystallization of the proteins occurs are well established.[23,24,31–34] In our experiments (see Figs. 10.5 and 10.6) the conditions were chosen such that after incubation of the biotin lipid-containing monolayer with fluoresceinated streptavidin, the monolayer was—after expansion to lower pressures—homogeneously fluorescent at any polarization of the excitation [Fig. 10.6(a)]. The lateral density in this phase was kept low enough that the streptavidin film was not ordered under these conditions. After compression into the fluid-solid coexistence of the matrix lipid DPPC, dark areas in the distribution of the streptavidin fluorescence appeared [Fig. 10.6(b)]. Obviously, the growing solid DPPC domains had excluded the fluorescent-labeled streptavidin. With the increasing fractional area of the solid lipid domains, this lead to an increase in the concentration of the streptavidin-biotin lipid complex in the remaining fluid DPPC matrix. Eventually this compression results in a solidification of the protein phase. When seen with linearly polarized

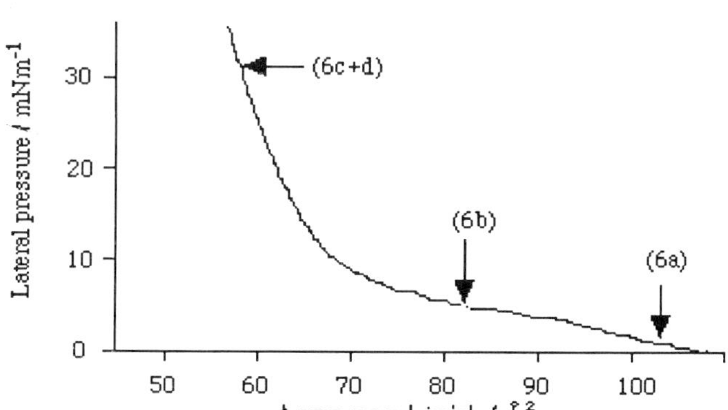

FIG. 10.5. DPPC film containing streptavidin-biotin lipid complexes at the air/water interface: pressure-area isotherm at 20°C (compression).

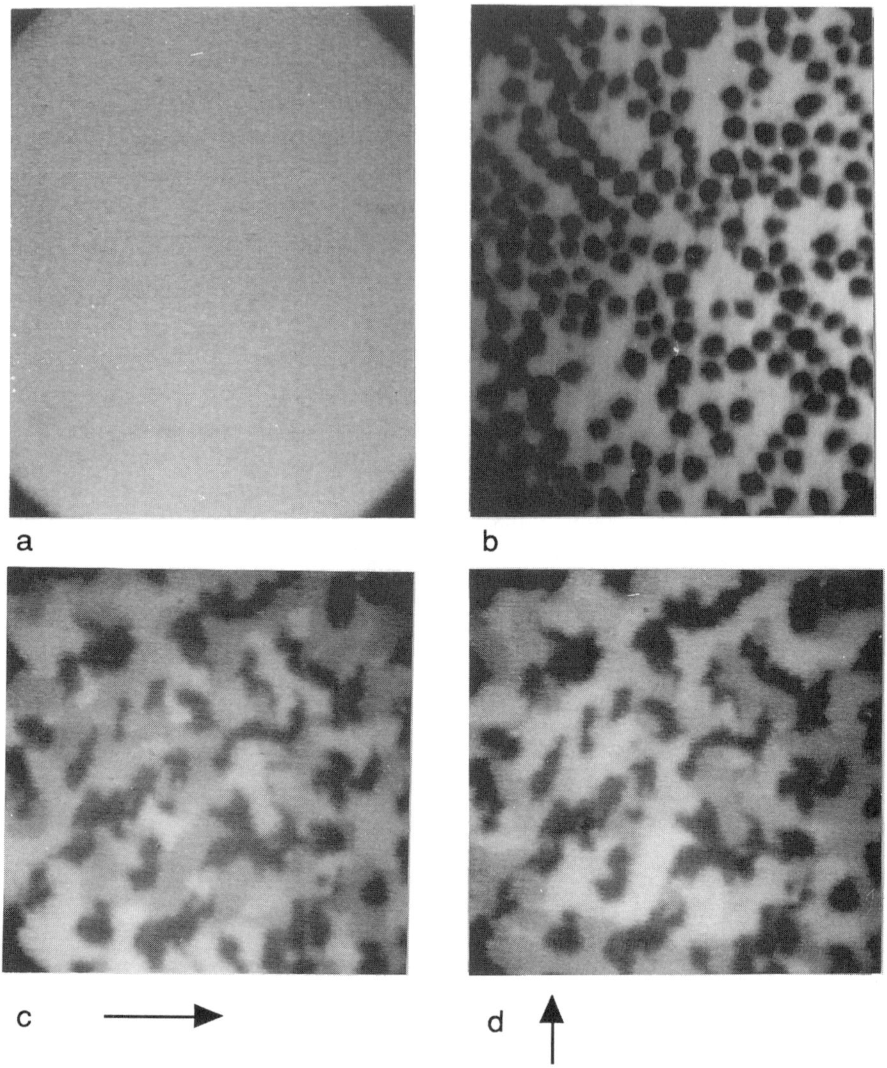

FIG. 10.6. Microfluorescence images of lipid-anchored streptavidin in a DPPC monolayer at different pressures. (a) At low pressures the lipid-anchored proteins are distributed homogeneously. (b) In the fluid-solid coexistence region, they are excluded from the solid phase of DPPC. (c) and (d) At high pressures the brightness of the fluorescent protein domains is a function of the excitation polarization (indicated with arrows in the lower images).

fluorescence excitation [Figs. 10.6(c) and (d)], the protein phase now shows a domain-like pattern where the relative intensity of the domains is a function of the orientation of the polarization (see arrows). This anisotropy means that the solid protein phase exhibits long-range order and is therefore likely to be a crystalline phase.[24] To our knowledge, this is the first example where the crystallization of a protein at the air/water interface had been directly induced by lateral compression. This crystallization process was found to be reversible: Expansion and compression lead to a comparable pattern.

For the formation of a recognition pattern, this last step of the crystallization of the receptors is not essential. The pattern may also consist of crystalline lipid domains in a fluid lipid-protein film. The finding that the presence of a small fraction of lipid-protein complexes does not drastically alter the thermodynamics of the matrix film is very important. It fits well with our model (Fig. 10.7) according to which the contribution of the proteins to the lateral interaction of the lipids should be limited due to the fact that the proteins are kept relatively far away from the surface. This in turn means that all the knowledge on the formation of a pattern in pure lipid films should—with minor modifications—be applicable to this type of system. One should be able to enrich such model receptors in the fluid phase of a broad variety of lipid films in the fluid-solid coexistence. One should also be able to design striped patterns as well as hexagonal superlattices. The crystalline lipid phases that are depleted from receptors would in all these cases serve as negative control areas.

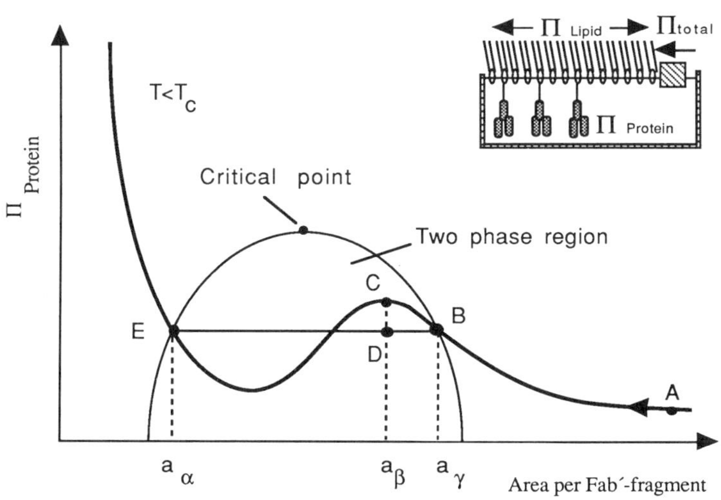

FIG. 10.7. Model for the gas-fluid coexistence of the lipid-protein film.

Several groups have shown that the mobility of LB films on solid supports depends sensitively on the coupling to the substrate. Lipid films in particular, which are transferred onto surfaces of long and ordered hydrocarbon chains like octadecyl trichlorosilane (OTS) films,[35] were shown immobile at room temperature.[36] Figure 10.8(a) shows in red fluorescence a phase-segregated Fab-lipid film after transfer onto a silicon wafer silanized with OTS. This pattern was stable at least for days. After blocking and incubation with haptenated BSA, which was additionally fluoresceinated, a congruent fluorescence pattern was measured in the green fluorescence [Fig. 10.8(b)]. A quantitative analysis of this binding assay (details are given in Ref. 10) resulted in an efficiency of the receptors of more than 60%. This means that the majority of the Fab′-lipids had retained the integrity of their binding sites throughout all procedures.

10.3. Conclusions

Our study shows that recognition patterns may be formed at the air/water interface by means of LB techniques. Inherent lateral self-organization mechanisms may be employed to design the shape and size of the pattern. The films may be transferred onto solid supports in such a way that the pattern is preserved and stable. Such films on solid support may be employed to detect certain molecules selectively. Whether or not this approach will be applicable to technologically relevant problems has yet to be shown.

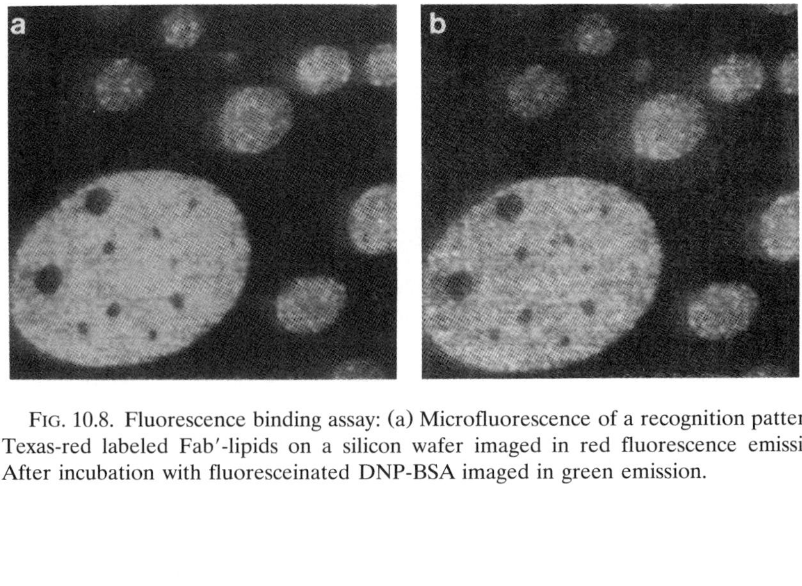

FIG. 10.8. Fluorescence binding assay: (a) Microfluorescence of a recognition pattern from Texas-red labeled Fab′-lipids on a silicon wafer imaged in red fluorescence emission. (b) After incubation with fluoresceinated DNP-BSA imaged in green emission.

Acknowledgments

This work was supported by the Deutsche Forschungsgemeinschaft and the Stiftung Volkswagenwerk. Helpful discussions with Mathias Lösche and Bob Gläser are acknowledged.

References

1. McConnell, H. M., Owicki, J. C., Parce, J. W., Miller, D. L., Baxter, G. T., Wada, H. G., and Pitchford, S. (1992). *Science* **257,** 1906–1912.
2. Hughes, R. C., Ricco, A. J., Butler, M. A., and Martin, S. J. (1991). *Science* **254,** 74–80.
3. Fischer, B., Egger, M., Heyn, S. P., and Gaub, H. E. (1993). *Langmuir* **9,** 136–140.
4. Stelzle, M., Weissmüller, G., and Sackmann, E. (1993). *J. Phys. Chem.* **97,** 2974–2981.
5. Schuhmann, W., Heyn, S. P., and Gaub, H. E. (1991). *Adv. Mater.* **3,** 388–391.
6. Welford, K. (1991). *Opt. Quant. Electron.* **23,** 1–27.
7. Ekins, R. P., and Thu, F. W. (1991). *Clin. Chem.* **37,** 1955–1967.
8. Norde, W. (1986). *Adv. Coll. Interface Sci.* **25,** 267–340.
9. Andrade, J. D., and Hlady, V. (1986). In: *Advances in polymer science.* Springer-Verlag, Berlin, Vol. 79, pp. 3–63.
10. Egger, M., Heyn, S. P., and Gaub, H. E. (1990). *Biophys. J.* **57,** 669–673.
11. Fodor, S. P. A., Read, J. L., Pirrung, M. C., Stryer, L., Tsai Lu, A., and Solas, D. (1991). *Science* **251,** 767–773.
12. Möhwald, H. (1990). *Ann. Rev. Phys. Chem.* **41,** 441–476.
13. Swalen, J. D., Allara, D. L., Andrade, J. D., Chandross, E. A., Garoff, S., Israelachvili, J., McCarthy, J. G., Murray, R., Pease, R. F., Rabold, J. F., Wynne, K. J., and Yu, H. (1987) *Langmuir* **3,** 932.
14. McConnell, H. M. (1991). *Ann. Rev. Phys. Chem.* **42,** 171–195.
15. Lösche, M., Sackmann, E., and Möhwald, H. (1983). *Ber. Bunsenges. Phys. Chem.* **87,** 848–852.
16. Weis, R. M., and McConnell, H. M. (1984). *Nature* **31,** 47–49.
17. Gaub, H. E., Moy, V. T., and McConnell, H. M. (1986). *J. Phys. Chem.* **90,** 1721–1725.
18. Moy, V. T., Keller, D. J., Gaub, H. E., and McConnell, H. M. (1986). *J. Phys. Chem.* **90,** 3198–3202.
19. Lösche, M., Helm, C., Mattes, H. D., Möhwald, H. (1985). *Thin Solid Films* **133,** 51–64.
20. McConnell, H. M., Keller, D., and Gaub, H. E. (1986). *J. Phys. Chem.* **90,** 1717.
21. Benvegnu, J. D., and McConnell, H. M. (1992). *J. Phys. Chem.* **96,** 6820–6824.
22. Seul, M., and Sammon, M. J. (1990). *Phys. Rev. Lett.* **64,** 1903–1906.
23. Blankenburg, R., Meller, P., Ringsdorf, H., and Salesse, C. (1989). *Biochem.* **28,** 8214–8221.
24. Darst, S. A., Ahlers, M., Meller, P., Kubalek, E. W., Blankenburg, R., Ribi, H., Ringsdorf, H., and Kornberg, R. D. (1991). *Biophys. J.* **59,** 387–396.
25. Egger, M., Heyn, S. P., and Gaub, H. E. (1992). *BBA* **1104,** 45–54.
26. Heyn, S. P., Tillmann, R. W., Egger, M., and Gaub, H. E. (1990). *J. Biochem. Biophys. Meth.* **22,** 145–158.
27. Heyn, S. P., Egger, M., and Gaub, H. E. (1990). *J. Phys. Chem.* **94,** 5073–5078.
28. Egger, M., Ohnesorge, F., Weisenhorn, A., Heyn, S. P., Drake, B., Prater, C. B., Gould, S. A. C., Hansma, P., and Gaub, H. E. (1990). *J. Struct. Biol.* **103,** 89–94.
29. Weisenhorn, A. L., Drake, B., Prater, C. B., Gould, S. A. C., Hansma, P. K., Ohnesorge, F., Egger, M., Heyn, S. P., and Gaub, H. E. (1990). *Biophys. J.* **58,** 1251–1258.

30. Crank, J. (1976). *The Mathematics of Diffusion*. Clarendon Press, Oxford.
31. Schmitt, F.-J., Weisenhorn, A. L., Hansma, P. K., and Knoll, W. (1992). *Thin Solid Films* **210 / 211,** 666–669.
32. Schmitt, F.-J., and Knoll, W. (1991). *Biophys. J.* **60,** 716–720.
33. Leckband, D. E., Israelachvili, J. N., Schmidt, F.-J., Knoll, W. (1992). *Science* **255,** 1419–1421.
34. Vaknin, D., Als-Nielsen, J. A., Piepenstock, M., and Lösche, M. (1991). *Biophys. J.* **60,** 1545–1552.
35. Maoz, R., and Sagiv, J. (1984). *J. Colloid Interface Sci.* **100,** 465–496.
36. Merkel, R., Sackmann, E., and Evans, E. (1989). *J. Phys. France* **50,** 1535–1555.

Electronic Properties

Photoinduced Electron Transfer in Monolayer Assemblies and Its Application to Artificial Photosynthesis and Molecular Devices

MASAMICHI FUJIHIRA

Tokyo Institute of Technology,
Department of Biomolecular Engineering,
4259 Nagatsuta, Midori-ku, Yokohama 226, Japan

11.1. Introduction

In biosystems, molecules organize spontaneously into complex functional entities with cooperating components of molecular dimensions. For example, well-organized molecular assemblies in biological membranes play an important role in the photosynthetic processes of plants and bacteria.[1] In 1984, the atomic structure of the *reaction center* (RC) of a purple bacterium *Rp. viridis*, i.e., the basic machinery for the primary processes of photosynthesis, was determined using x-ray diffraction[2] (Fig. 11.1). Immediately after this first structural study of the RC, almost the same structure was confirmed, again using x-ray crystallographic measurements for the RC of the other bacterium *Rb. sphaeroides*.[3,4] All RCs in plants and bacteria are complexes containing protein subunits and donor-acceptor molecules. The subunits are generally called *L*, *M*, and *H*, signifying their apparent relative molecular weights light, middle, and heavy, respectively. The electron donor is a special pair of chlorophyll molecules. Two similar electron transport chains lead away from the special pair, and each chain consists of a monomeric chlorophyll, a pheophytin, and a quinone. These centers span an inner membrane in the plant or bacterial cell, and their donor-acceptor complexes perform charge separation that creates a potential gradient across the membrane. The positive hole created in the special pair is reduced by an electron from

ISBN 0-12-523485-6

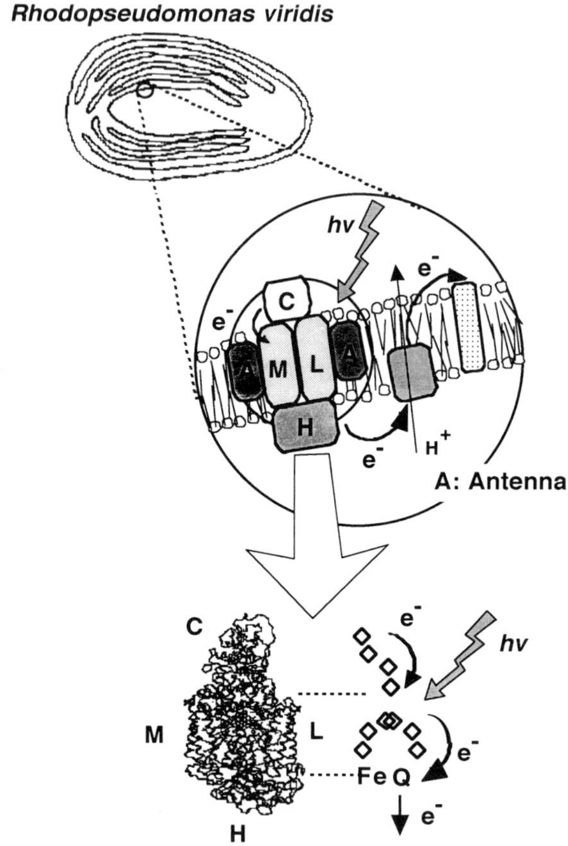

FIG. 11.1. A schematic diagram of the photosynthetic reaction center of *Rhodopseudomonas viridis*.

water and cytochrome in the plants and the bacteria, respectively. Before the charge separation can take place at any RC, solar energy must be harvested by light-absorbing antenna pigments and transmitted to the RC.[5] In green plants, the charge separation occurs in two different RCs working in series. The results of these primary processes are reduced products (NADPH), which appear on one side of the thylakoid membrane, and oxygen, which appears on the other.[6] The four-electron oxidation of water to molecular oxygen is catalyzed by the oxygen-evolving complex containing manganese. NADPH together with ATP synthesized during the primary process is used further in a series of dark reactions by which CO_2 is converted into useful fuels.

To design artificial photosynthetic molecular systems[7–10] for solar energy conversion, it is of great interest to mimic the elaborate molecular machinery for the light harvesting and the charge separation processes. Thus, it is essential to consider the structure and function of the asymmetric spatial arrangement of electron donors and acceptors in the charge separation unit across the biological membrane (Fig. 11.2). To create the electrochemical potential gradient across the biological membrane, the

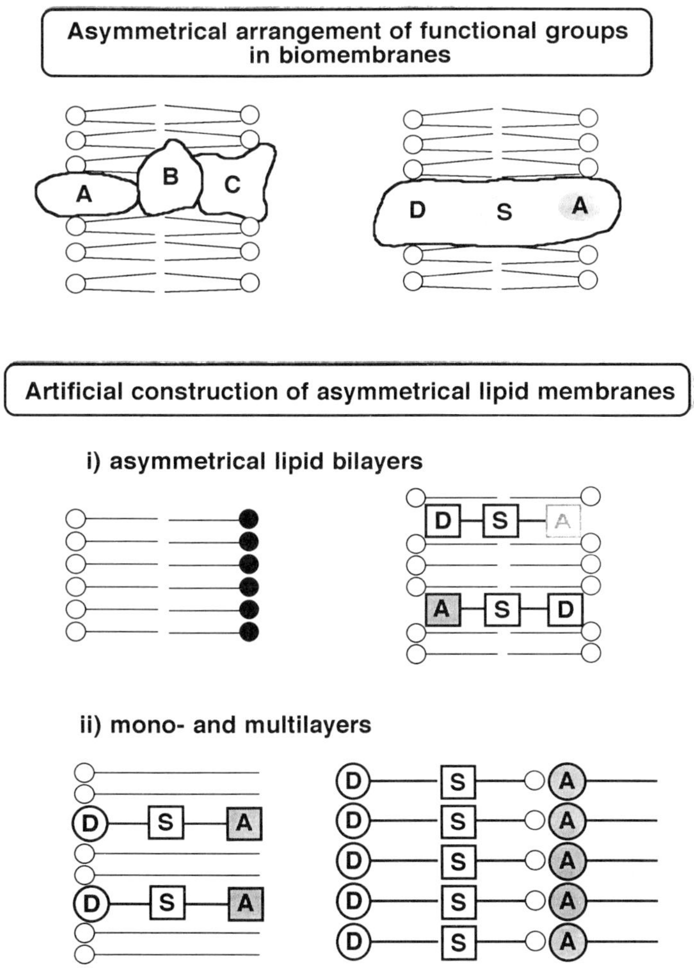

FIG. 11.2. Asymmetrical arrangement of functional groups in biomembranes and artificial construction of asymmetrical lipid membranes: (a) asymmetric lipid bilayers and (b) mono- and multilayers built by the LB method.

unidirectional arrangement of the donor-acceptor molecules is crucial. It is interesting to note here that Kuhn proposed the concept of the light-driven electron pump[7] about two decades ago without detailed knowledge of RCs. His idea played an important role and was our guiding principle. Consideration of the barrier height based on positive hole transfer through the valence band (or HOMO) of the intervening alkyl chain from the excited sensitizer to the electron donor was the only minor modification that we should have added to his electron pump model. Among the possible techniques that might enable the construction of artificial asymmetric membranes, two seemed to be the most promising when we started this project in 1983: (1) the fabrication technique for asymmetrical lipid bilayers[11] and (2) the Langmuir-Blodgett (LB) technique for monolayer or complexed heterogeneous multilayers.[8,9,12] Nonamphiphilic triad molecules embedded in a symmetric bilayer cannot be oriented unidirectionally without making half of them inactive by the redox reaction with the species added only in the solution located in one side of the symmetric membrane[13] [Fig. 11.2(a)]. Considering the possibility of redistribution into a symmetric membrane by flip-flop motion of molecules in the asymmetric bilayer, we concluded that the LB film is likely to be the most appropriate artificial material by which spatial arrangements of the various functional moieties across the film can be constructed readily at atomic dimensions. In a series of studies, we have attempted to simulate the elemental processes of photosynthesis by taking advantage of LB monolayer assemblies.

In this chapter, developments from our group in the past 10 years are reviewed with emphasis on the following subjects:

1. The construction of charge separation units by the use of A/S/D multilayered systems and A-S-D triad monolayer systems [see Fig. 11.2(b)], where A, S, and D stand for an electron acceptor, a sensitizer, and an electron donor moiety, respectively.
2. The increase of photoelectric conversion efficiencies in terms of the electron transfer kinetics, i.e., ΔG^0 dependence, the electrical double-layer effect on ΔG^0 dependence, and distance dependence.
3. The simulation of the primary process by the use of a mixed monolayer of triad and antenna molecules.
4. The construction of reaction center compartments by making use of two-dimensional (2-D) phase separation in a mixed monolayer of hydrocarbon (HC) and fluorocarbon (FC) amphiphilic molecules studied using the atomic force microscope (AFM), the friction force microscope (FFM), and the scanning surface-potential microscope (SSPM). The application of phase separation to future molecular and biomolecular optoelectronic devices is also discussed.

5. Artificial photosynthesis with multilayered LB films consisting of the charge separation unit and the catalytic layers for multielectron transfer reactions, such as CO_2 reduction and H_2O oxidation, in series.

11.2. Construction of Charge Separation Units with LB Films

11.2.1. A/S/D Multilayered Systems

Since the introduction of the molecular diode idea by Aviram and Ratner[14] in 1974 following the previously described light-driven electron pump proposed by Kuhn[7] in 1972, a variety of molecular and biomolecular devices have been proposed both theoretically and conceptually.[15,16] The realization of these devices, however, has not been attained readily by the use of practical material. For example, Mezger and Panetta[17] have made an effort to realize the molecular diode proposed by Aviram and Ratner. To the best of our knowledge, the first A/S/D molecular photodiode based on Kuhn's idea[7] of the light-driven electron pump was fabricated with the molecularly ordered film on a gold optically transparent electrode (AuOTE) prepared by the LB method.[18] Before this work, the group of Kuhn had developed a variety of photoelectric LB devices including one or two electronically active layers as described in Ref. 8 and references therein. In some of their devices, the electrode acts as D or A. The structure and function of the molecular photodiode of A/S/D are shown in Figs. 11.3(a) and (b), where hydrophilic parts and hydrophobic units are indicated by circles and squares, respectively. Three functional compounds tend to orient regularly in the heterogeneous LB films due to their amphiphilic properties.

Another interesting and fascinating application of LB films is their use as controlled-thickness spacers or "distance keepers."[8,9] Therefore, the distances between the three functional moieties, i.e., A, S, and D, can be closely controlled at known values. The electron transfer process in such molecular assemblies is free from any complication due to diffusion. Kuhn,[8] Möbius,[9] and others[19] have studied the distance dependence of the rate of photoinduced electron transfer in LB films. Their observed dependence agrees in a qualitative sense with the other experimental[20–23] and theoretical[24] results, which are described by the following equations[24]:

$$k = k(r)\exp(-\Delta G^*/RT) \tag{11.1}$$

$$k(r) = k_0 \exp\{-\beta(r - r_0)\} \tag{11.2}$$

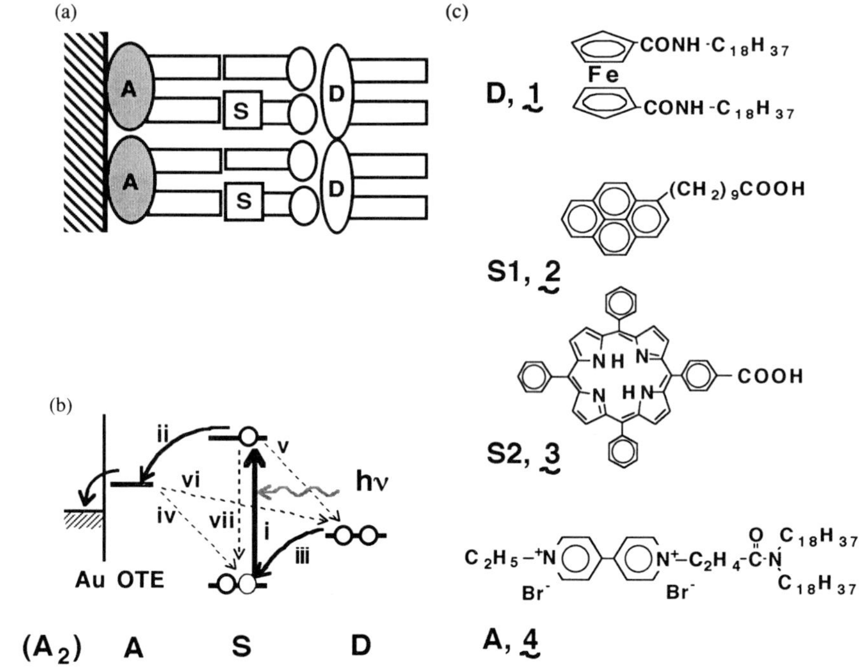

FIG. 11.3. Molecular photodiode with heterogeneous A/S/D LB film on AuOTE: (a) structure, (b) energy diagram, and (c) structural formulas of A, S, and D amphiphiles.

where k is the first-order rate constant; $k(r)$ the preexponential factor of the rate constant; ΔG^* the free-energy barrier of reaction; r_0 the value of distance r at which $k(r)$ equals some preassigned value of k_0, i.e., 1×10^{13} s^{-1}; and β is the parameter of the distance dependence of the rate. Kuhn,[8] however, explained his observed smaller slope, i.e., a smaller β value, by assuming the adiabatic case due to the rigid environment. He also tried to explain the temperature dependence, the effect of changes in the protein, and the unidirectionality in the photosynthetic reaction centers by his approach.[25]

In addition to the distance dependence, the effect of the standard free-energy difference ΔG^0 for the electron transfer process[24,26,27] is another important factor in determining the rate of electron transfer. The relationship of Eq. (11.3) was first introduced by Marcus where the free-energy barrier for the reaction, i.e., ΔG^* in Eq. (11.1), is given[24] in terms of ΔG^0 and the reorganization energy λ:

$$\Delta G^* = (\lambda/4)(1 + \Delta G^0/\lambda)^2 \qquad (11.3)$$

The changes in bond lengths of the reactants and the changes in solvent orientation coordinates in the electron transfer process are related[24] to λ. In Marcus's original theory, the motion of the nuclei was treated classically. There have been several attempts[24,27] to treat the nuclear coordinates quantum mechanically and to modify the equation for the energy gap ΔG^0. In connection with the design of the proper energy diagram for the molecular photodiode, the inverted region, where the rate decreases with an increase in a large excess of $-\Delta G^0$, predicted by Eq. (11.3) is most important. The presence of the inverted region has been confirmed experimentally by the use of internal electron transfer systems with rigid spacers and the electron transfer in solutions.[28–32]

Keeping the distance and ΔG^0 dependence in mind, we considered how to design a better molecular photodiode. In Fig. 11.3(b), the energy diagram of the A/S/D molecular photodiode is depicted as a function of distance across the LB film. If the forward processes indicated by arrows with solid lines are accelerated, and the backward processes with dashed lines are retarded by setting the distances and the energy levels appropriately, the photoinduced vectorial flow of electrons can be achieved. Namely, the acceleration by setting ΔG^0 equal to λ is assumed for the forward electron transfer processes ii and iii, while the retardation, as a consequence of the inverted region, is assumed for the back-electron transfer processes iv and v. Once an electron-hole pair is separated successfully, the recombination of the pair across the large separation by LB film (process vi) is hindered by the distance dependence.

The three kinds of functional amphiphilic derivatives used for the first A/S/D-type molecular photodiode[18] are shown in Fig. 11.3(c) together with porphyrin sensitizer **3** used later. By depositing these three amphiphiles on AuOTE, as shown in Fig. 11.3(a), and by the use of the resulting electrode as a working electrode in a photoelectrochemical cell,[20] the photoinitiated vectorial flow of electrons was achieved and detected as photocurrents. The AuOTE is a metal electrode and hence does not by itself possess a rectifying ability as does a semiconductor electrode.[20] Yet, in spite of the inability of the substrate electrode to rectify, the photocurrent had opposing directions depending on the spatial arrangement of A/S/D or D/S/A.[33] The direction was in accordance with the energy profile across LB films in Fig. 11.3(b). Much higher photocurrents were observed[18,33] for stacks of multilayers of each component, for example, in the form of A,A,A/S,S,S,S/D,D,D and D,D,D/S,S,S,S/A,A,A. The directions of the photocurrents also agreed with those we expected for these multilayered systems. An amphiphilic porphyrin derivative **3** was also used as a sensitizer.[33]

In such heterogeneous LB films, the long alkyl chains intervened between the A and S and between the S and D moieties. As a result, part of the excited sensitizers were deactivated by the emission of photons [Fig. 11.3(b), vii] without quenching by electron transfer. To cope with this problem, polyimide LB films consisting of A, S, and D units were used for constructing more efficient molecular photodiodes in collaboration with Kakimoto, Imai, and their coworkers.[34,35] They had reported the preparation and properties of polyimide LB films.[36] Since polyimide LB films have no long alkyl spacer between the layers (monolayer thickness, 0.4 to 0.6 nm), electrons should be more readily transferred. For example, a submicroampere order of photocurrents was observed[34] for AuOTEs that were coated with six layers of aromatic polyimide itself as A, two layers of a porphyrin derivative with aliphatic polyimide as S, and six layers of a ferrocene derivative with aliphatic polyimide as D. These magnitudes are about 10 times larger than those for photodiodes with conventional LB films. Later, the hole carrier property of the D layer was further improved by the use of triphenylamine in place of ferrocene as the D moiety.[35]

Other than our group, recently Isoda and his coworkers[37] have investigated molecular devices based on biological electron transport systems using molecular heterojunctions. In addition, Sakamoto and coworkers reported very recently an A/S/D-type photocell fabricated by electropolymerization.[38]

11.2.2. A-S-D Triad Monolayer Systems

Another approach to shortening the distances between the functional moieties is the use of unidirectionally oriented amphiphilic triad monolayers.[18,39] Each triad contained an A, S, and D moiety as its functional subunits. Other groups also studied the two-step photodriven charge separation and back-electron transfer reactions of triad molecules of the S-A_1-A_2 or A-S-D types.[29,40–43] They succeeded in retarding charge recombination in homogeneous solution, but did not attempt to orient these triad molecules in one direction and thus to conduct a direct photoelectric conversion. In our first amphiphilic triad **5** (SAD C11 in Fig. 11.4),[18] A, S, and D corresponded to viologen, pyrene, and ferrocene moieties, respectively. The viologen moiety is hydrophilic, whereas the pyrene and ferrocene moieties are hydrophobic. Subunits A and S were linked with a C_6 alkyl chain, while subunit D was linked to subunit A with a longer C_{11} alkyl chain. Later, a modified triad **6** (SAD C16 in Fig. 11.4) was synthesized[39] in order to improve the balance of the two distances between A and S and between S and D. In Fig. 11.4 the structures of the "linear" type A-S-D triads and their reference compounds, A-S, S-D, and S-, are also shown. The reason we synthesized linear type triads is described later.

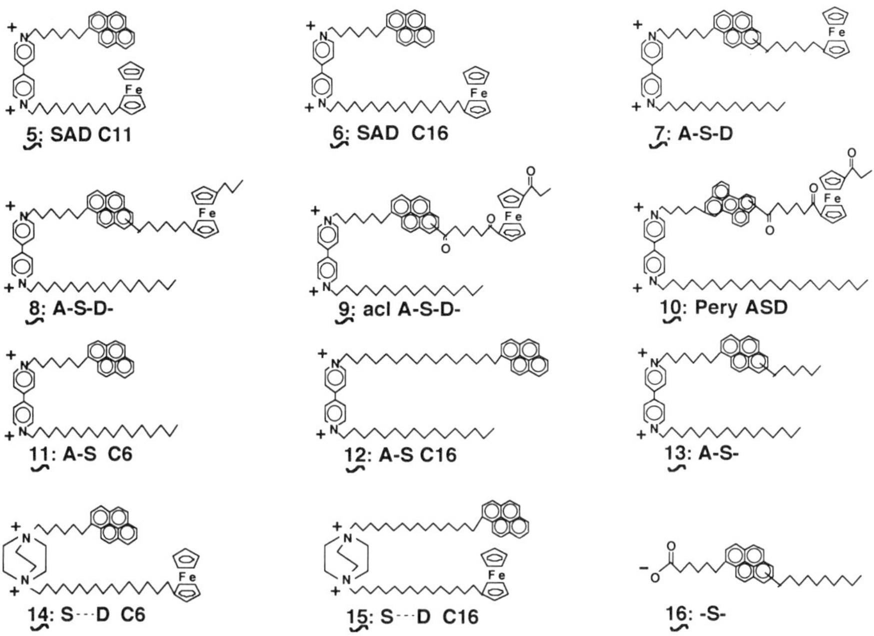

FIG. 11.4. Structural formulae of folded-type S-A-D and linear-type A-S-D triads and their reference A-S, S-D, and -S- compounds.

Due to their amphiphilic properties, the A, S, and D moieties were considered to be arranged spatially in this order, owing to the difference in length of two alkyl chains, from the electrode to the electrolyte solution perpendicularly as vizualized in Fig. 11.5(b). This arrangement was confirmed by observing anodic photocurrents, whose direction was in accordance with the photoinduced vectorial flow of electrons expected from the energy diagram of the oriented triad, as depicted in Fig. 11.5(d). From the shape illustrated in Fig. 11.5(b), S-A-D is called the "folded"-type triad. The photodiode function of these folded-type S-A-D triads was studied in detail[39] in terms of (1) the wavelength of the incident photons, (2) the surface pressure applied for monolayer deposition, (3) the applied electrode potentials, (4) the distances between A and S and between S and D, and (5) the role of the D moiety.

The photocurrent spectra in terms of the formal quantum efficiency[20] were recorded for the mixed monolayers [triad **5** and arachidic acid (AA) (1 : 2)], which were deposited on AuOTEs at the two different surface pressures indicated by P and Q in Fig. 11.5(a). The most important and interesting point was that the photocurrent increased by a factor of about

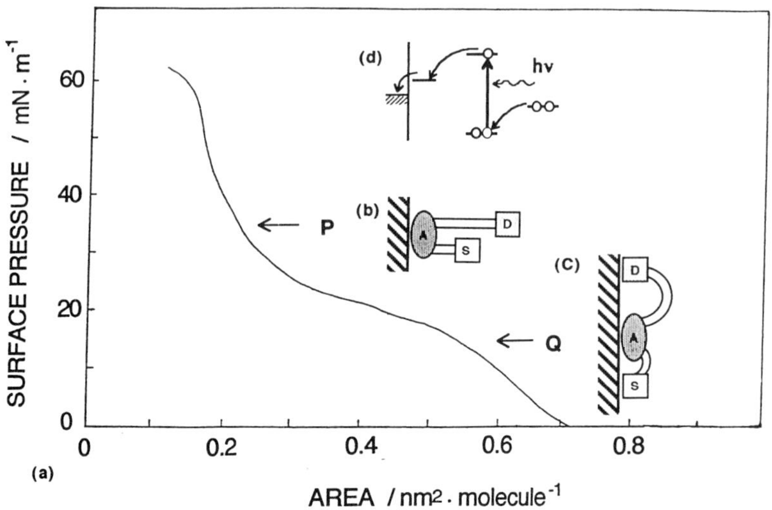

FIG. 11.5. Surface pressure-area isotherm (a) of a mixed monolayer of triad **5** with arachidic acid (1 : 2); schematic representation of (b) oriented and (c) nonoriented triad **5**; and (d) the corresponding energy diagram.

20 when the surface pressure for film deposition was changed only from 15 (at *Q*) to 35 mN/m (at *P*). The increase in the surface concentration by compression from *Q* to *P* was insufficient to explain such a dramatic increase in the photocurrent. Rather, the nonoriented structure of triad **5** and the unfavorable relative location of the A, S, and D moieties [Fig. 11.5(c)] in the monolayer at the low surface pressure (*Q*) might be responsible for the small photocurrent. By contrast, the much larger photocurrents observed at the high surface pressure (*P*) support the more favorable orientation of the folded-type S-A-D triad molecule postulated in Fig. 11.5(b).

A higher efficiency of triad **6** than that of triad **5** was also observed in spite of almost the same cross-sectional areas of these triads. The higher efficiency might be attributed to a better matching in concurrent electron transfer reactions between A and S and between S and D owing to an improved balance between the A-S and S-D distances. Another amphiphilic compound **11** (A-S C6) without the D moiety in Fig. 11.4 confirmed a positive contribution of the D moieties in triads **5** and **6**.

Very recently, Wang and coworkers[44] synthesized a new type of folded-type triads, porphyrin-viologen-carbazole, and studied the photoelectrochemical properties of SnO_2 electrodes coated with LB films of these triads for photoelectric conversion.

To improve the orientation of triad molecules in the monolayer, the "linear"-type A-S-D triads in Fig. 11.4 were synthesized.[45] The mixed monolayer of one of the linear-type A-S-D triads (**7**) with behenic acid exhibited a much higher photocurrent than did that of the folded-type S-A-D triads. This indicates that a more ideal spatial arrangement of the A, S, and D moieties was attained for the linear triad molecules in the mixed monolayer.

11.3. Photoelectric Conversion Efficiencies in Terms of the Electron Transfer Kinetics

11.3.1. ΔG^0 Dependence

As described in Section 11.2, the kinetics of photoinduced multistep electron transfer plays a crucial role in efficiencies of the molecular photodiodes. To clarify the energy gap and the distance dependence of the photoinduced electron transfer in heterogeneous LB films or triad monolayers, nanosecond and picosecond laser photolysis experiments were carried out.

To measure each forward electron transfer rate constant for processes ii or iii in Fig. 3(b) in the photoinduced multistep electron transfer in the A/S/D and D/S/A LB films, the luminescence decay curves were recorded for sensitizers such as pyrene (**2**) and $Ru(bpy)_3^{2+}$ (**17**) derivatives confined as one monolayer in LB films. These films also contained other monolayers of acceptor or donor amphiphiles, which were deposited apart from the sensitizer monolayer by a fixed distance as shown in Fig. 11.6. The structures of the sensitizers and those of amphiphilic acceptors and donors are shown in Figs. 11.3 and 11.7. In addition to the difference of 1.0 V in the oxidation potentials between the excited pyrene and the $Ru(bpy)_3^{2+}$ derivatives, the redox potentials of four types of acceptors (**4**, **18**–**20**) ranged widely up to 1.8 V. This enabled us to examine the energy gap ΔG^0 dependence of the photoinduced electron transfer rate of the A/S LB films.[46] It was concluded from the results that the possibility of a Marcus-type inverted region exists at highly negative ΔG^0, although it is not definite yet because of the limited data.

11.3.2. Electrical Double-Layer Effect on ΔG^0 Dependence

To examine the energy gap law for S/D LB films, four kinds of amphiphilic ferrocene derivatives (**21**–**24** in Fig. 11.7) were synthesized to serve as electron donors with different standard redox potentials E^0's and

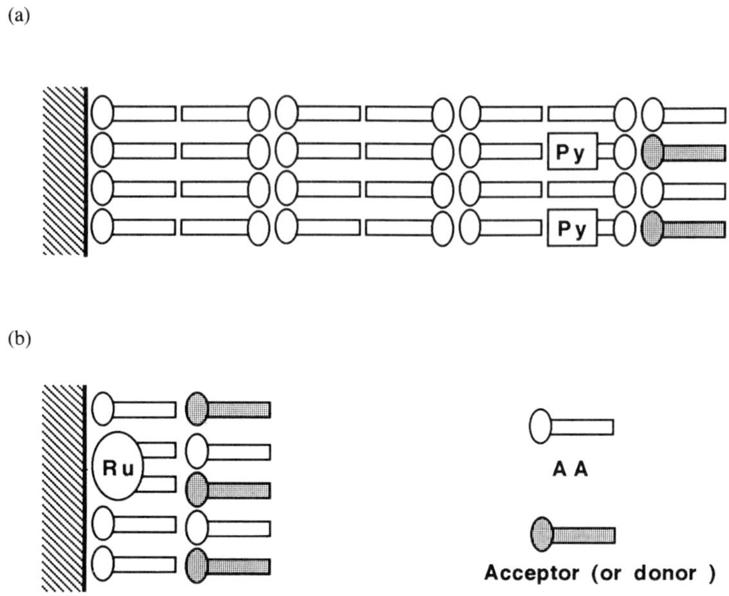

FIG. 11.6. Heterogeneous LB films used for kinetic study of photoinduced electron transfer: (a) pyrene decanoic acid (**2**), and electron acceptor A (or electron donor D), and (b) RuC19 (**17**), and A (or D).

with the same alkyl chain spacer.[47] The comparison of the energy gap ΔG^0 dependence of the reductive quenching of the $Ru(bpy)_3^{2+}$ derivative **17** (RuC19) with the ferrocene derivatives for three different systems, i.e., the LB films, the micellar,[48] and the solution systems, implies that the electrical potential difference between the hydrophilic headgroups in LB films has to be taken into account for estimation of the effective energy gap ΔG^0. Namely, an efficient photoinduced electron transfer quenching happens even in a S/D LB film in which the reaction was expected to be uphill ($\Delta G^0 > 0$) on the basis of half-wave potentials and thus too slow to be detected. The result contradicted also with that of the corresponding solution system.

The effect of the inner potential difference in the electrical double layer was further studied by using three kinds of amphiphilic ferrocene derivatives, AFc (**21**), CFc (**25**), and NFc (**26**), with an anionic, a cationic, and a nonionic headgroup, respectively[49] (Fig. 11.7). All the redox potentials of three ferrocene derivatives were about 0.9 V versus SCE and more positive than the reduction potential of excited $Ru(bpy)_3^{2+}$ of 0.6 V versus SCE. The luminescence decay curves for RuC19 in three types of S/D systems

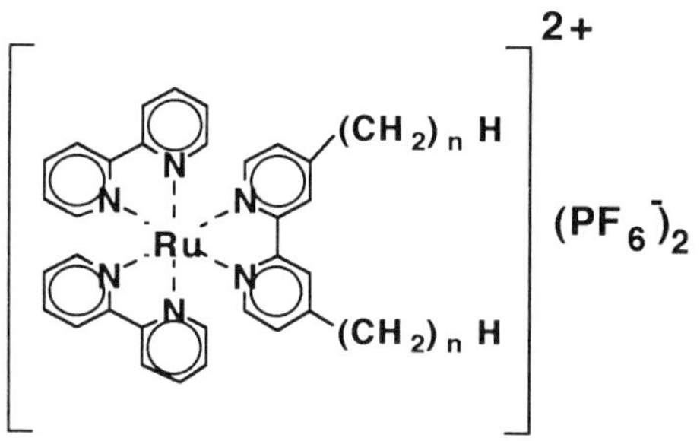

17 Ru C19 (n = 19)

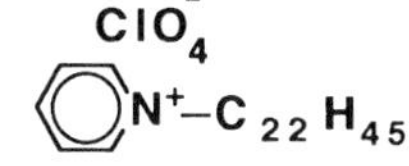

18

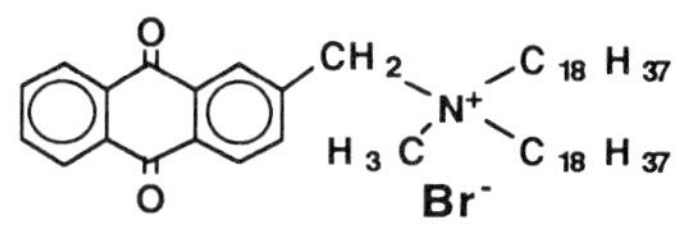

19

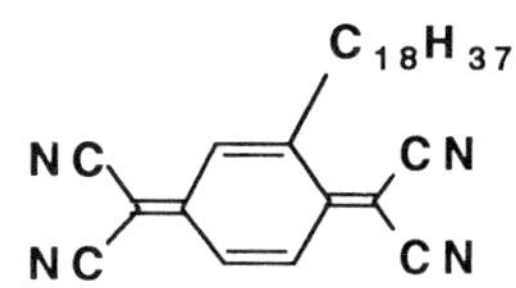

20

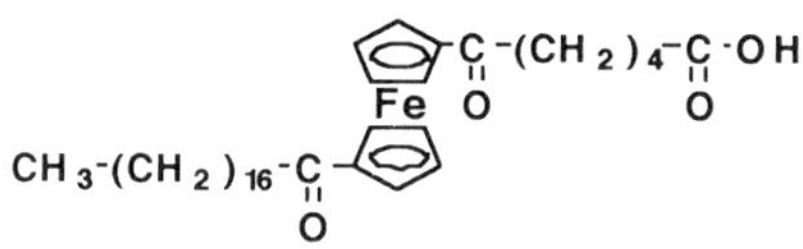

21

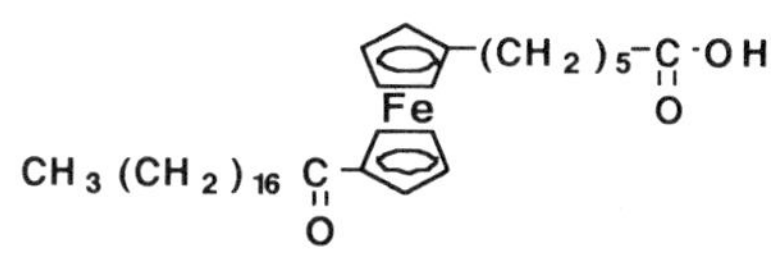

22

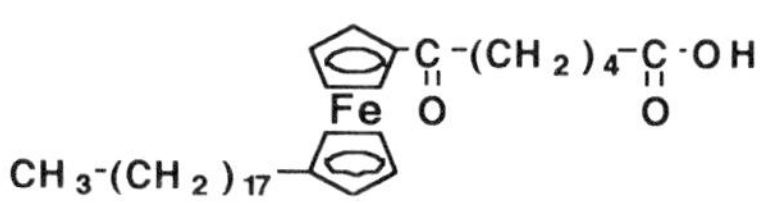

23

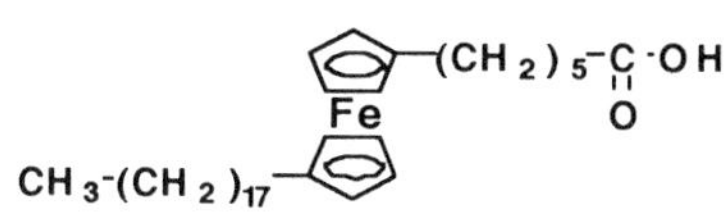

24

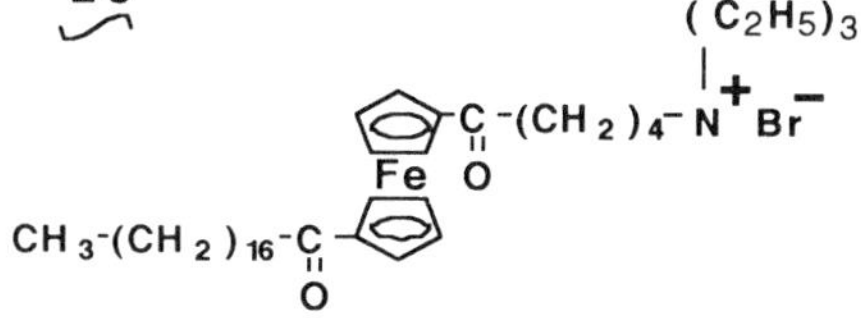

25

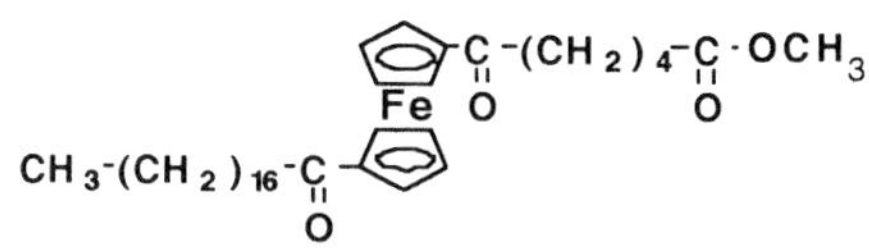

26

FIG. 11.7. Structural formulas of amphiphilic $Ru(bpy)_3^{2+}$ (**17**) sensitizer and electron acceptor (**18–20**) and donor (**21–26**) amphiphilic derivatives used in the kinetic study.

were recorded together with that in a reference LB film in which the pure monolayer of AA was deposited in place of the ferrocene donor layer.[49] Only the decay curve for S/AFc showed a fast decay component, while the other two curves are very similar to that for the reference LB film[49] and showed almost a single exponential decay without appreciable quenching.

The change in sign of the headgroup charge was expected to vary signs of the electrical double layer and thus the effective energy gap ΔG^0. The inner potential difference[50] $\Delta\phi$ will readily become a few hundred millivolts as observed for monolayers at the air/water interface[51] and for micellar surfaces.[52] If this potential difference is taken into account, appreciable electron transfer quenching, observed specifically in the RuC19/anionic ferrocene derivative system, is rationalized by the effective ΔG^0_{LB} in LB films in the following:

$$\Delta G^0_{\mathrm{LB}} = \Delta G^0 + \mathrm{F}\Delta\phi \tag{11.4}$$

where F is the Faraday constant. The numeral 2 was introduced erroneously in the second electrical potential term of a corresponding equation in Refs. 47, 48, and 49 (in such a way as $\Delta G^0_{\mathrm{LB}} = \Delta G^0 + 2F\Delta\phi$). It is noteworthy that the electrostatic potential effect was proposed to explain why the photochemical reaction follows the L pathway in the photosynthetic reaction center with almost C_2 symmetry.[4]

11.3.3. Distance Dependence

To increase the response time and efficiency of the molecular photodiode consisting of folded-type S-A-D triad molecules, it is important to know the distance and orientation dependence of the electron transfer quenching[19–23] in A-S and S-D monolayers. For this purpose, A-S [Fig. 11.8(a)] and folded-type S-D [Fig. 11.8(b)] amphiphiles with an alkyl chain spacer of different carbon numbers were prepared. The luminescence decays of these LB films were measured with picosecond laser photolysis.[53] The linear decrease in $\log k_q$ with an increase in carbon number was observed for the A-S monolayers [Fig. 11.8(c)], while the regular dependence of $\log k_q$ on the alkyl chain lengths could not be obtained for folded-type S-D systems. This result indicates that the alkyl chains in the A-S dyad molecules are extended in their monolayers and therefore the distance between A and S can be controlled by changing the carbon number; this is not the case for the folded-type S-D dyads. Folding of the alkyl chain in the S-D dyad in the monolayer illustrated in Fig. 11.8(d) may account for the nonideal result. The linear-type -S-D dyads are now synthesized. As is clear from Fig. 11.8(c), the slopes for the straight lines for the A-S monolayers are dependent on the dilution of the dyad with

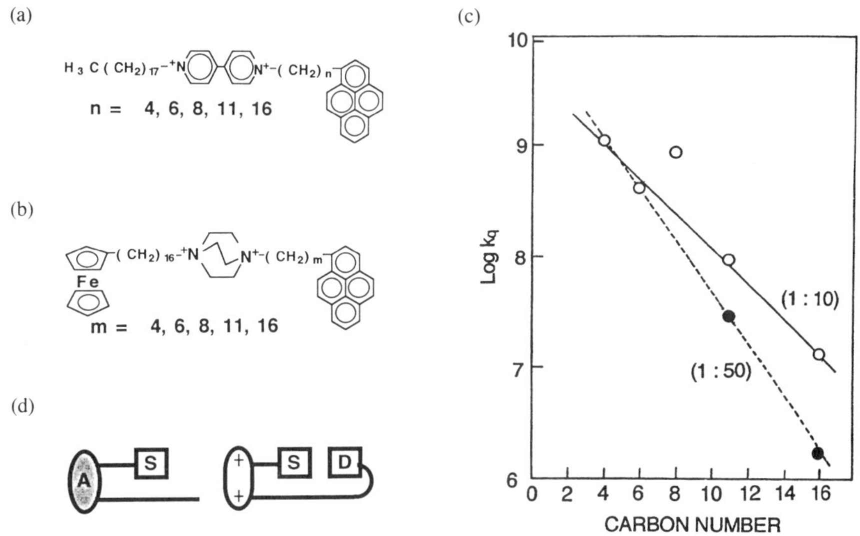

FIG. 11.8. Structural formulas of (a) A-S and (b) S-D compounds used to study the effect of alkyl chain lengths on kinetics of electron transfer quenching in mixed monolayers of A-S (or S-D) and AA. (c) Logarithms of k_q as a function of alkyl chain length in carbon number where the ratios indicate the molar ratios between A-S and AA in mixed monolayers. (d) Schematic representation of an extended A-S structure in contrast with a folded S-D structure.

arachidic acid. The lower slope ($\beta = 3.0\ \text{nm}^{-1}$), observed for 1 : 10 dilution than that ($\beta = 4.4\ \text{nm}^{-1}$) observed for 1 : 50 dilution indicates clearly that, in addition to the intramolecular quenching of the excited S-D dyad, the intermolecular electron transfer quenching by the accepting moieties of the surrounding dyads cannot be neglected even for the 1 : 10 dilution. Of course, the contribution of the intermolecular quenching became more important with the increase in the carbon number. For the 1 : 50 dilution, the contribution from the intermolecular quenching may be negligibly small. Then, the slope obtained can be compared with those of other intramolecular systems determined experimentally in dilute solutions[21–23] and calculated theoretically.[24,54,55] In particular, the difference in the slopes between A-S- and -S-D systems is interesting in terms of the super exchange mechanism via vacant orbitals (i.e., LUMO) and filled orbitals (i.e., HOMO), respectively. These data for the extended single alkyl chains will also be useful for understanding of electron exchange couplings in terms of through bond and through space.[54,55] The electron exchange couplings give us a deeper understanding of the distance dependence in

Eq. (11.2) and play an important role in long-distance electron transfer in proteins.[56,57]

11.4. Simulation of the Primary Process by the Use of a Mixed Monolayer with Triad and Antenna Molecules

In the next step, we simulated the light harvesting and succeeding charge separation processes by means of a monolayer assembly consisting of synthetic antenna pigments and triad molecules, as illustrated in Fig. 11.9.[58,59] For the light-harvesting (H) antenna pigments, an amphiphilic pyrene derivative was used.[60] For the amphiphilic linear triad molecule, a perylene moiety, as the S unit, and viologen and ferrocene moieties, as the A and the D units, respectively, were used. The structures of the antenna **16** and the triad **10** molecule are also shown in Fig. 11.4. Because of the overlap of the emission spectrum of the antenna pyrene and the absorption spectrum of the sensitizer perylene moiety of the triad [Fig. 11.10(a)], light energies harvested by the antenna molecules were efficiently transferred to the sensitizer moiety of the triad via the Förster mechanism shown schematically in Fig. 11.9(c). Then, the excitation energy of the

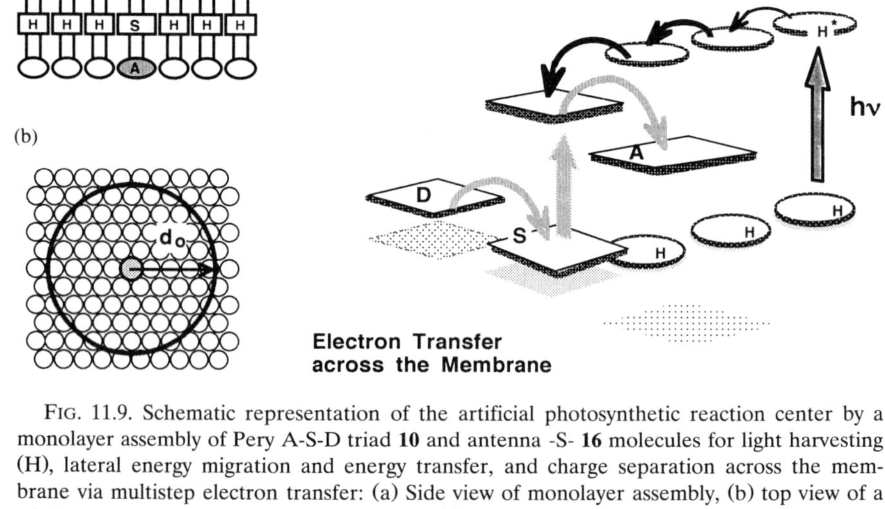

FIG. 11.9. Schematic representation of the artificial photosynthetic reaction center by a monolayer assembly of Pery A-S-D triad **10** and antenna -S- **16** molecules for light harvesting (H), lateral energy migration and energy transfer, and charge separation across the membrane via multistep electron transfer: (a) Side view of monolayer assembly, (b) top view of a triad surrounded by antenna molecules, and (c) energy diagram for photoelectric conversion in a monolayer assembly.

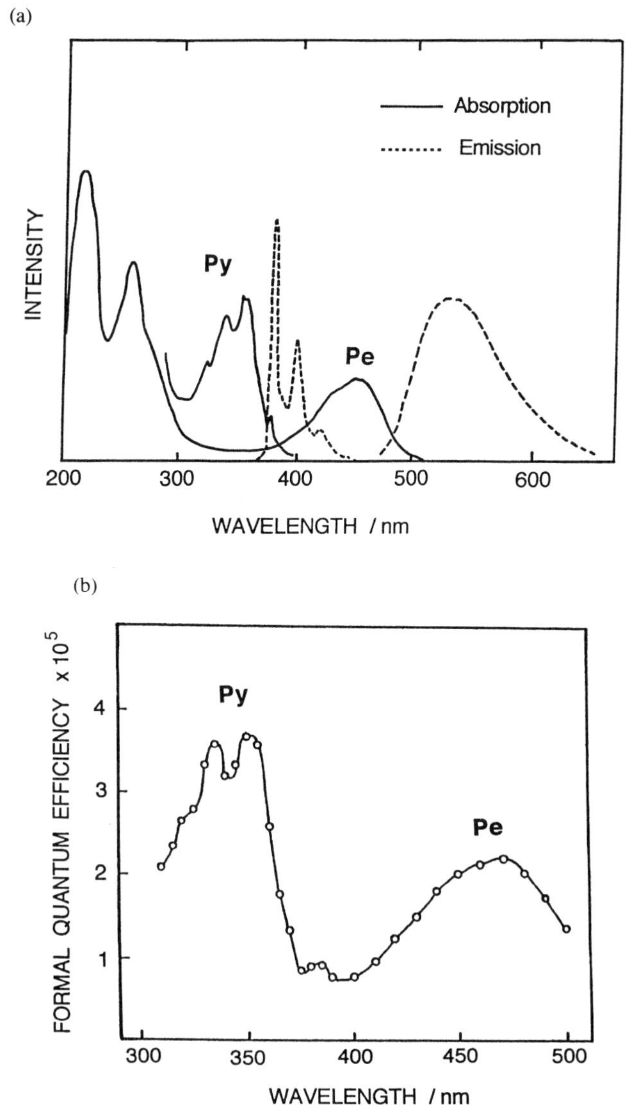

FIG. 11.10. (a) UV and visible absorption and emission spectra of antenna pyrene **16** and perylene triad **10** derivatives in ethanol and (b) photocurrent spectrum of the mixed monolayer of **10** and **16** (1 : 4) deposited on AuOTE.

perylene moiety should be converted to electrical energies via multistep electron transfer across the monolayer as described earlier.

Figure 11.10(b) shows[58,59] the photocurrent spectrum of the mixed monolayer of the triad and the antenna with a molar ratio of 1 : 4. Maxima of anodic photocurrents at about 350 and 470 nm are found. These correspond to the adsorption maxima of pyrene and acylated perylene. The anodic direction of the photocurrent agrees with the energy diagram and the orientation of the triad shown in Fig. 11.9. In contrast, a negligible photoresponse was obtained with the pure antenna monolayer. The result indicates that charge separation in the triad molecules was initiated by light absorption both with the perylene sensitizer itself and with the pyrene antennas followed by the energy transfer.

11.5. Construction of Nanodomains in Monolayers by the Use of Phase Separation

11.5.1. The Conceivable Approaches to "Real Molecular Devices"

The molecular photodiode shown in Fig. 11.5(b) was expected to exhibit the photodiode function by itself, i.e., by single molecule. So far, however, the molecular photodiodes developed in our laboratory have the molecular dimensions in their thickness (z axis), while the scale in the lateral directions (x-y plane) was in the macroscopic dimensions such as 1×1 cm^2. Thus, if our molecular devices are made in the molecular dimensions even in the lateral directions and in addition if access to the devices in the molecular dimensions is available for electron or photon communication, the resulting molecular devices will be suited to be called *real molecular devices*.

To attain this final goal, we must solve the following problems:

1. How to make the lateral dimensions of our devices on the molecular or nanoscales.
2. How to connect the real molecular devices with practical wires or some other means for electronic and optoelectronic communications from the external circuit.

Several approaches are conceived in connection with the first question. These are the following:

1. The construction of nanodomains in mixed monolayers by taking advantage of potential self-assembling properties of the constituent amphiphiles for phase separation[61–65]

2. The use of the tips of scanning probe microscopes for molecular manipulation[66,67]
3. The compartmentalization of the macroscopic-scale monolayer assembly into a vast number of nanoscale divisions by the use of AFM tips for wearing out nanoscale boundary lines[62,68,69]
4. Nanolocalized syntheses of active amphiphiles by the use of STM and photon STM (an optical fiber with a nanoscale tip end) tips for the electrochemical and photochemical syntheses.[70-72]

The approaches to the second question are as follows:

1. The use of a STM tip for connecting molecular electronic devices with the external circuit for direct current (electrons) leading (i.e., electronic contact mode)[73]
2. The use of a SSPM tip for connecting molecular photodiodes with the external circuit for photovoltage (localized surface potential) reading (i.e., electronic noncontact mode)[63]
3. The use of a photon STM tip for nanoscale local photoillumination of molecular optoelectronic devices.

In this chapter, we describe the approaches 1 and 3 for the first problem and approaches 1, 2, and 3 for the second problem. These approaches have been studied recently in our laboratory. Some of them have been studied in collaboration with the group of Professor Güntherodt at the University of Basel.

11.5.2. The Study of Phase Separation with AFM and FFM

As shown in Fig. 11.1, the RCs in the biological membranes in the plant or bacterial cell look like "islands" made of protein in the "sea" made of the bilayer lipid membrane. The structure of protein islands in a lipid sea of biological membranes (shown in Fig 11.11) was proposed initially by Singer and Nicolson[74] and is called a *fluid mosaic model*. The fluidity of the membrane, i.e., the diffusional motion of the membrane protein, is not necessarily important, but is rather undesirable for the localization model of the planer molecular devices. However, the model is very suggestive for compartmentalization of an assembled molecular functional unit such as an artificial A-S-D triad charge separation unit surrounded by antenna molecules, as shown in Fig. 11.9(b). The membrane proteins are self-assembled via three-dimensional folding of the polypeptide chain and, if necessary, gathers to form a complex—like the RC consisting of the subunits in Fig. 11.1—which acts as a functional unit isolated in the lipid sea. Isolating or constructing a single triad-antenna assembly [Fig. 11.9(b)]

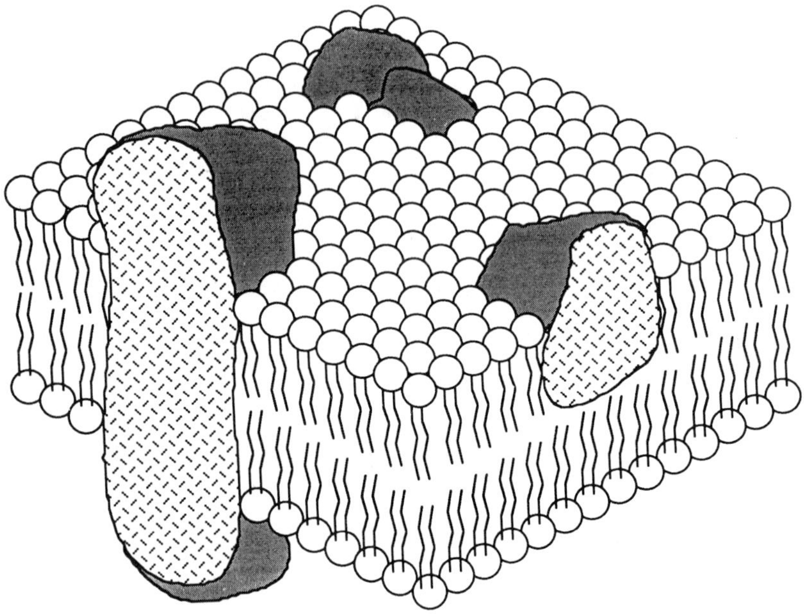

FIG. 11.11. Fluid mosaic model of biological membranes.

as an artificial counterpart of the natural RC is our strategy to cope with the first problem.

FC compounds are known to act as oil repellents as well as water repellents.[75] This means that FCs will be phase-separated from HC compounds. Concentration of the FC amphiphiles at the surface of the mixed LB films of HC and FC amphiphiles was studied by the angular dependence of ESCA measurements.[76] Many studies have already demonstrated that the STM[77] and AFM[78] can effectively image organic thin films and surfaces.[79-81] Therefore, in collaboration with the Basel group, we tried to observe the phase separation in the mixed monolayer[62] and bilayer[61] of the mixture of perhydrocarbon and partially fluorinated amphiphiles with AFM and FFM.[82-84] Figure 11.12 shows AFM and FFM images[62] of the mixed monolayer of AA and $C_9F_{19}C_2H_4OC_2H_4COOH$ (**27**) (1 : 1) polyion complexed with poly 4-methyl-vinylpyridinium cations[85,86] deposited on oxidized Si(100). The phase separation is obvious both from the AFM and FFM images. The detailed study reveals that the "island" surface consists of HC chains, whereas the "sea" surface, i.e., the continuous phase, consists of FC chains. The size and shape of the islands depend on the chain lengths of the amphiphiles[62] and the subphase composition and

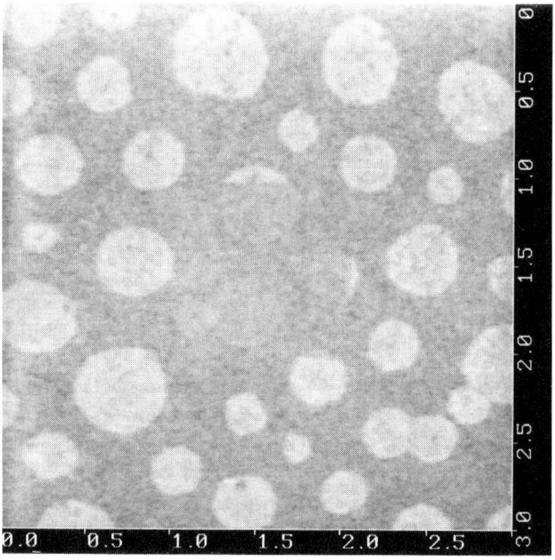

(a)

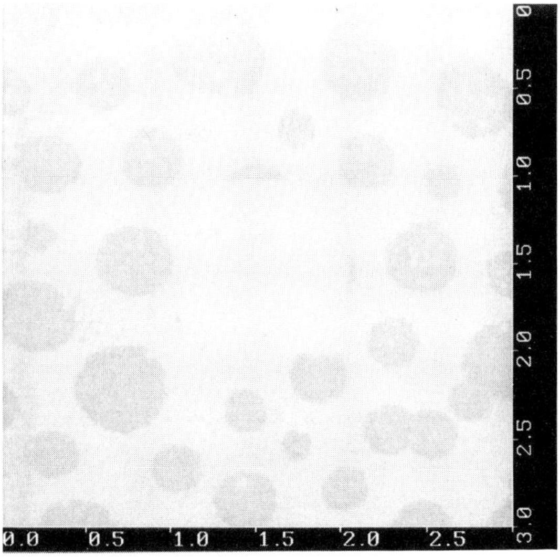

(b)

FIG. 11.12. AFM and FFM of a mixed monolayer of **27** and AA (1 : 1) deposited on oxidized Si(100) ($3 \times 3\ \mu m^2$): (a) topography with step height of 2.4 nm and (b) friction force map. Islands located in a central part are removed by narrow scans ($1 \times 1\ \mu m^2$) under a higher force.

pH.[65] The size can be changed readily from micron to submicrons by changing types of cationic polymers. The chemical assignment of the surfaces between the islands and the sea was also confirmed[65] by the comparison between AFM and fluorescence microscopic images of the mixed monolayer of **16** and **27**.

11.5.3. The Study of Phase Separation with SSPM

The surface dipole moments of the FC and the HC chains are known to be significantly different from each other.[51] From macroscopic measurements it is well known that surface potentials or *contact potential differences* (CPDs) are highly material dependent and are related to work functions[87] and surface dipole moments.[87–90] This prompted us to measure local surface potential distribution in the nanoscale to confirm the chemical differentiation of the phase-separated surfaces in a different manner. The high-resolution potentiometry by electrostatic force microscopy[91–93] or the high lateral resolution CPD measurements (Kelvin probe force microscopy[94,95]) was devised already for simultaneous topographic and CPD measurements. The method was further modified and called scanning Maxwell stress microscopy (SMM)[96] or scanning surface-potential microscopy (SSPM).[63] Figure 11.13 shows a schematic diagram of our setup for SSPM. The electrostatic force on a conducting tip held close to a conducting surface is given by

$$F = -(1/2)V^2(\partial C/\partial z) \tag{11.5}$$

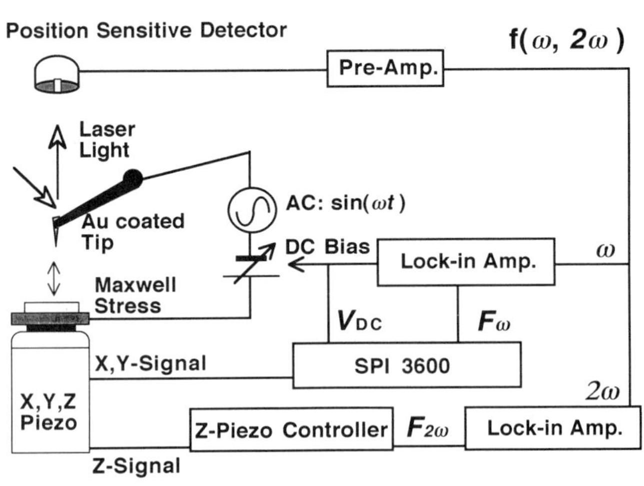

FIG. 11.13. Schematic diagram of a SSPM.

where V denotes the potential difference between sample surface and tip.[63,91,94,96] If the tip potential is sinusoidally driven by an external source so as to give $V = V_{DC} + V_{AC}\sin(\omega t)$, there appear to be three force components, i.e., DC, ω, and 2ω components, each of which gives information on the electrostatic properties of the surface and the surrounding medium. If the frequency ω is much less than the resonance frequency of the cantilever, the amplitude of vibration of tip A will be given by

$$A = S\{V_{DC}^2 + 2V_{DC}V_{AC}\sin(\omega t) + (V_{AC}^2/2)[1 - \cos(2\omega t)]\} \quad (11.6)$$

where S is the calibration factor relating potential difference to amplitude of vibration and includes the elastic response of the tip (i.e., the elastic constant of the cantilever k) as well as $(\partial C/\partial z)$:

$$S = -(1/2k)(\partial C/\partial z) \quad (11.7)$$

If the 2ω component is used for the z piezocontroller so as to keep $-S(V_{AC}^2/2)$ constant as shown in Fig. 11.13, the amplitude of the ω component, $2S(V_{DC}V_{AC})$ is proportional to V_{DC} (i.e., CPD) because of the constancy of S and V_{AC}. In other words, the ratio of these terms gives $-4V_{DC}/V_{AC}$ and V_{AC} is known as an experimental parameter; thus V_{DC} can be calculated from the ratio. However, when frequencies ω and 2ω are not far from the resonance frequency of the cantilever, we have to take into account the effect of the resonance on the amplitudes at ω and 2ω to calculate the corrected ratio.[97] The calculated V_{DC} (i.e., CPD) is displayed directly on the x-y frame. Instead of that, if the amplitude of the ω component is used as a feedback signal for the DC bias so as to nullify itself as shown in Fig. 11.13, then the applied DC bias is equal to $-V_{DC}$ (i.e., CPD) itself, and the value of V_{DC} thus obtained is also displayed.

Since the z signal, i.e., the applied voltage to the z piezo, is changed so as to keep S constant in Fig. 11.13 as described earlier, the z signal also can be displayed in an x-y frame. At this point in the argument, it is interesting to note what kind of physical property is kept constant under the condition of S constant. If we assume that the capacitance can be approximated by the parallel condenser, S will be described as follows:

$$\begin{aligned} S &= +(1/2k)(\varepsilon_r\varepsilon_0 A/z^2) \\ &= +(1/2k)(C/z) \end{aligned} \quad (11.8)$$

where ε_r is the relative permittivity of dielectric material sandwiched by the condenser electrodes, ε_0 the vacuum permittivity, A the area of the condenser, and z the distance between the electrodes. When the metal coated tip scans a sample surface consisting of two different metal domains without organic film coatings, e.g., gold stripes on chromium substrate,[63] the tip-sample separation (z spacing) can be kept constant from Eq. (11.8) under the constant S. This results from the fact that in this case the dielectric material is always air ($\varepsilon_r = \varepsilon_a$) between the electrodes and thus the capacitance is independent of the tip position in the x-y plane as long as the z-spacing is kept constant. In other words, the z signal image in this case should be exactly the same as the topography given by AFM. This expectation was confirmed experimentally.[97]

On the contrary, the dielectric consists of more than two layers of insulators (including the air gap) in the case of the mixed HC and FC monolayer deposited on an atomically flat conducting substrate.[61,62] In the latter case, the dielectric constant and the thickness of the film change from place to place, i.e., the thickness and dielectric constant of the island part are different from those of the sea part as shown in Fig. 11.14. Depending on the structures of the mixed monolayer, the following relations for the change in z under constant S, Δz, will be obtained.[97]

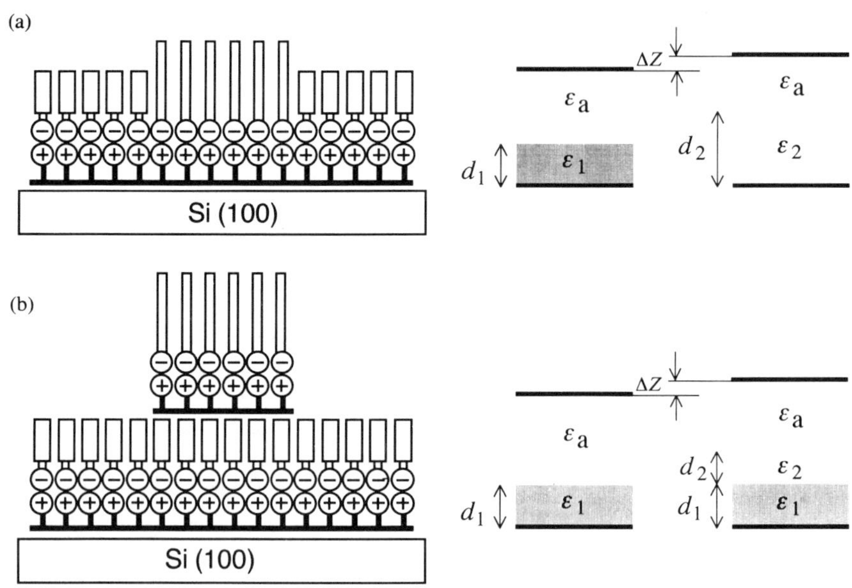

FIG. 11.14. Schematic diagrams of phase-separated HC and FC mixed monolayers on Si(100) and corresponding parallel plate condenser models between a substrate Si(100) surface covered with a mixed monolayer and a gold vapor-deposited tip: (a) side-by-side structure and (b) on-top structure.

For the side-by-side structure shown in Fig. 11.14(a):

$$\Delta z = d_2(1 - \varepsilon_a/\varepsilon_2) - d_1(1 - \varepsilon_a/\varepsilon_1) \tag{11.9}$$

For the on-top structure shown in Fig. 11.14(b),

$$\Delta z = d_2(1 - \varepsilon_a/\varepsilon_2) \tag{11.10}$$

where ε_a is the relative permittivity of air of 1.0005, d_1 and d_2 are the thicknesses of the monolayers with different relative permittivities ε_1 and ε_2, respectively, for the side-by-side structure, and for the on-top structure d_2 is the thickness of the monolayer with ε_2 on top of the first monolayer with d_1 and ε_1. As is clear from Eqs. (11.9) and (11.10), Δz can be related finally to the relative permittivities ε_1 and ε_2, because d_1 and d_2 can be determined independently from AFM experiments for both structures.

As described, the spacing between the tip and the conducting surface is not kept strictly constant when the 2ω component, i.e., S, is kept constant. Nevertheless, the distance can be regarded to be almost constant, because the total distance is usually more than 10 nm and Δz is small, because of the small values of the relative permittivities of the usual LB films. This was also confirmed experimentally. For example, on a sample with a surface of oxidized Si(100) covered with a mixed monolayer of perhydrobehenic acid (BA, $C_{21}H_{43}COOH$) and partially fluorinated carboxylic acid **27** (1 : 1) polyion complexed with cationic polymer, poly(4-vinyl-N-methylpyridinium), Δz observed between the island and the sea was about 0.3 to 0.6 nm and much smaller than the height difference of about 2 nm obtained by AFM under the constant force mode. On the basis of the on-top model, ε_2 was calculated[64,97] to be about 1.3. The on-top model was previously inferred from the experimental finding[62] of the incredible step height of 2.4 nm between the island and the sea of the mixed monolayer of **27** and AA. The step height for the side-by-side model is estimated to be about 0.5 nm from the difference in molecular lengths.[61] The side-by-side model was also contradicted by the observation that the bottom level exposed by shearing the more deformable island area is at the same height as the sea level.[62]

Figure 11.15(a) shows a $4.6 \times 5.0\ \mu m^2$ top-view AFM image of a vapor-deposited gold film surface on a glass plate covered with a polyion complexed mixed monolayer of $C_9F_{19}COOH$ (**28**) and AA.[63] But the image was almost the same as an AFM image observed for the same gold surface uncovered with the mixed monolayer. This indicates that the surface roughness of the gold surface was too large to see the topography of the mixed monolayer itself. In contrast with this, the difference in heights between the hydrocarbon islands and the FC sea is clearly seen in an AFM image for the same mixed monolayer deposited on an atomically flat

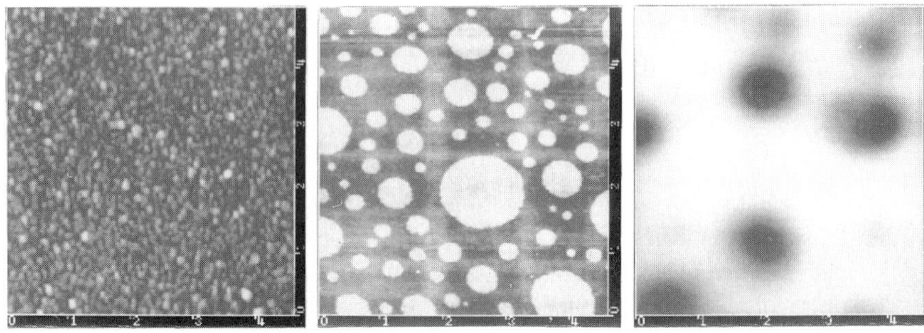

(a) (b) (c)

FIG. 11.15. (a) AFM and (c) SSPM images of a gold film covered with a polyion complexed monolayer of **28** and AA (1 : 1) (~ 4.6 × 5.0 μm^2) and (b) an AFM image of the same mixed monolayer on oxidized Si(100).

hydrophilic Si(100) surface as shown in Fig. 11.15(b). The image was similar to those observed for a mixed monolayer of **27** and AA as shown in Fig. 11.12(a). Figure 11.15(c) shows a top view of a SSPM image observed on the same surface for the AFM image of Fig. 11.15(a). In spite of the roughness of the surface as shown in Fig. 11.15(a), an image with clear contrast of the surface potential difference between the HC islands and the FC sea is discernible. A comparison of the images of Figs. 11.15(b) and (c) shows clearly that even islands smaller than 1 μm in diameter can be seen in the SSPM image. The surface potential difference between the islands and the sea can be related to the difference in the surface dipole moments due to the terminal CH_3- and CF_3- groups.[51] The macroscopic surface potential difference between the pure polyion complexed monolayers of **28** and AA deposited on the same gold films was measured to be about 0.5 V. The observed difference between the island and the sea area, however, was much less than this difference. Again the difference was interpreted well by taking account of the on-top model. The detailed structure of the mixed monolayers shown in Figs. 11.12 and 11.15 are now under intensive investigation.[64,97]

In addition to the clear distinction between the surfaces of the islands and the sea, a difference in the types of built-up films, i.e., X, Y, and Z layers, can be readily distinguished by the SSPM. This is because dipole moments perpendicular to the solid support are canceled in every bilayer of Y-type multilayers, in contrast to the additive contribution of the dipole moments in X- and Z-type multilayers. Between X- and Z-type multilayers, the direction of the oriented dipoles is opposite. Also, the localized transformation of the oriented dipoles, for example, from X- to Y-type due

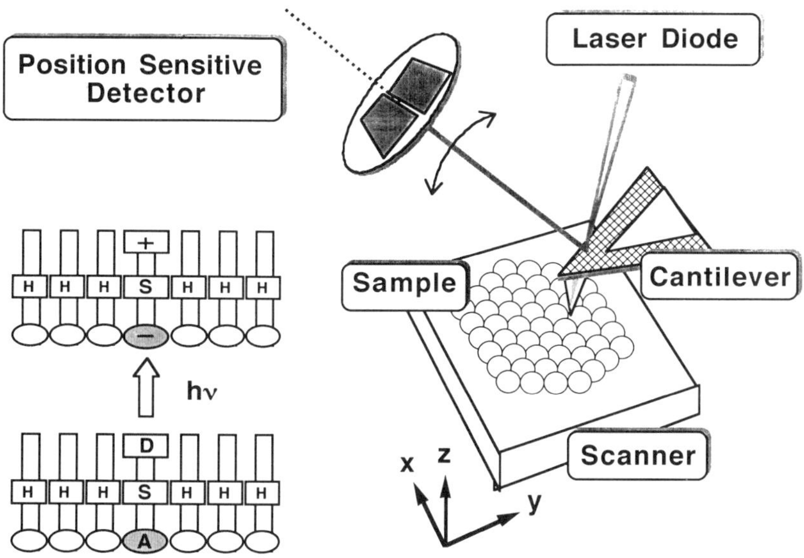

FIG. 11.16. Detection of a photoinduced change in surface potentials of an island of a mixed monolayer of A-S-D triad and antenna H molecules surrounded by an FC sea.

to "overturning,"[88] should be observed by the change in the distribution of surface potentials as a function of time elapsed after the deposition. These aspects of the structure of LB films are also now under study.[97]

11.5.4. Application of 2-D Phase Separation and Compartmentalization by AFM to Molecular Devices

FFM[61,62] and SSPM[63,64,97] have been used to distinguish between different chemical species on the different surfaces of the phase-separated monolayer assemblies, as described earlier. In addition to their capability of local analysis, for example, the SSPM can be used to read information written into localized domains. In Fig. 11.16, detection by the tip of the SSPM (a noncontact mode) of the charge-separated state of an A-S-D triad surrounded by antenna molecules in the form of a single isolated island upon photoirradiation[98] is illustrated. A bit of information in this case is stored in each island. Current SSPM can also be used to study the photoresponse of the natural RCs in the form of monolayer assemblies.[99,100] In the latter case, each RC can play the role of a bit. The injected localized charge into the surface of an insulator has been detected in the noncontact mode.[101]

The high lateral resolution in the illumination can be accomplished with near-field optical microscopic[72,102–107] irradiation with the sharpened tip of an optical fiber of the photon STM. Figure 11.17 illustrates that the photoresponse of a single oriented triad molecular photodiode is to be observed in a contact mode—via tunneling current—with a photon STM tip with a protruded STM metal tip. The tunneling through a single molecule is attempted by use of a STM tip.[73] Note that the electron transfer reaction proceeding in the individual molecule of the RC or the triad looks like one molecular process, but a large number of vibrational modes of the polypeptide chains or of the LB assemblies are involved through the solvent organization energy. In other words, the single molecular device such as the RC and the triad molecule surrounded by a large number of antenna molecules should exhibit clearly its molecular photodiode function without suffering from the serious noise of thermal fluctuation. Thus, if the process is repeated many times during the observation, the time averaging of the process of individual molecules may afford a meaningful result. How many events in a molecule or how many molecules in an event are necessary as a definite signal source is another interesting subject for future study.[108]

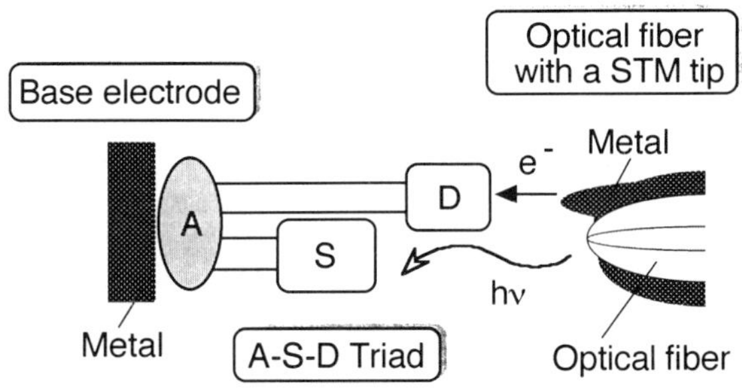

(a)

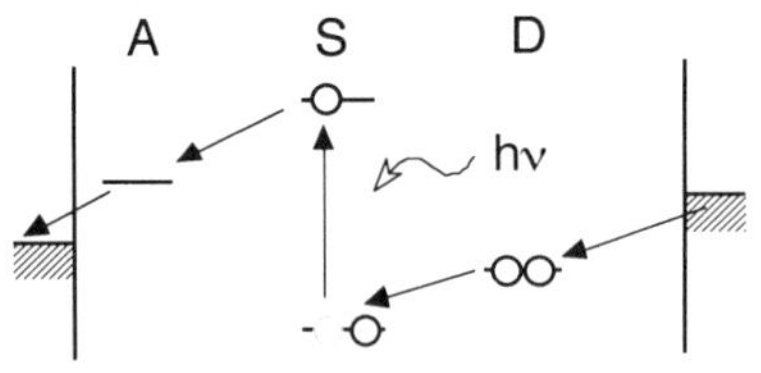

(b)

FIG. 11.17. Assumed photocurrent measurements of a single oriented triad molecule illuminated with a photon STM tip with a protruded STM metal tip: (a) structure and (b) energy diagram.

The surfaces of the islands of the mixed monolayer of HC and FC amphiphiles can be removed readily by the AFM tip under relatively weak force, while the surface of the sea part was difficult to wear with the same tip under higher forces.[62] Figure 11.18 shows that a squared depression in one of the islands, which was created by wearing after several narrower scans with a higher load, was observed by a wide scan with the same AFM tip under the lowest forces. The level of the square dug with the preceding narrow scan was almost the same as that of the sea, as already shown in Fig. 11.12(a). Not only the levels, but also the friction[62] [Fig. 11.12(b)] and the surface potential[64] observed in the square were in correspondence with those of the sea, respectively, as described previously.

The ease of wear of the HC of islands as described enabled us to perform nanometer lithography.[62] The dramatic difference in etching rate was observed between covered and uncovered surfaces with LB films.[68] In addition to use as nanometer lithography, the wear can be used to compartmentalize a macroscale monolayer assembly into a vast number of nanometer-scale divisions. Recently, we also succeeded in making a pattern with HC and FC domains[109] created by the combination of lithography and chemical modification with silanization reagents.[20]

Preferential site-selective adsorption of tobacco mosaic virus (TMV) on the well-characterized mixed monolayer of HC and FC described earlier[61–64] was studied recently by Frommer *et al.*[110] Adsorption occurs preferentially on the FC plane, at the juncture between HC and FC domains, i.e., a step site. This definitive assignment of binding loci is made possible by the highly localized resolution of a noncontact imaging mode in which the probe vibrates about 1 to 2 nm above the sample surface, avoiding actual physical contact between probe and sample. The experimental finding that the step site functions as the actual location for the adsorption of TMV suggested to us the usefulness of the step site between the two domains with different levels as "a line with a molecular dimension." These molecular lines can be made readily by wearing LB films with the use of nanolithography by an AFM tip.

11.6. Artificial Photosynthesis with Multilayered LB Films

The object of this section is to construct an artificial photosynthetic system using a complex LB film consisting of two or three types of LB film stacked in a series as shown in Fig. 11.19. One of them has a photoinduced charge separation function as described in Sections 11.2, 11.3, and 11.4, and the others possess catalytic activities for oxidation and reduction

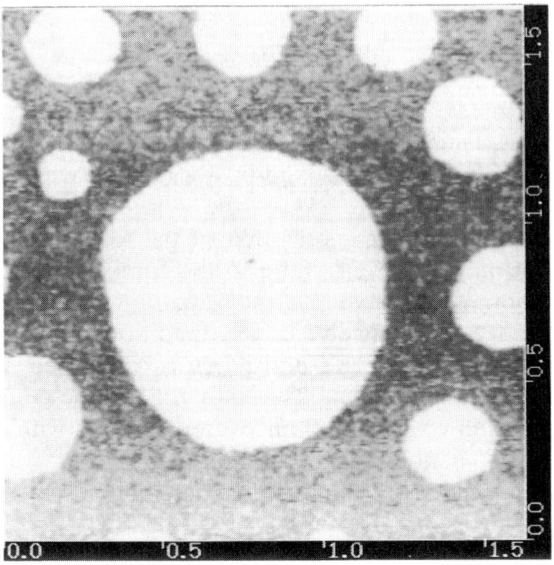

(a)

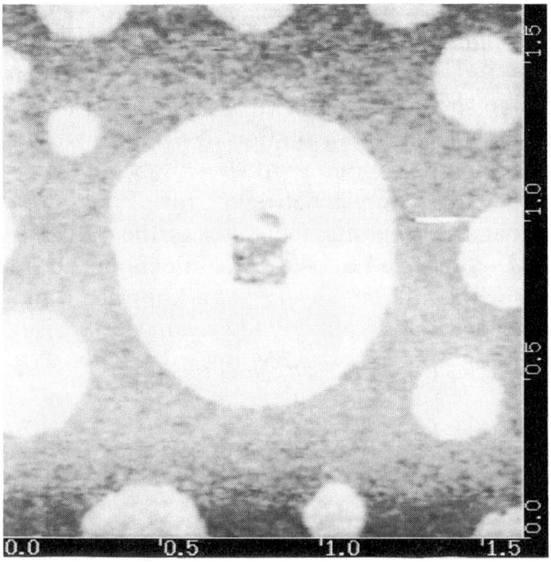

(b)

FIG. 11.18. AFM image of an artificially created hole on an island of a mixed monolayer of **27** and $C_{23}H_{47}COOH$ (1 : 1) ($1.6 \times 1.7\ \mu m^2$): (a) before and (b) after wearing.

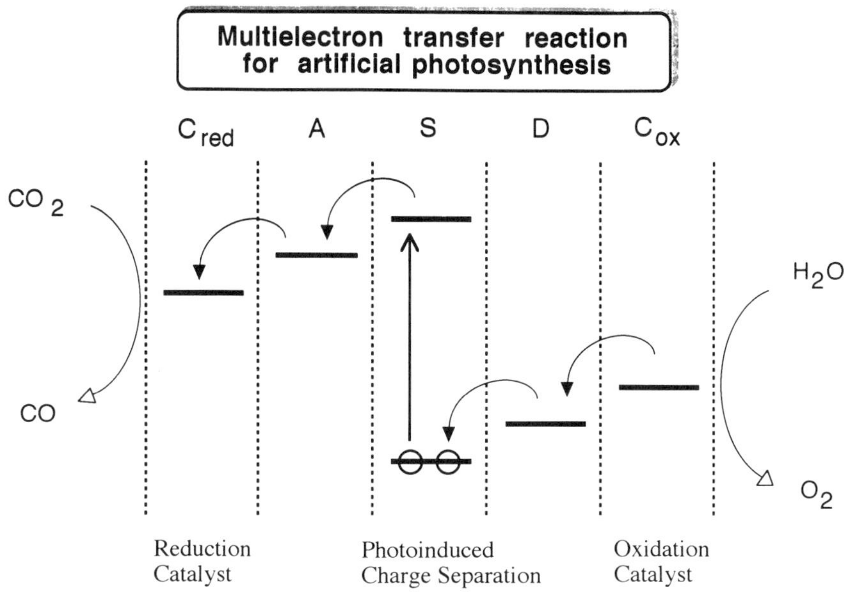

FIG. 11.19. Artificial photosynthetic system using a complex LB film consisting of a charge separation A/S/D (or A-S-D) layer sandwiched between two redox catalysis monolayers.

reactions achieved by the separated holes and electrons, respectively. The electrocatalytic activity of LB films of amphiphilic nickel cyclams[111] was examined for the reduction of CO_2, since nickel cyclam is known[112] to possess high catalytic activity and selectivity to CO. Two amphiphilic Ni(II) complexes with the long-alkyl-substituted 1,4,8,11-tetraazacyclotetradecane (cyclam) shown in Fig. 11.20 were synthesized, i.e., nickel(II) tetrakis(N-hexadecyl)cyclam **29** (Ni-R_4-cyclam) and nickel(II) N-hexdecylcyclam **30** (Ni-R_1-cyclam). Glassy carbon disk electrodes were modified with LB films

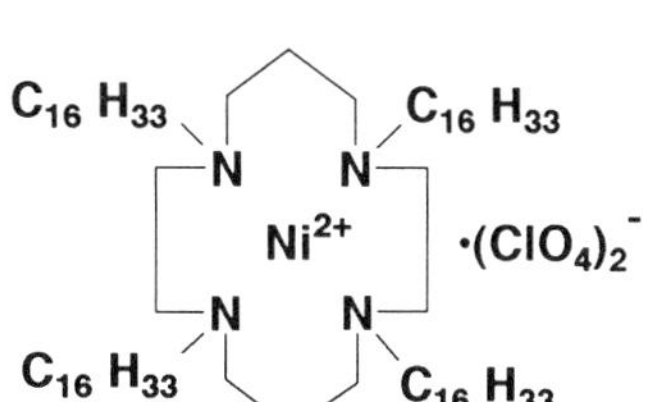

29 Ni-R_4-Cyclam

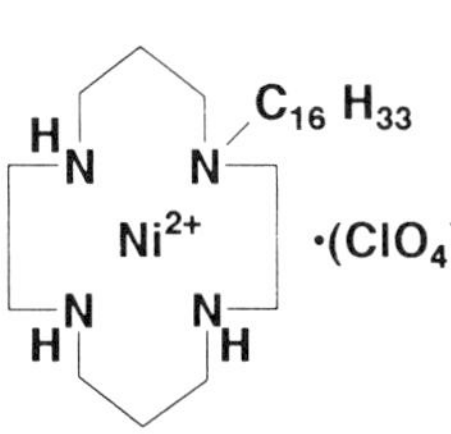

30 Ni-R_1-Cyclam

FIG. 11.20. Structural formulas of amphiphilic Ni(II) complexes with N-long alkyl substituted cyclams.

of these amphiphilic nickel complexes and the electrocatalytic activities of these electrodes were examined by *cyclic voltammetry* (CV)[113] in aqueous solutions saturated with N_2 or CO_2. The redox waves of the LB films of the alkyl derivatives of nickel(II) cyclam observed under N_2 were shifted in the positive direction relative to that of nickel(II) cyclam in a homogeneous solution. Under CO_2, the increase in the reductive current of these electrodes compared with that under N_2 was observed at more negative potentials than -1.3 V versus SCE in an aqueous solution with a constant pH value. This result shows that the LB films of these complexes possess sufficient electrocatalytic activity to reduce CO_2 despite their minimal amounts.

In addition to the electrochemical study of the LB modified electrodes, the electrocatalytic reduction of CO_2 on mercury with nickel(II) cyclam itself was studied in detail[114] by CV, polarography, and electrocapillarity in aqueous solutions, with and without the catalyst under N_2, CO, and CO_2. The effect to the catalytic activity and the stability of the nickel complexes of the short-alkylated cyclams as determined by the position and the number of the introduced alkyl groups to cyclam were also investigated.[115] The products of the electrocatalytic reduction of CO_2 with nickel(II) complexes of cyclam and the alkylated cyclams were analyzed *in situ* during their CV by modified differential electrochemical mass spectrometry (DEMS), originally developed by Wolter and Heitbaum.[116] Here "differential" means a time derivative of the total quantity of the product of interest, and thus is proportional to the electrolytic current. The first electrochemical mass spectrometer for the accumulated product being proportional to the electrolytic charge was invented about two decades ago by Bruckenstein and Gadde.[117] In our DEMS apparatus, an amalgamated gold mesh electrode is placed in contact with a nonwetting polytetrafluoroethylene inlet membrane, one side of which contacts the solution being electrolyzed and the other side contacts the high-vacuum system of a mass spectrometer.[118] CO and H_2 were shown to be the major products with these nickel(II) cyclams at low and high negative potentials, respectively.

The conventional DEMS apparatuses including the one just discussed were equipped with electrochemical cells with stationary solution. When a volatile reactant such as CO_2 is dissolved in the electrolyte, however, the reactant in the vicinity of the electrode surface will be depleted by the mechanisms such as (1) the exhaust of the reactant by evacuation and (2) the degassing by concomitantly evolving gases, i.e., H_2 and O_2 due to electrolysis of water, in addition to (3) the consumption of the volatile reactant by the electrochemical reaction. To cope with these problems, a new type of DEMS with a stationary gas-permeable gold electrode in a rotational flow produced by a rotating rod was devised.[119,120] The electrochemical characteristics of a stationary disk electrode placed in a rotational flow have been studied already.[121,122] This type of DEMS has a much simpler structure and a higher sensitivity in comparison with related equipment.[123,124]

Derivatization of ruthenium complexes for catalytic CO_2 reduction with long-alkyl chains is now being attempted[125] in addition to the cyclam derivatives.

11.7. Conclusions

With the near completion of the study on the reduction of CO_2 with LB monolayer catalysts, we started the study of the amphiphilic catalysts, which can oxidize water to O_2 by positive holes generated with the LB assembly for charge separation. Our final goal is the construction of LB film molecular assemblies for artificial photosynthesis as schematically shown in Fig. 11.21. We are now modifying the new DEMS rotating rod through which a laser beam passes to illuminate complexed LB films with a structure as shown in Fig. 11.21 deposited on a porous water-repellent membrane.[126] We hope to attain this goal in the near future. In addition to

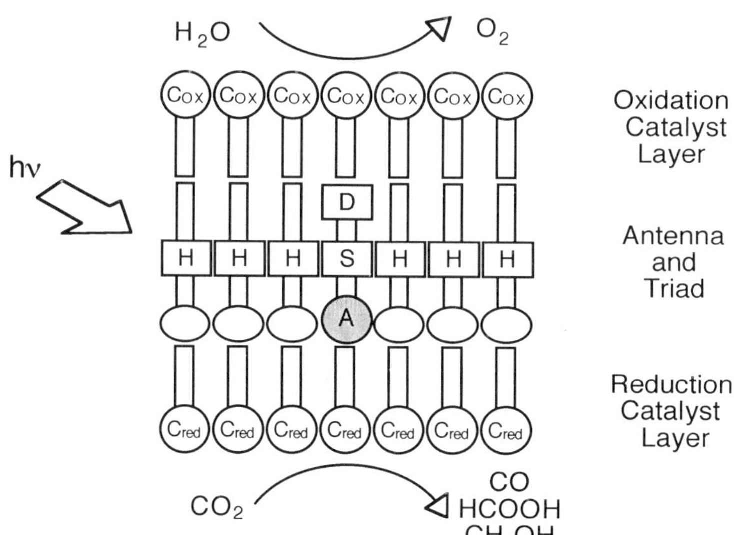

FIG. 11.21. Schematic structure of LB film molecular assembly for artificial photosynthesis.

the photoinduced charge-separation molecular devices, we are also developing a photoinduced proton transfer device[127] in connection with the proton pump of bacteriorhodopsin.[128]

Acknowledgments

I would like to thank my colleagues in our laboratory who collaborated in this research. I also would like to express my thanks to Rene Overny, Ernst Meyer, Jane Frommer, and Prof. H.-J. Güntherodt, the University of Basel, for their collaboration in the study of the mixed monolayer of HC and FC amphiphiles with AFM and FFM. Professor H. Kuhn, Max-Planck-Institute in Göttingen, and Dr. Abraham Ulman, Eastman Kodak Company, have been kind enough to read the manuscript and offer their suggestions for its improvement. This research was supported by a grant-in-aid for scientific research 61470076, that of new program 03NP0301, and those on priority areas 63604534 and 63612506, and that for developmental scientific research 04555195 from the Ministry of Education, Science, and Culture and by a grant from the Nissan Science Foundation.

References

1. Stryer, L. (1988). *Biochemistry*, 3rd ed. Freeman, New York.
2. Deisenhofer, J., Epp, O., Miki, K., Huber, R., and Michel, H. (1984). *J. Mol. Biol.* **180,** 385; (1985). *Nature* **318,** 618.
3. Chang, C. H., Schiffer, M., Tiede, D., Smith, U., and Norris, J. R. (1985). *J. Mol. Biol.* **186,** 201; Norris, J. R., and Raghavan, M. (1991). In: *Photochemical Conversion and Storage of Solar Energy* (E. Pelizzetti, and M. Schiavello, eds.). Kluwer Academic Publishers, Dordrecht, p. 141.
4. Norris, J. R., and Schiffer, M. (1990). *C & EN News* July 30, 22.
5. Mimuro, M. (1992). In: *Dynamics and Mechanisms of Photoinduced Electron Transfer and Related Phenomena* (N. Matage, T. Okada, and H. Masuhara, eds.). Elsevier, Amsterdam, p. 551.
6. Amesz, J., ed. (1987). *Photosynthesis*, Elsevier, Amsterdam.
7. Kuhn, H. (1972). *Chem. Phys. Lipids* **8,** 401.
8. Kuhn, H. (1979). *J. Photochem.* **10,** 111; (1983). *Thin Solid Films* **99,** 1; (1986). *Proc. Robert A. Welch Foundation* **30,** 339.
9. Möbius, D. (1978). *Ber. Bunsenges. Phys. Chem.* **82,** 848; (1981). *Acc. Chem. Res.* **14,** 63.
10. Fujihira, M. (1992). In: *Dynamics and Mechanisms of Photoinduced Electron Transfer and Related Phenomena* (N. Mataga, T. Okada, and H. Masuhara, eds.). Elsevier, Amsterdam, p. 453.
11. Takagi, M., Azuma, K., and Koshimoto, U. (1965). *Ann. Rep. Biol. Works, Fac. Sci. Osaka Univ.* **13,** 107; Fendler, J. H. (1982). *Membrane Mimetic Chemistry*. John Wiley, New York.

12. Kuhn, H., Möbius, D., and Bücher, H. (1972). In: *Techniques of Chemistry* (A. Weissberger, and B. Rossiter, eds.). Vol. I, *Physical Methods of Chemistry*, Part IIIB. Wiley Interscience, New York.
13. Seta, P., Bienvenue, E., Moore, A. L., Mathis, P., Bensasson, R. V., Liddell, P., Pessiki, P. J., Joy, A., Moore, T. A., and Gust, D. (1985). *Nature* **316,** 653; Lymar, S. V., Parmon, V. N., and Zamarev, K. I. (1991). In: *Topics in Current Chemistry, Vol. 159, Photoinduced Electron Transfer III* (J. Mattay, ed.). Springer Verlag, Berlin, p. 1.
14. Aviram, A., and Ratner, M. A. (1974). *Chem. Phys. Lett.* **29,** 277.
15. Carter, F. L. ed. (1982). *Molecular Electronic Devices*. Marcel Dekker, New York; Carter, F. L., ed. (1987). *Molecular Electronic Devices II*. Marcel Dekker, New York.
16. Carter, F. L., Siatkowski, R. E., and Wohltjen, H., eds. (1988). *Molecular Electronic Devices*. North-Holland, Amsterdam.
17. Metzger, R. M., and Panetta, C. A. (1987). In: *Molecular Electronic Devices II* (F. L. Carter, ed.). Marcel Dekker, New York, p. 5.
18. Fujihira, M., Nishiyama, K., and Yamada, H. (1985). *Thin Solid Films* **132,** 77.
19. Hsu, Y., Penner, T. L., and Whitten, D. G. (1992). *J. Phys. Chem.* **96,** 2790; Mooney, W. F., and Whitten, D. G. (1986). *J. Phys. Chem.* **108,** 5712.
20. Osa, T., and Fujihira, M. (1976). *Nature* **264,** 349; Fujihira, M., Ohishi, N., and Osa, T. (1977). *Nature* **268,** 226; Fujihira, M., Kubota, T., and Osa, T. (1981). *J. Electroanal. Chem.* **119,** 379; Fujihira, M. (1986). In: *Topics in Organic Electrochemistry* (A. J. Fry, and W. E. Britton, eds.). Plenum, New York, p. 255.
21. Calcaterra, L. T., Closs, G. L., and Miller, J. R. (1983). *J. Am. Chem. Soc.* **105,** 670; Closs, G. L., Calcattera, L. T., Green, N. J., Penfield, K. W., and Miller, J. R. (1986). *J. Phys. Chem.* **90,** 3673; Miller, J. R. (1987). *Nouv. J. Chim.* **11,** 83.
22. Warman, J. M., De Haas, M. P., Paddon-Row, M. N., Cotsaris, E., Hush, N. S., Oevering, H., and Verhoeven, J. W. (1986). *Nature* **320,** 615; Verhoeven, J. W. (1986). *Pure Appl. Chem.* **58,** 1285; Zamaraev, K. I., and Khairutdinov, R. F. (1992). In: *Topics in Current Chemistry, Vol. 163, Photoinduced Electron Transfer IV* (J. Mattay, ed.). Springer Verlag, Berlin, p. 1.
23. Leland, B. A., Joran, A. D., Felker, P. M., Hopfield, J. J., Zewail, A. H., and Dervan, P. B. (1985). *J. Phys. Chem.* **89,** 5571.
24. Marcus, R. A., and Sutin, N. (1985). *Biochim. Biophys. Acta* **811,** 265.
25. Kitzing, E. V., and Kuhn, H. (1990). *J. Phys. Chem.* **94,** 1699; Kuhn, H. (1986). *Phys. Rev.* **A34,** 3409.
26. Marcus, R. A. (1965). *J. Chem. Phys.* **43,** 2654; Siders, P., and Marcus, R. A. (1981). *J. Am. Chem. Soc.* **103,** 748.
27. Ulstrup, J., and Jortner, J. (1975). *J. Chem. Phys.* **63,** 4358.
28. Miller, J. R., Calcaterra, L. T., and Closs, G. L. (1984). *J. Am. Chem. Soc.* **106,** 3047; Miller, J. R., Beitz, J. V., and Huddleston, R. K. (1984). *J. Am. Chem. Soc.* **106,** 5057.
29. Wasielewski, M. R., Niemczyk, M. P., Svec, W. A., and Pewitt, E. B. (1985). *J. Am. Chem. Soc.* **107,** 5562.
30. Mataga, N. In book cited in Ref. 5, p. 3; Asahi, T., and Mataga, N. (1991). *J. Phys. Chem.* **95,** 1956; Ohno, T., Yoshimura, A., Mataga, N., Tazuke, S., Kawanishi, Y., and Kitamura, N. (1989). *J. Phys. Chem.* **93,** 3546; Mataga, N., Kanda, Y., Asahi, T., Miyasaka, H., Okada, T., and Kakitani, T. (1988). *J. Chem. Phys.* **127,** 239.
31. Gould, I. R., and Farid, S. (1992). *J. Phys. Chem.* **96,** 7635; Gould, I. R., Ege, D., Moser, J. E., and Farid, S. (1990). *J. Am. Chem. Soc.* **112,** 4290; Gould, I. R., Ege, D., Mattes, S. L., and Farid, S. (1987). *J. Am. Chem. Soc.* **109,** 3794.
32. Suppan, P. (1992). In: *Topics in Current Chemistry, Vol. 163, Photoinduced Electron Transfer IV* (J. Mattay, ed.). Springer Verlag, Berlin, p. 95.

33. Fujihira, M., Nishiyama, K., and Yoneyama, H. (1987). *J. Chem. Soc. Jpn.* 2119.
34. Nishikata, Y., Morikawa, A., Kakimoto, M., Imai, Y., Hirata, Y., Nishiyama, K., and Fujihira, M. (1989). *J. Chem. Soc. Chem. Commun.* 1772.
35. Nishikata, Y., Fukui, S., Kakimoto, M., Imai, Y., Nishiyama, K., and Fujihira, M. (1992). *Thin Solid Films* **210 / 211,** 296.
36. Suzuki, M., Kakimoto, M., Konishi, T., Imai, Y., Iwamoto, M., and Hino, T. (1986). *Chem. Lett.* 395; Kakimoto, M., Suzuki, M., Konishi, T., Imai, Y., Iwamoto, M., and Hino, T. (1986). *Chem. Lett.* 823; Nishikata, Y., Kakimoto, M., Morikawa, A., Kobayashi, I., Imai, Y., Hirata, Y., Nishiyama, K., and Fujihira, M. (1989). *Chem. Lett.* 861.
37. Isoda, S., Nishikawa, S., Ueyama, S., Hanazato, Y., Kawakubo, H., and Maeda, M. (1992). *Thin Solid Films* **210 / 211,** 290; Isoda, S., Maeda, M., Miyasaka, H., and Mataga, N. (1991). *Chem. Phys. Lett.* **182,** 379; Isoda, S. (1992). *FED J.* **2,** 59.
38. Sakamoto, K., Aramaki, K., and Nishihara, H. (1993). *Chem. Lett.* 659.
39. Fujihira, M., and Yamada, H. (1988). *Thin Solid Films* **160,** 125.
40. Nishitani, S., Kurata, N., Sakata, Y., Misumi, S., Karen, A., Okada, T., and Mataga, N. (1983). *J. Am. Chem. Soc.* **105,** 7771; Mataga, N., Karen, A., Okada, T., Nishitani, S., Kurata, N., Sakata, Y., and Misumi, S. (1984). *J. Phys. Chem.* **88,** 5138.
41. Moore, T. A., Gust, D., Mathis, P., Mialoeq, J. C., Chachaty, C., Bensasson, R. V., Land, E. J., Doizi, J. C., Liddel, P. A., Lehman, W. R., Nemeth, G. A., and Moore, A. L. (1984). *Nature* **307,** 630.
42. Gust, D., and Moore, T. A. (1992). *Adv. Photochem.* **16,** 1; Gust, D., and Moore, T. A. (1991). In: *Topics in Current Chemistry, Vol. 159, Photoinduced Electron Transfer III* (J. Mattay, ed.). Springer Verlag, Berlin, p. 103.
43. Osuka, A., Nakajima, S., Maruyama, K., Matage, N., Asahi, T., Yamazaki, I., and Nishimura, Y. In book cited in Ref. 5, p. 550.
44. Wang, X. D., Zhang, B. W., Bai, J. W., Cao, Y., Xiao, X. R., and Xu, J. M. (1992). *J. Phys. Chem.* **96,** 2886.
45. Fujihira, M., and Sakomura, M. (1989). *Thin Solid Films* **179,** 471.
46. Fujihira, M., Nishiyama, K., and Aoki, K. (1988). *Thin Solid Films* **160,** 317.
47. Kondo, T., Yamada, H., Nishiyama, K., Suga, K., and Fujihira, M. (1989). *Thin Solid Films* **179,** 463; Kondo, T., and Fujihira, M. (1990). *Kobunshi Ronbunshu* **47,** 921.
48. Suga, K., Fujita, S., Yamada, H., and Fujihira, M. (1990). *Bull. Chem. Soc. Jpn.* **63,** 3369.
49. Kondo, T., and Fujihira, M. (1991). *Chem. Lett.* 191; Kondo, T., Yanagisawa, M., and Fujihira, M. (1991). *Electrochim. Acta* **36,** 1793; Fijihira, M., Yanagisawa, M., Kondo, T., and Suga, K. (1992). *Thin Solid Films* **210 / 211,** 265.
50. Bockris, J. O'M., and Reddy, A. K. N. (1970). *Modern Electrochemistry.* Plenum, New York.
51. Vogel, V., and Möbius, D. (1985). *Thin Solid Films* **132,** 205; (1988). *Thin Solid Films* **159,** 73; (1988). *J. Colloid Interface Sci.* **126,** 408.
52. Wolff, C., and Grätzel, M. (1977). *Chem. Phys. Lett.* **52,** 542; Grätzel, M. (1989). *Heterogeneous Photochemical Electron Transfer.* CRC Press, Florida.
53. Yamada, H., Ohtani, H., and Fujihira, M. Unpublished work; Fujihira, M. (1991). In: *Photochemical Processes in Organized Molecular Systems* (K. Honda, ed.). Elsevier, Amsterdam, p. 463.
54. Beratan, D. N., Onuchic, J. N., and Hopfield, J. J. (1985). *J. Chem. Phys.* **83,** 5325; Beratan, D. N., Betts, J. N., and Onuchic, J. N. (1992). *J. Phys. Chem.* **96,** 2852.
55. Kuki, A. (1991). In: *Structure and Bonding. 75. Long Range Electron Transfer in Biology* (G. Palmer, ed.). Springer Verlag, Berlin, p. 49.
56. Electron Transfer in Biology and the Solid State. In: *Advances in Chemistry Series, Vol. 226* (M. K. Johnson, R. B. King, D. M. Kurtz, Jr., C. Kutal, M. L. Norton, and R. A. Scott, eds.). American Chemical Society, Washington, 1990.

57. Beratan, D. N., Onuchic, J. N., and Gray, H. B. (1990). In: *Electron Transfer Reactions in Metallo-proteins, Metal Ions in Biological Systems, Vol. 27* (H. Sigel, and A. Sigel, eds.), Marcel Dekker, New York, p. 97.
58. Fujihira, M. (1988). *2nd International Symposium on Bioelectronic and Molecular Electronic Devices R&D Association for Future Electron Devices,* December 12–14, 1988, Fujiyoshida, Japan, p. 35; Fujihira, M., Sakomura, M., and Kamei, T. (1989). *Thin Solid Films* **180,** 43.
59. Fujihira, M. (1990). *Mol. Cryst. Liq. Cryst.* **183,** 59.
60. Fujihira, M., Kamei, T., Sakomura, M., Tatsu, Y., and Kato, Y. (1989). *Thin Solid Films* **179,** 485.
61. Overney, R. M., Meyer, E., Frommer, J., Brodbeck, D., Lüthi, R., Howald, L., Güntherodt, H.-J., Fujihira, M., Takano, H., and Gotoh, Y. (1992). *Nature* **359,** 133.
62. Meyer, E., Overney, R. M., Brodbeck, D., Howald, L., Lüthi, R., Frommer, J., Güntherodt, H.-J., Fujihira, M., Takano, H., and Gotoh, Y. (1992). *Thin Solid Films* **220,** 132.
63. Fujihira, M., Kawate, H., and Yasutake, M. (1992). *Chem. Lett.* 2223.
64. Fujihira, M., Kawate, H., Takano, H., Gotoh, Y., Overney, R. M., Meyer, E., Frommer, J., and Güntherodt, H.-J. Unpublished work.
65. Fujihira, M., Kamei, T., Takano, H., and Gotoh, Y. Presented at *9th International Symposium on Surfactants in Solution*, June 10–15, 1992, Varna, Bulgaria.
66. Schneiker, C., and Hameroff, S. (1988). In: *Molecular Electronic Devices* (F. L. Carter, R. E. Siatkowski, and H. Wohltjen, eds.), North-Holland, Amsterdam, p. 69.
67. Rohrer, H. (1992). *Ultramicroscopy* **42–44,** 1.
68. Fujihira, M., and Takano, H. Unpublished work.
69. Chi, L. F., Eng, L. M., Graf, K., and Fuchs, H. (1992). *Langmuir* **8,** 2255.
70. Liu, Z. F., Hashimoto, K., and Fujishima, A. (1990). *Nature* **347,** 658.
71. Fujihira, M. (in press). *ACS Advances in Chemistry 240, Symposium on Molecular and Biomolecular Electronics* (R. R. Birge, ed.). American Chemical Society, Washington.
72. Fujihira, M. (1992) *Materials Research Society Spring Meeting Symposium Proc. Vol. 277, Macromolecular Host-Guest Complexes: Optical, Optoelectronic, and Photorefractive Properties and Applications* (S. A. Jenekhe, ed.), MRS, Pittsburgh, p. 47.
73. Pomeranz, M., Aviram, A., McCorkle, R. A., Li, L., and Schrott, A. G. (1992). *Science* **255,** 1115.
74. Singer, S. J., and Nicolson, G. L. (1972). *Science* **175,** 720.
75. Hato, M., Fujihira, M., Nakayama, H., and Shinoda, K. (1970). *Nippon Kagaku Zasshi* **91,** 448.
76. Higashi, N., Kunitake, T., and Kajiyama, T. (1986). *Macromolecules* **19,** 1362.
77. Binnig, G., Rohrer, H., Gerber, C., and Weibel, E. (1982). *Appl. Phys. Lett.* **40,** 178.
78. Binnig, G., Quate, C. F., and Gerber, C. (1986). *Phys. Rev. Lett.* **56,** 930.
79. Frommer, J. (1992). *Angew. Chem. Int. Ed.* **31,** 1298.
80. Rabe, J. P. (1992). *Ultramicroscopy* **42–44,** 41.
81. Heckl, W. M. (1992). *Thin Solid Films* **210 / 211,** 640.
82. Mate, C. M., McClelland, G. M., Erlandsson, R., and Chiang, S. (1987). *Phys. Rev. Lett.* **59,** 1942.
83. Meyer, G., and Amer, N. M. (1990). *Appl. Phys. Lett.* **57,** 2089.
84. Marti, O., Colchero, J., and Mlynek, J. (1990). *J. Nanotechnology* **1,** 141.
85. Fujihira, M., and Gotoh, Y. (1992). In: *Nanostructures Based on Molecular Materials* (W. Göpel, and C. Ziegler, eds.). VCH Publishers, Weinheim, p. 177.
86. Umemura, J., Hishiro, Y., Kawai, T., Takenaka, T., Gotoh, Y., and Fujihira, M. (1989). *Thin Solid Films* **178,** 281.

87. Parsons, R. (1954). In: *Modern Aspects of Electrochemistry, Vol. 1* (J. O'M. Bockris, ed.). Butterworths, London, p. 103.
88. Gaines, Jr., G. L. (1966). *Insoluble Monolayers at Liquid-Gas Interfaces*. John Wiley, New York.
89. Roberts, G., ed. (1990). *Langmuir-Blodgett Films*. Plenum Press, New York.
90. Ulman, A. (1991). *Ultrathin Organic Films*. Academic Press, San Diego.
91. Martin, Y., Abraham, D. W., and Wickramasinghe, H. K. (1988). *Appl. Phys. Lett.* **52,** 1103.
92. Abraham, D. W., Martin, Y., and Wickramasinghe, K. (1988). *Proc. SPIE* **897,** 191.
93. Weaver, J. M. R., and Abraham, D. W. (1991). *J. Vac. Sci. Technol.* **B9,** 1559.
94. Nonnenmacher, M., O'Boyle, M. P., and Wickramasinghe, H. K. (1991). *Appl. Phys. Lett.* **58,** 2921.
95. Nonnenmacher, M., O'Boyle, M. P., and Wickramasinghe, H. K. (1992). *Ultramicroscopy* **42–44,** 268.
96. Yokoyama, H., Saito, K., and Inoue, T. (1992). *Mol. Electronics. Bioelectronics* **3,** 79; Inoue, T., and Yokoyama, H. (1992). In: *4th International Symposium on Bioelectronic and Molecular Electronic Devices*, Nov. 30–Dec. 2, 1992, Miyazaki, Japan, p. 36.
97. Fujihira, M., and Kawate, H. Unpublished work.
98. Fujihira, M., Kawate, H., and Sakomura, M. Unpublished work.
99. Tasuda, Y., Hirata, Y., Sugino, H., Kumei, M., Hara, M., Miyake, J., and Fujihira, M. (1992). *Thin Solid Films* **210 / 211,** 733.
100. Hirata, Y., Nukanobu, K., Hara, M., Asada, Y., Miyake, J., and Fujihira, M. (1992). *Chem. Lett.* 2277.
101. Stern, J. E., Terris, B. D., Mamin, H. J., and Rugar, D. (1988). *Appl. Phys. Lett.* **53,** 2717; Saurenbach, F., and Terris, B. D. (1992). *IEEE Trans. Indust. Appl.* **28,** 256.
102. Synge, E. H. (1928). *Phil Mag.* **6,** 356.
103. Ash, E. A., and Nicholls, G. (1972). *Nature* **237,** 510.
104. Lewiss, A., Isaacson, M., Harootunian, A., and Muray, A. (1984). *Ultramicroscopy* **13,** 227; Lieberman, K., Harush, S., Lewis, A., and Kopelman, R. (1990). *Science* **247,** 59.
105. Pohl, D. W., Denk, W., and Lanz, M. (1984). *Appl. Phys. Lett.* **44,** 651.
106. Fischer, U. Ch. (1985). *J. Vac. Sci. Technol.* **B3,** 386.
107. Betzig, E., Trautman, J. K., Harris, T. D., Weiner, J. S., and Kostelak, R. L. (1991). *Science* **251,** 1468; Betzig, E., and Trautman, J. K. (1992). *Science* **257,** 189.
108. Ulman emphasized, however, in p. 422 in Ref. 90 that electronic operations will not be conducted by individual molecules because of the uncertainty principle. This can be understood also if both the expected percent of back reaction and the acceptable error associated with a logic operation are concerned. On the other hand, open and close of a channel in a cell membrane with one biomolecule was readily measured as the single channel ionic current. [See, for example, Sackmann, B., and Neher, E., eds. (1983). *Single-Channel Recording*. Plenum, New York.]
109. Fujihira, M., Morita, Y., and Kawate, H. Unpublished work.
110. Frommer, J., Lüthi, R., Meyer, E., Anselmetti, D., Overney, R., Güntherodt, H.-J., and Fujihira, M. Unpublished work.
111. Hirata, Y., Suga, K., and Fujihira, M. (1989). *Thin Solid Films* **179,** 95; Akiba, U., Nakamura, Y., Suga, K., and Fujihira, M. (1992). *Thin Solid Films* **210 / 211,** 381.
112. Beley, M., Collin, J. P., Ruppert, R., and Sauvage, J. P. (1986). *J. Am. Chem. Soc.* **108,** 7461.
113. Bard, A. J., and Faulkner, L. R. (1980). *Electrochemical Methods Fundamentals and Applications*. John Wiley & Sons, New York.
114. Fujihira, M., Hirata, Y., and Suga, K. (1990). *J. Electroanal. Chem.* **292,** 199.

115. Fujihira, M., Nakamura, Y., Hirata, Y., Akiba, U., and Suga, K. (1991). *Dennki Kagaku* **59,** 532.
116. Wolter, O., and Heitbaum, J. (1984). *Ber. Bunsenges. Phys. Chem.* **88,** 2.
117. Bruckenstein, S., and Gadde, R. R. (1971). *J. Am. Chem. Soc.* **93,** 5941.
118. Hirata, Y., Suga, K., and Fujihira, M. (1990). *Chem. Lett.* 1155.
119. Fujihira, M., and Noguchi, T. (1992). *Chem. Lett.* 2043.
120. Fujihira, M., and Noguchi, T. (1993). *J. Electroanal. Chem.* **347,** 457.
121. Hamada, S., Itoh, M., Matsuda, H., and Yamada, J. (1978). *J. Electroanal. Chem.* **91,** 107.
122. Bucur, R. V., Mecea, V., and Bartes, A. (1977). *Electrochim. Acta* **22,** 499.
123. Tegtmeyer, D., Heindrichs, A., and Heitbaum, J. (1989). *Ber. Bunsenges. Phys. Chem.* **93,** 201.
124. Wasmus, S., Cattaneo, E., and Vielstich, W. (1990). *Electrochim. Acta* **35,** 771.
125. Kondo, T., Kitamura, T., and Fujihira, M. (1992). *Abstract of 9th International Conference on Photochemical Conversion and Storage of Solar Energy*. Beijing, China, August 23–28, 1992, p. 189.
126. Fujihira, M., Noguch, T., and Oyamada, M. Presented at *Abstract of 183rd Meeting of the Electrochemical Society*, Honolulu, Hawaii, May 16–21, 1993.
127. Fujihira, M., and Niidome, Y. (1992). *Thin Solid Films* **210 / 211,** 378.
128. *Primary Processes* (1987). In: *Photobiology, Springer Proceedings in Physics 20* (T. Kobayashi, ed.). Springer Verlag, Berlin.

Photoelectric Behavior of Bacteriorhodopsin Thin Films at the Solid/Liquid Interface

TSUTOMU MIYASAKA

Fuji Photo Film Co., Ltd., Ashigara Research Laboratories, Minamiashigara, Kanagawa, Japan 250-01

12.1. Introduction

The photoelectric properties of organic molecules have been studied extensively using their thin dry films placed at the interface of conductive electrodes. Early studies included photoconductivity measurements on photographic sensitizing dyes by Nelson[1] who extended his target dyes to chlorophylls to infer the energy conversion mechanism of photosynthesis.[2] There have since been a number of other studies on the photovoltaic properties of synthetic[3] and biological[4] dyes. In electrochemistry, photographic dye sensitization has been applied by exploiting various sensitizing dyes at the interface of a conductive or semiconductive electrode and a liquid electrolyte.[5] Eventually, dye-sensitized photoelectrochemical cells were proposed as a potential model for photosynthetic energy conversion.[6,7] These dry and wet photocells bearing a light-sensitive thin organic film produce a photocurrent due to electron transfer across the solid/solid or solid/liquid interface. This interface serves as an effective potential barrier to rectify the electron flow. As long as the chromophore of a molecule under electronic excitation is made to contact an electrode or a medium (liquid, gas, etc.) containing an electron donor or acceptor, the possibility arises that the chromophore undergoes electronic oxidation or reduction as a result of carrier exchange. This is the essential reason why photoelectric conversion using organic dyes more or less carries the risk of photodegradation of the chromophore, often caused by photooxidation involving singlet oxygen.

One rare exception to such cases is likely to be bacteriorhodopsin (bR), a sort of photosensitive chromoprotein that contains retinal as a chro-

ISBN 0-12-523485-6

mophore in its protein cavity. Because retinal forms a stable Schiff base bound with the protein and its reactions with oxygen diffused into the cavity are perfectly inhibited, this protein shows unusual stability against light. Further, bR is not a redox-active molecule and its reaction site, retinal, is well protected by the surrounding close-packed helices of protein. This is not the case for the photosynthetic reaction centers, other important photosensitive proteins; they perform redox reactions involving the chromophore (chlorophyll) and surrounding donors and acceptors, thus allowing for the possibility of oxidative degradation of the chromophore under unfavorable reaction conditions. In the bR molecule, light energy is first converted to a geometrical change of the chromophore and the cavity, which subsequently provides a chemical energy to transport an ion (proton). The photoelectric response of bR is obtained as a result of charge displacements accompanying such a geometrical change inside the molecule. This chapter describes some of the unique properties that we can elicit from bR and its thin film at interfaces as a result of its photoelectric behavior.

12.2. Photoelectric Response Caused by bR Photocycle

As a protein analog of the visual pigment rhodopsin, bR (MW ~ 26,000) is produced by a halobacterium and isolated in the form of fragments of purple membrane (PM),[8] which contains two-dimensional (2-D) crystals of oriented bR molecules. The biological role of bR is to transport protons from the intracellular to the extracellular side of a closed cell. This proton pumping establishes a pH gradient necessary for adenosine triphosphate (ATP) synthesis with which the bacterium can manage respiration while experiencing a lack of oxygen. To continue photorespiration, bR performs a unique cyclic reaction (photocycle). Although the retina's rhodopsin is decomposed (bleached) by photoreaction, bR is always regenerated after each photocycle and can thus survive under prolonged irradiation of solar energy. Figure 12.1 shows a simplified scheme of the photocycle, which comprises a series of intermediates, a light-produced K state and subsequent thermal states (L–O).[9,10] The quantum efficiency of the photocycle has been estimated[9] to be about 0.7. This is of a level similar to that of visual rhodopsin.

Proton pumping occurs during the formation of the M species in the time range of 40 μs to 1 ms, driven by a conformational change of protein, which changes a pK_a of the retinal-enveloping cavity.[11,12] This is a charge transfer process that contributes to photoelectric conversion. In previous

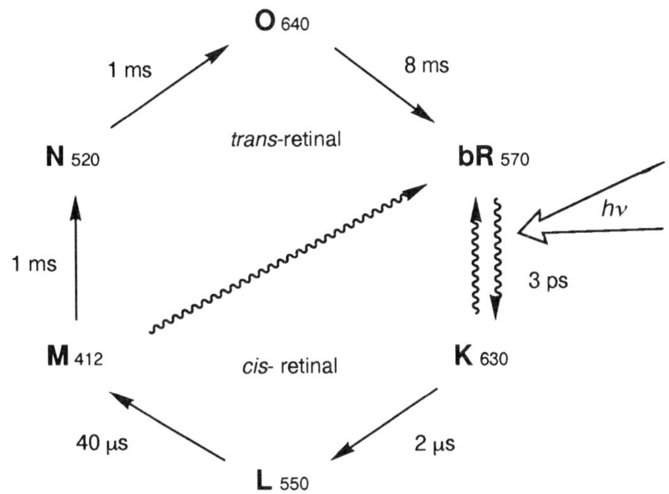

FIG. 12.1. Photocycle of bR comprising a series of intermediates K, L, M, N, and O. Wavy arrows represent photochemical processes. Solid arrows represent the thermal decay of a species. Subscripts denote the maximum absorption wavelength of a species. The photoexcited short-lived J state before the formation of K has also been identified, but is not included here.

methods, which employed a bR-mounted film as a membrane separating two electrolyte solutions,[13] light-induced rapid electric signals are obtained that are often accompanied by a slow response component due to proton transfer across the membranes; protons are pumped out to the solution phase and accumulate on the membrane surface, causing a change in electrochemical potential. For the purpose of photoelectric conversion, however, the process of proton pumping is not regarded as able to make a key contribution because the rate of carrier (proton) transfer is too slow (diffusion-limited) to cause a rapid and efficient electrical response on an external electrode.

Of central importance for the photoelectric process is the initial step of forming the K state where retinal undergoes rapid isomerization from the *trans* to the *cis* form on a picoseconds time scale.[12] This rapid conformational change brings about a displacement of charge, that is, a several angstrom shift of the positive charge localized on the protonated Schiff base of retinal (bound to lysine216 residue), as shown in Fig. 12.2. As a consequence, a vectorial change in the dipole moment occurs in the retinal-enclosing cavity. This induces a rapid electric response on an electrode adjacent to bR molecules, detected as a voltage change or an electric current (photocurrent).

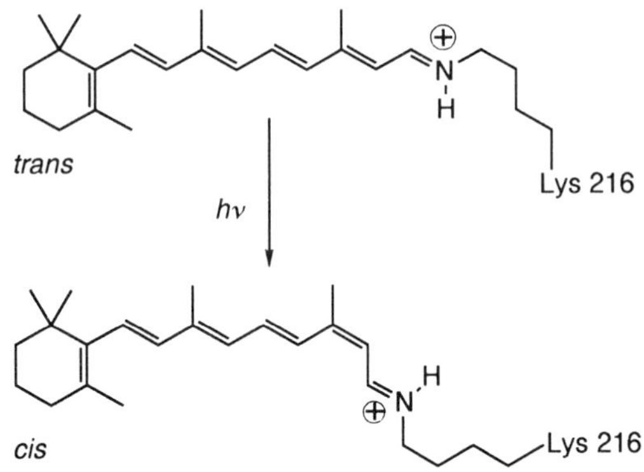

FIG. 12.2. Isomerization of the protonated retinal Schiff base bound within the bR cavity. Displacement of positive charge takes place by the movement of the Schiff base in the process of bR to K conversion.

A similar electrostatic response can also be obtained with thin films of synthetic dyes, such as azobenzene derivatives,[14] which undergo photoisomerization. Such an electrostatic response can be detected without direct contact between molecules and electrodes. For bR, this type of photoelectric response was first measured by Nagy[15] in 1978. Nagy used dry films of oriented PMs deposited on a conductive glass and separated from a counterelectrode by an air layer. Varo[16] later studied the photovoltaic effects of electrodeposited PM films by vacuum depositing a counterelectrode on the PM film. Both methods required a unidirectional orientation of PM to elicit efficient photoresponse. Furuno *et al.*[17] employed a Langmuir-Blodgett (LB) method to obtain oriented PM films and measured transient photocurrents due to charge displacements of bR. Many other photoelectric measurements have been conducted, including dry and wet systems, as reviewed by Trissl.[18] Of particular importance is the fact that under optimized conditions regarding the cell structure and electric setup for measurements, the rise time of the bR-induced photoelectric response reaches the time range of several picoseconds.[19,20] This shows that the response actually comes from the very initial step of the photocycle forming the K intermediate. From a practical standpoint, it is noteworthy that bR holds such a high potential for rapid response, something that is normally difficult to realize with existing solid-state photodiodes.

The direction of the bR photoelectric response is rectified according to the molecular orientation of the PM film. Recent time-resolved studies[21,22] indicate that the initial rapid photocurrent ascribed to the charge displacement caused by bR-to-K isomerization takes place in the direction opposite to that of proton pumping (from the carboxyl terminal to the amino terminal side of bR). Subsequent slower and reverse components of photocurrents (10^{-5} to 10^{-3} s) can be explained as arising from the process of proton pumping synchronized with a three-dimensional (3-D) structural change of the protein. Vectorially opposite components of displacement photocurrents have also been confirmed[23] for the other bacterial rhodopsin analog, halorhodopsin, which pumps Cl^-.

12.3. Photoelectric Behavior of bR at Solid/Liquid Interfaces

Although photoelectrochemistry has previously exploited many photosensitive dyes as able sensitizers at the electrode/liquid electrolyte interface, there has been no attempt to utilize bR in such a mode, since bR is apparently not an able electron donor or acceptor that performs redox interactions with an electrode. The merit of using bR molecules at such an interface is therefore assumed to be an alteration or improvement of the charge displacement effects of bR by exposing the molecule to a certain potential gradient at the electrochemical interface. We have investigated[24] for the first time the photoelectrochemical behavior of bR according to the method used by Miyasaka *et al.*[25] for light energy conversion with a thin film of chlorophyll molecules. To immobilize bR molecules at the electrode/electrolyte interface, LB films of PM were prepared by an improved method,[24] which enables a homogeneous dispersion of PM fragments on the water surface without causing dissolution of PM into the aqueous subphase. The LB technique for PM was attempted earlier by Hwang *et al.*[26] and was applied for a photoelectric measurement by Furuno *et al.*[17] who assumed, based on the direction of electric response, that PM on the water surface tends to orient with the cytoplasmic (carboxyl terminal) side toward the subphase. We have chosen as a substrate of a PM LB film an SnO_2 conductive electrode that has a hydrophilic surface suitable for depositing PM and yields low background redox currents due to oxygen and water reduction. Monolayers of PM were built up on the SnO_2 surface by the modified horizontal deposition method,[27] which forms Y-type (alternately stacked) LB films.

A photoelectrochemical cell was prepared comprised of an aqueous solution of a supporting electrolyte (KCl), counterelectrode (Pt), reference electrode (SCE), and the SnO_2 electrode coated with a PM LB film. Upon visible light irradiation to the PM film, transient photocurrents were generated in the absence of redox species in the electrolyte. The electrode potential applied to SnO_2 (versus SCE) significantly affected the magnitude of the photocurrent, but the direction of photocurrents was always maintained as cathodic (electron flow from the electrode to electrolyte side) regardless of the direction of applied potentials (electric fields).[28] This strong rectification of photocurrents is very characteristic of bR film placed at the SnO_2/electrolyte interface.

A significant finding concerning the charge-displacement photocurrent produced at the electrode/electrolyte interface is that it is of high efficiency and it responds differentially to a change in the light intensity with a good quantitativity to all ranges of intensity. A clear differential-type photocurrent of the order of 10^{-8} A can be obtained from a single PM monolayer under illumination with a 150-W xenon arc lamp. A typical response profile is exhibited in Fig. 12.3. A stepwise change in the intensity of incident light causes photocurrents that respond in the time-differential mode, showing a rapid relaxation to the background zero level when the intensity settles to a constant level. The magnitude of displacement photocurrents far exceeds those previously obtained with a dry thicker LB film (50 layers) of PM (10^{-10} A).[17] The quantum efficiency of the photocurrent (for the transient peak) was measured to be of the order of 10^{-2}, which is still low compared to the efficiency of the photocycle but of a significantly higher level as an artificial conversion system in comparison

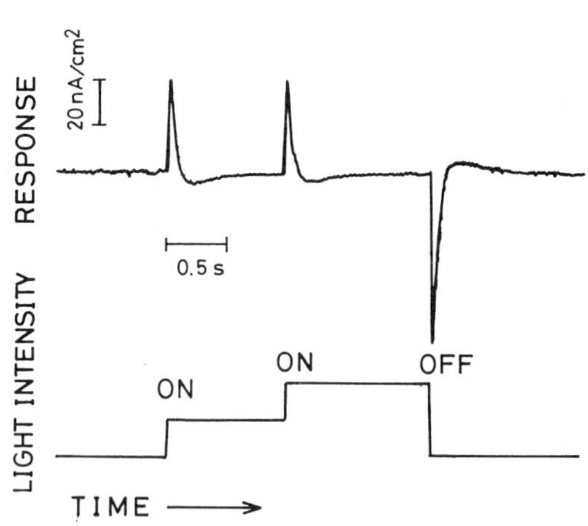

FIG. 12.3. Photocurrent response of a thin PM film at the SnO_2/electrolyte interface. The time course of a change in incident light intensity is shown at the bottom. The response is differential against changes in intensity.

to the previous bR-based systems.[29] Such a quantum conversion (from photon quanta to electrons) using previous dry sandwich photovoltaic cells is inevitably of low efficiency because of the intrinsically high electric resistance of PM samples. In contrast, relatively large photocurrents have been obtained from sandwich cells using PM dipped in wet aqueous gels.[30] The high conversion efficiency of our system is apparently due to low internal electric resistance of the cell endowed by the electrolyte (ionic conductor), which soaks the molecules of bR. It has been established that the charge-displacement current induced on an electroconductive surface is inversely proportional to the distance between the molecule under reaction and the conductive surface.[31] In our method, the spatial gap between bR molecules and the SnO_2 electrode is filled with a conductive aqueous electrolyte, which can directly capture the charge-displacement effect in bR.

The mechanism of photocurrent generation and its rectification is discussed taking the effect of the electrochemical interface into account. First, that the origin of the differential photocurrent is invariably bR is evident[24] from the action spectra that follow the absorption of bR peaking at around 560 nm. An important aspect of the photocurrent is its strong rectification in the cathodic direction. Further, the magnitude of photocurrent tends to increase with an increase in cathodic polarization of the electrode.[28] Negative bias applied to the electrode thus assists the generation of photocurrent. As for SnO_2 and other oxide electrodes, it is known that the surface of an electrode in contact with an aqueous solution is negatively charged due to dissociated hydroxyl groups, the extent of which determines the surface potential of electrodes as a function of the pH of the electrolyte. In neutral solutions, the surface of the SnO_2 is almost totally dissociated, resulting in a negatively charged layer, which, as a consequence of forming an electrostatic double layer, establishes a strong electric field across the SnO_2/electrolyte interface. In view of this, the current rectification can be interpreted as a field effect at the SnO_2/electrolyte interface on the charge displacement of bR molecules. From immuno-gold labeling experiments on the side characteristics of PM, we found that the orientation of PM fragments spread on the water surface is substantially random.[32] Assuming a random orientation of PM (bR) present on the SnO_2 surface, there should exist components of charge displacement that oppose the direction of the observed cathodic flow of photocurrent. However, such reverse components of displacement are presumably suppressed by the interfacial electrostatic field, and the electrode may be responding solely to the charge displacements that occur, and possibly are enhanced, in the direction along the field. This situation is sketched in Fig. 12.4, where the positive charge in a bR molecule means a

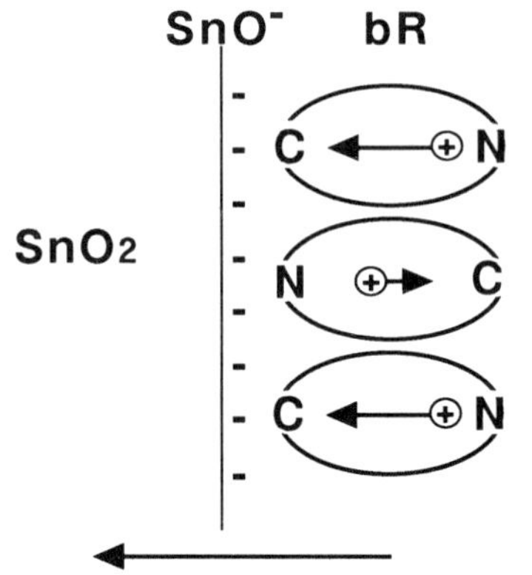

FIG. 12.4. Scheme showing a proposed mechanism for charge-displacement photocurrent at the SnO_2/electrolyte interface. The C and N marked in the bR molecule represent the carboxyl- and amino-terminal sides of bR, respectively. Molecules of bR are in contact with a negatively charged surface of SnO_2, which gives the electrostatic field shown by the long arrow. The arrow inside the molecule represents the vector of positive-charge displacement, which is suppressed in strength for a molecule oriented in such a direction that the charge displacement occurs opposing the direction of the electric field.

statistical distribution of charge under a stationary state of the photocycle given by a certain light intensity.

Several phenomena support this mechanism. First, photoresponse is weakened when measured on a gold electrode that forms a least oxide layer. Second, cathodic photocurrents still occur even when an electrode is illuminated that bears a sample of a PM film with its substantial orientation opposing the direction producing the cathodic response, i.e., orientation directing the extracellular side to the electrode.[33] Third, photocurrent tends to be saturated in magnitude when the number of PM monolayers built up on the electrode is increased. This tendency is shown in Fig. 12.5, which plots the peak photocurrent versus the number of layers of PM LB film, i.e., film thickness (one PM is approximately 4 nm thick). The result implies that the contribution of bR to charge-displacement current is limited to the PM layer positioned close to the electrode surface where the field effect becomes more significant.

We recently elucidated[33] that the origin of cathodic photocurrent is the transfer of positive charge occurring from the extracellular to the intercellular side of bR, opposite to the direction of proton pumping, being initiated by the *trans*-to-*cis* photoisomerization of bR in the photocycle.

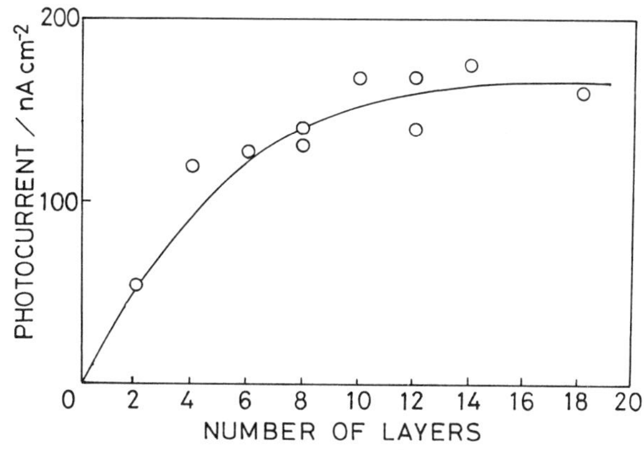

FIG. 12.5. Magnitude of bR-induced charge-displacement photocurrent as a function of the number of PM monolayers. PM LB films were built up on the SnO_2 electrode surface and their photocurrents were measured under cathodic polarization of the electrode potential (around −0.5 V versus SCE) in a neutral electrolyte solution (0.1 M KCl).

The response time measured under pulse excitation, with minimized resistance and capacitance in external circuits, was much shorter than a microsecond, showing that the response is not associated with the proton pumping.

12.4. Device Application for Optical Sensing and Image Processing

Rapid and clear differential responsivity obtained with the bR-immobilized cell is advantageous for potential applications in the construction of a visually functioning light sensor because differential responsivity plays a fundamental role in visual information processing. From a practical standpoint, the advantage of using bR-based photoelectrochemical cells is again emphasized in the respect that the bR-induced photocurrents are entirely of a nonfaradaic type, causing no electrochemical consumption of ambient materials. Although the present cell takes an electrochemical structure, it should actually be characterized as an ion-conductive diode having a photosensitive rectifying interface. We have converted the cell into a more compact sandwich structure, similar to liquid crystal devices, comprising junctions of SnO_2/PM film/electrolyte layer/counterelectrode (Au).[34] The fluid electrolyte solution was displaced here with a thin layer (200 to 300 μm) of an aqueous polymer gel containing 4% carboxylmethyl-chitin and 2 M KCl (pH 7). A gold-sublimated glass plate was used as a counterelectrode, which, with insertion of a Teflon film spacer, seals the conductive gel layer. This solid/liquid junction photocell generated differential photocurrents that were cathodically rectified and, in contrast to the electrochemical system, much less influenced by the applied bias between two electrodes. Experiments have been carried out to assess the light-sensing ability of the photocell. As the result of quantum conversion, a linear relationship between the photocurrent (output electrons) and light intensity (incident photons) was confirmed that covers an intensity range of more than three orders of magnitude.[34] Also confirmed is that the output response is a function of the rapidity of the change in light intensity, i.e., time differential of the intensity (I), dI/dt, showing an increase with increased dI/dt.

Regarding potential applications of the bR-based photocell for visual information processing, an image sensor has been fabricated by means of pixelized light sensor. A PM LB film (14 layers) was coated on a patterned indium-tin-oxide (ITO) electrode, which bears a network of 256 ITO square pixels (1.3×1.3 mm). For parallel signal transactions, each pixel has an independent wire (ITO pattern) led to the edge terminals of the photocell where external circuits are connected. The bR-induced photocurrents from the pixel network were converted in the voltage mode and amplified to output finally to an LED display device comprising 256 picture units.[35] With this system optical images received by the bR pixels are simultaneously displayed on the LED panel. Image sensing experiments were carried out by projecting various optical images on the receptor plane of the sensor by means of a commercial video projector. Immediately it was shown that our bR-based image sensor rapidly detects a moving optical target while ignoring still objects. A typical example is demonstrated in Fig. 12.6, which shows a hand movement captured by the sensor. The sensor outputs no image signals during the time the hand stops movement [Fig. 12.6(a)], but once a movement starts it outputs optical information, rapidly responding to any light intensity change caused by optical scanning. As shown in Fig. 12.6, the image sensor can simultaneously extract the profile (edge components) of a moving image, thus working as a motion sensor and an edge detector at the same time.

Motion-sensing and edge-detecting functions are typical of the biological photoreceptors in human and animal vision systems. Biological edge detection is performed in retinal ganglion cells where differential electric signals are produced for subsequent image processing (recognition, memory, etc.) in the cerebral neurons. A typical application of the edge detection ability is pattern recognition. We have studied the edge detection from some letter images using the 256-pixel bR-based image sensor as an artificial photoreceptor. As a simulation of vibrational eye movements, a

(a)

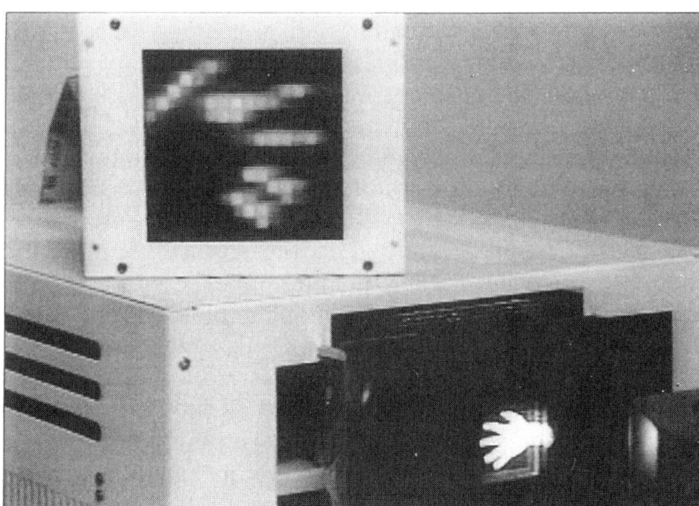

(b)

FIG. 12.6. Detection of a hand motion by the 256-pixel bR-based photoreceptor. The image of a moving hand was projected to the photoreceptor (lower right) to perform real-time image processing, as displayed in the panel. The photoreceptor ignores still images when the hand stops its motion (a) but it begins detecting the profile of the hand image at the moment the hand starts jiggling (b). On this sensing system moving objects are selectively detected in the edge-enhanced form. Reprinted with permission from Miyasaka, T., and Koyama, K., *Applied Optics*, Copyright © 1993, The Optical Society of America.

phenomenon requisite to visual perception, a letter image was projected to the bR-based photoreceptor under oscillation of the optical axis. When oscillation was performed in the vertical and horizontal directions, the photoreceptor could actually extract the horizontal and vertical edge components, respectively, from an alphabetical letter image.[35] Vectorially selective extraction of edge components could also be done by this method, which is a key process to letter recognition.

These experiments corroborate that the bR-based optical device possesses a unique photoelectric responsivity, which enables rapid preprocessing of optical information. It is to be emphasized that such a preprocessing ability (differential responsivity) is realized at the material level by immobilizing bR at the solid/liquid interface, which is well suited for eliciting charge displacement effects and also for keeping bR in its high activity under accommodation in an aqueous medium.

12.5. Conclusions

For photoelectric conversion, charge displacements of a bR molecule are most effectively captured as an electric current at the interface of a conductive electrode and aqueous electrolyte solution. Light quanta are converted to electron flow (photocurrent) at a relatively high quantum efficiency with a linear relationship between the two quantities over a wide range of light intensity. Using a thin LB film of bR, photocurrents are well rectified at the oxide electrode (SnO_2, etc.)/electrolyte interface owing to the interfacial static field attributed to the surface charge. Under this rectification, bR shows a characteristic differential responsivity to a change in light intensity, which is a key function in biological photoreceptors. An artificial photoreceptor (image sensor) was fabricated using a bR-immobilized pixelized electrode, which was proved to function as a motion sensor with an edge detection capability. The bR-based photoreceptor can thus perform rapid preprocessing of visual information at the material level, i.e., without the assistance of computational circuitry and hardware.

The exceptionally high stability of bR against light is noteworthy for future potential applications to optical devices. Bacteriorhodopsin owes its stability to the photocycle, which keeps out degradative side reactions and is driven solely by the internal conformational change (retinal isomerization) without involving reactions with external redox species. In fact, bR maintains its photoelectric function over a long period of time even when the molecule is brought into such a powerful redox medium as the electrode/electrolyte interface.

Acknowledgments

The author wishes to acknowledge Mr. Koichi Koyama and Mr. Naoto Yamaguchi who have investigated orientation of PM films. The author also thanks Dr. Isamu Itoh for his useful and informative insights.

References

1. Nelson, R. C. (1956). *J. Opt. Soc. Am.* **46,** 10–13.
2. Nelson, R. C. (1957). *J. Chem. Phys.* **27,** 864–867.
3. Meier, H. (1974). *Organic Semiconductors.* Verlag Chemie, Weinheim; Morel, D. L., Ghosh, A. K., Feng, T., Stogryn, E. L., Purwin, P. E., Shaw, R. F., and Fishman, C. (1978). *Appl. Phys. Lett.* **32,** 495–497; Sakai, K., Saito, M., Sugi, M., and Iizima, S. (1985). *Jap. J. Appl. Phys.* **24,** 865–869.
4. Tang, C. W., and Albrecht, A. C. (1975). *Nature* 254–259, 507; Dodelet, J. P., Le Brech, J., and Leblanc, R. M. (1979). *Photochem. Photobiol.* **29,** 1135–1145.
5. Meier, H., Albrecht, W., Tschirwitz, U., Zimmerhackl, E., and Geheeb, N. (1977). *Ber. Bunsenges. Physik. Chem.* **81,** 592–598; Osa, T., and Fujihira, M. (1976). *Nature* **264,** 349–350; Giraudeau, A., Fan, F.-R. F., and Bard, A. (1980). *J. Am. Chem. Soc.* **30,** 5137–5142; Tributsch, H., and Calvin, M. (1971). *Photochem. Photobiol.* **14,** 95–112; Miyasaka, T., Watanabe, T., Fujishima, A., and Honda, K. (1980). *Photochem. Photobiol.* **32,** 217–212.
6. Kampas, F. J., Yamashita, K., and Fajer, J. (1980). *Nature* **284,** 40–42; Fujihira, M., Sakomura, M., and Kamei, T. (1989). *Thin Solid Films* **180,** 43–50.
7. Miyasaka, T., Watanabe, T., Fujishima, A., and Honda, K. (1979). *Nature* **277,** 638–640.
8. Oesterhelt, T., and Stoeckenius, W. (1974). *Method Enzymol.* **31,** 667–678.
9. Birge, R. R., Cooper, T. M., Lawrence, A. F., Masthay, M. B., Vasilakis, C., Zhang, C.-F., and Zidovetzki, R. (1989). *J. Am. Chem. Soc.* **111,** 4063–4074.
10. Birge, R. R. (1990). *Ann. Rev. Phys. Chem.* **41,** 683–733.
11. Sheves, M., Albeck, A., Friedman, N., and Ottolenghi, M. (1986). *Proc. Natl. Acad. Sci. USA* **83,** 3262–3266.
12. El-Sayed, M. A. (1992). *Acc. Chem. Res.* **25,** 279–286.
13. Hong, F. T. (1994). *ACS Adv. Chem. Ser.* **235,** 531–560; Dancshazy, Z., and Karvaly, X. (1976). *FEBS Lett.* **72,** 136–138; Singh, K., Korenstein, R., Lebedeva, H., and Caplan, S. R. (1980). *Biophys. J.* **31,** 393–401.
14. Majima, Y., Naruse, H., and Iwamoto, M. (1992). *Thin Solid Films* **210 / 211,** 82–85.
15. Nagy, K. (1978). *Biochem. Biophys. Res. Commun.* **85,** 383–390.
16. Varo, G. (1981). *Acta Biol. Acad. Sci. Hung.* **32,** 301–310.
17. Furuno, T., Takimoto, K., Kouyama, T., Ikegami, A., and Sasabe, H. (1988). *Thin Solid Films* **160,** 145–151.
18. Trissl, H.-W. (1990). *Photochem. Photobiol.* **51,** 793–818.
19. Trissl, H.-W., Gartner, W., and Leibl, W. (1989). *Chem. Phys. Lett.* **158,** 515–518.
20. Simmeth, R., and Rayfield, G. W. (1990). *Biophys. J.* **57,** 1099–1101.
21. Groma, G. I., Szabo, G., and Varo, G. (1984). *Nature* **308,** 557–558.
22. McIntosh, A., and Boucher, F. (1991). *Biochim. Biophys. Acta* **1056,** 149–158.
23. Der, A., Szaraz, S., and Keszthelyi, L. J. (1992). *Photochem. Photobiol. B: Biol.* **15,** 299–306.
24. Miyasaka, T., and Koyama, K. (1991). *Chem. Lett.* 1645–1648.

25. Miyasaka, T., Watanabe, T., Fujishima, A., and Honda, K. (1978). *J. Am. Chem. Soc.* **100,** 6657–6665.
26. Hwang, S.-B., Korenbrot, J. I., and Stoeckenius, W. (1977). *J. Membr. Biol.* **36,** 115–135.
27. Miyasaka, T., Maekawa, Y., and Koyama, K. (1989). *Thin Solid Films* **180,** 73–83.
28. Miyasaka, T., and Koyama, K. (1992). *Thin Solid Films* **210 / 211,** 146–149.
29. Kononenko, A., Lukashev, E., Chamorovsky, S. K., Maximychev, A. V., Timashev, S. F., Chekulaeva, L. N., Rubin, A. B., and Paschenko, V. (1987). *Z. Biochim. Biophys. Acta* **892,** 56–67.
30. Liu, S. Y. (1990). *Biophys. J.* **57,** 943–950.
31. Iwamoto, M., and Majima, Y. (1988). *Jpn. J. Appl. Phys.* **27,** 721–275.
32. Yamaguchi, N., Jimbo, Y., Arai, M., and Koyama, K. (1993). *FEBS Lett.* **324,** 287–292.
33. Koyama, K., Yamaguchi, N., and Miyasaka, T. (1994). *Science* **265,** 762–765.
34. Miyasaka, T., Itoh, I., and Koyama, K. (1992). *Science* **255,** 342–344.
35. Miyasaka, T., and Koyama, K. (1993). *Appl. Opt.* **32,** 6371–6379.

Hole-Burning Spectroscopy of Dye-Doped Langmuir-Blodgett Films

M. ORRIT and J. BERNARD

Centre de Physique Moléculaire Optique et Hertzienne, C.N.R.S. et Université Bordeaux I, 351 Cours de la Libération, F-33405 Talence Cedex, France

13.1. Introduction

Optical methods are traditionally applied to the study of Langmuir-Blodgett (LB) films. Langmuir and Blodgett were the first to prepare ultrathin films of fatty acid soaps, to transfer them to solid substrates, and to investigate optically their thickness and refraction indices.[1,2] Later, Kuhn and colleagues used spectroscopic methods extensively to study LB films.[3] They investigated the orientation of dye molecules in the films by linear dichroism both in transmission and reflection,[4] and examined the influence of aggregation on absorption and luminescence spectra.[5] They also devised elegant optical experiments to test Förster's theory of energy and electron transfer, and to study the spontaneous emission of a molecule in the vicinity of an interface.[3] More recently, the potential of the LB technique to assemble molecules in a planned way drew the attention of many researchers. This led to numerous fundamental studies of LB films by the various methods of physical chemistry,[6] among them optical spectroscopy.[7–11]

The field of optics and spectroscopy has been revolutionized by the advent of tunable lasers. Nonlinear optical properties of matter are studied in a number of laboratories; LB films are particularly interesting systems for nonlinear optics because of the possibility that active molecules can be organized and noncentrosymmetric arrangements can be prepared for second harmonic generation[11–15] as can novel materials for three-wave

ISBN 0-12-523485-6

frequency mixing.[16] Photochemical and photophysical processes are studied with short light pulses in a variety of systems, recently joined by LB films.[7,8]

The present chapter is concerned with other types of optical experiments in which lasers are indispensable, namely, high-resolution spectroscopy. The high-resolution techniques use intense laser waves with very well-defined frequencies to select and drive efficiently the desired transitions of atoms, ions, or molecules, often isolated in vapors or molecular beams. High-resolution techniques can also be applied to solids (such as LB films), provided their temperature is low enough.[17] At low temperatures, the absorption band of a solid solution consists of a broad superposition of the much narrower homogeneous lines of individual solute molecules, shifted at random by microscopic disorder in the surrounding solvent cage: The absorption band is inhomogeneously broadened. By illuminating the sample with a monochromatic laser, only resonant molecules are excited. After repeated transitions through the excited state, permanent changes eventually occur either in the molecular structure or in the structure of the solvent cage: A narrow photochemical or photophysical hole has been burnt in the absorption spectrum (Fig. 13.1). Since the system is kept at low temperature, these changes are practically permanent on the timescale of the measurement. Other methods have been developed to display narrow homogeneous structures in inhomogeneous bands, such as the spectroscopy of single molecules,[18] but the hole-burning method is the most useful because it is easy and general.[19]

Hole-burning spectroscopy provides information about the homogeneous linewidth, which is the inverse phase lifetime.[17] This width is therefore sensitive to the lifetime of the excited state and provides information about energy transfer, for example, in biological systems.[20-22] The hole also probes the dynamics (closely related to the structure) of the surrounding matrix, both at short times[23] (dephasing from residual thermal motion in the nanosecond range) and at longer times[24,25] (spectral diffusion from motion on timescales from a few nanoseconds to a second and beyond).

A hole is also a narrow spectral feature that will be much more sensitive to external perturbations than the inhomogeneous band in which it is burned (the ratio of inhomogeneous to homogeneous widths is often larger than 10,000). Even modest electric[26-28] or magnetic[29] fields, mechanical pressure,[30] or stresses can broaden and shift spectral holes.

Finally, hole-burning has interesting potential applications to the optical storage and processing of information,[19] as was recently demonstrated by experiments on holographic storage[31] and computation on stored data using combined changes in the frequency and applied electric field.[32]

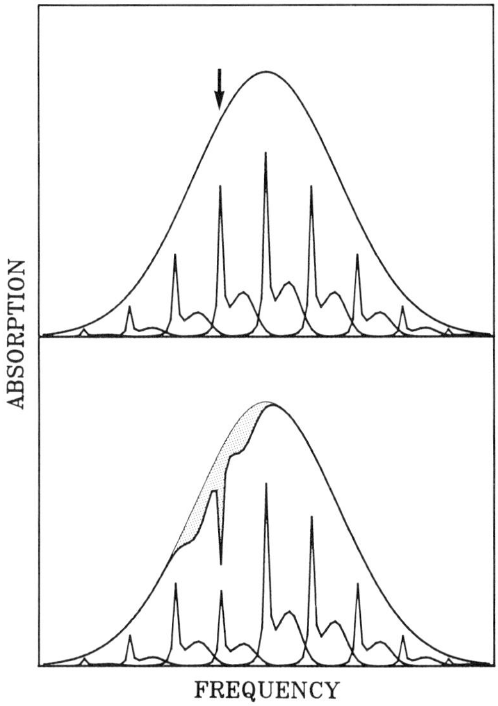

FIG. 13.1. Schematic diagram of an inhomogeneously broadened absorption band, arising from the superposition of the homogeneous absorption bands of single molecules (top). The homogeneous absorption band consists of a sharp zero-phonon line and of a broad phonon wing at higher frequency. If we illuminate at a given frequency (arrow), the resonant molecules are selectively excited and may undergo irreversible transformations: Some absorption strength has been removed from the spectrum (shaded area on the bottom diagram) around the irradiation frequency; a permanent hole has been burned in the absorption spectrum. Hereafter, only the narrow zero-phonon part of the hole is considered.

Since the time of an early study on perylene included in a Cd-arachidate multilayer,[33] little work had been devoted to hole burning in LB films. A few years ago, we started to apply hole burning to the study of dye-doped LB films.[34] Our main results and conclusions since then are summarized in this chapter, which should stimulate further work in this field. As the example of our work will show, hole burning can be used to characterize LB films, but the specific features of LB films such as their partially ordered structure, important surface effects, and the possibility to organize them in a planned way make them interesting matrices for hole burning and its potential applications.

13.2. Experimental Setup

The experimental techniques and procedures in high-resolution spectroscopy may not be familiar to all readers of this chapter. Therefore, to allow other authors to reproduce our results, the experimental setup is described in some detail, with emphasis on the features we deliberately devised for the study of LB films.

13.2.1. Sample Preparation

Most of the LB films used in this study were Cd arachidate assemblies, doped with minute concentrations of dye (typically 1 dye molecule for 500 matrix molecules), which did not alter significantly their surface pressure-area isotherms. They were prepared and transferred according to the standard conditions[3] (surface pressure of 25 mN/m, with a transfer speed of a few centimeters per minute) in the laboratory of Dr. Möbius in the Max-Planck-Institute at Göttingen, Germany. The dye molecules used in this study were chosen for their spectral range of absorption (580 to 640 nm, the tuning range of our laser) and for their good fluorescence yield, in order to detect holes in the fluorescence excitation spectrum. The non-amphiphilic dye molecules resorufin (from Aldrich) and porphine (from Sigma) are commercially available. The amphiphilic cyanine S20 was synthetized by Sondermann some 20 years ago in Göttingen. In the case of the study of amphiphilic porphyrin[35,36] in Section 13.3.1, the samples were prepared in Moscow by Samoilenko. The concentration of the dye was higher in these samples (1 : 100) and allowed intermolecular interaction effects to appear.

All multilayers were deposited according to the Y procedure (contacts between hydrophilic and hydrophobic parts excluded), which gives the maximum stability for strongly amphiphilic molecules. Due to the low dye concentration, the number of dye molecules in a single dye-doped monolayer would have been too small for a reasonable signal-to-noise ratio. Therefore, we used a stack of about 30 thin glass slides (see Fig. 13.2) covered on both sides by the LB film to improve the signal. The slides were tilted to Brewster incidence angle to ensure the same illumination (linearly polarized) to all molecules.

Multilayers and monolayers with higher dye concentrations were characterized at room temperature by conventional absorption, reflection,[37] and fluorescence spectroscopy before hole burning. In all cases, broad bands were obtained as expected for inhomoeneously broadened spectra. The overlap between absorption and emission 0–0 bands was good, a necessary condition to observe spectral holes with reasonable contrast.

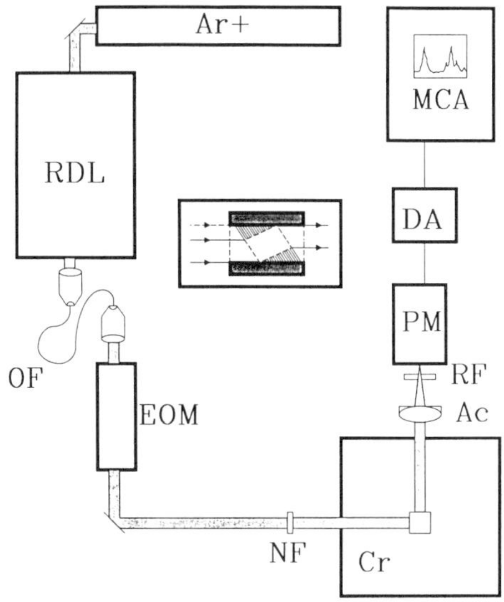

FIG. 13.2. Diagram of the setup used for hole-burning experiments. The high-resolution light source is a cw ring dye laser (RDL) pumped by an argon ion laser. The light beam is spatially filtered by a single-mode optical fiber (OF) and power stabilized by an electro-optical modulator (EOM). The beam, attenuated by a neutral filter (NF), illuminates the sample in the cryostat (Cr). The fluorescence from the sample is collected by a lens (Ac), ridded from stray exciting light by a red-pass glass (RF) and detected as photocounts by the photomultiplier tube (PMT) and discriminator amplifier (DA). The excitation spectrum is stored in a multichannel analyzer as the excitation frequency is scanned. The sample holder is sketched in the middle of the figure, with a number of glass or silica plates covered on both sides by the LB film under study. The plates are tilted at the Brewster incidence angle to avoid multireflections.

The experiments on the Stark effect[38,39] were done using a multilayer sandwiched between a transparent ITO electrode and an opaque aluminum electrode. The electrical contact on the aluminum was established using a thin (50-μm-diameter) gold thread with a droplet of silver lacquer.

13.2.2. Optical Setup

The diagram of the setup is shown in Fig. 13.2. The detection method chosen is fluorescence excitation because of its high sensitivity. The fluorescence signal appears on a dark background and is therefore best suited to samples with low optical densities, such as diluted LB films. (We estimate the optical density of our dilute monolayer and multilayer samples to be less than 0.01.) Detecting shallow holes in transmission spectra would require a modulation of the signal and lock-in detection, as we did in the case of the electric field effect.[39]

The excitation source was a commercial high-resolution ring dye laser (Coherent Radiation Model 699-21). The laser dye's fluorescence band is narrowed by a birefringent filter and two intracavity Fabry-Pérot etalons. An additional improvement of the resolution is obtained by locking the laser's frequency on an external etalon. In this way, short-term frequency jitter is reduced to about 1 MHz (3.10^{-5} cm^{-1}). In the study of most LB films, such a high resolution was superfluous. Cheaper continuous or pulsed lasers with broader lines (10^{-3} cm^{-1} or broader) could also be used to obtain valuable information, with the advantage of a larger scanning interval (ours was limited to 1 cm^{-1}).

The power needed for hole-burning experiments is usually very small, in the range of some microwatts or milliwatts per square centimeter, depending on the burning mechanism. This is similar to the power needed in those bulk systems where hole burning proceeds from hydrogen bond rearrangements.[40,41] Therefore, we assume this to be the most efficient burning mechanism for polar or ionic dyes imbedded in the hydrophilic part of the layer.

In hole-burning experiments, one compares absorption (or excitation) spectra before and after burning. Therefore, it is important to have very stable illumination conditions throughout the experiment. In fluorescence excitation measurements, signal fluctuations can arise from intensity fluctuations and from geometrical jitter of the exciting beam. To eliminate the latter effect, we filtered spatially the exciting beam by means of a birefringent single-mode optical fiber. Both ends of the fiber were cut at the Brewster incidence angle to avoid feedback into the ring dye laser. After the fiber, the intensity of the geometrically stable beam was stabilized by an electro-optic modulator (a commercial noise-eater, model Lass II from Conoptics). Finally, the laser beam was attenuated by calibrated neutral density filters. Burning and reading sequences were driven by a home-built electronic device.

13.2.3. Cryostat and Detection

The sample was placed in a helium flow and bath cryostat with three windows for optical measurements (from TBT-Air Liquide). The temperature could be varied from 1.7 K in liquid helium to 20 K and beyond. Two features were important for our work in LB films. First, the output window was rather larger for an efficient collection of fluorescence. Second, it was possible to use rather bulky samples (a few cubic centimeters), corresponding to a maximal covered area of about 100 cm^2.

The fluorescent light was filtered by a red-pass colored glass (Schott RG630 for resorufin and S20 dyes; RG645 or RG665 for porphyrins) instead of a monochromator. In this way, nearly the total intensity of the emission could be detected, whereas the entrance slit, the grating, and the frequence selectivity of a monochromator would have entailed large light losses. The phototube (RCA 31034 A02) was a red-sensitive tube with an AsGa cathode, operated in the photon-counting mode for the detection of low light fluxes. The signal was usually lower than 300,000 counts per second. The electric pulses were amplified and shaped by an amplifier-discriminator (EG & G Model 1182). The fluorescence signal was stored as a function of excitation frequency in a multichannel analyzer (Tracor Northern 7200) and processed by an IBM-compatible personal computer.

13.3. The Hole Width and Its Temperature Dependence

Hole-burning spectroscopy gives direct information about photochemistry and photophysics in dense matter at low temperatures. This direction of research would gain from the use of LB films as matrices, because they allow us to organize and bring together reactant molecules in a planned way. In the work presented here, we did not touch this subject, but instead we tried to determine the intrinsic properties of the hole, mainly its width, as a function of control parameters. As mentioned in the introduction, the hole width is connected to the low-temperature dynamics of the matrix. Other hole properties, such as the depth, the burning kinetics,[36] or the spectral distribution of the photoproduct (antihole),[40] would give more insight into the burning mechanism.

13.3.1. Energy Transfer

The homogeneous width is necessarily larger than the inverse lifetime T_1^{-1} of the excited state (Heisenberg's limit). For short lifetimes, broadening by dephasing and spectral diffusion is negligible and the hole width (twice the homogeneous width) gives T_1 directly. Energy transfer in complex molecular systems often reduces the lifetime of the donor considerably: The transfer times can be deduced from hole-burning spectroscopy, as was done in some biological antenna systems and reaction centers.[20–22]

A few years ago, we studied LB films doped with fairly high concentrations of an amphiphilic porphyrin.[35] Here, energy can be transferred only between porphyrin donors and acceptors within the inhomogeneous band (Fig. 13.3). It is much easier for a donor molecule in the center of the band to find acceptors than for a molecule in the red wing. Therefore, holes are

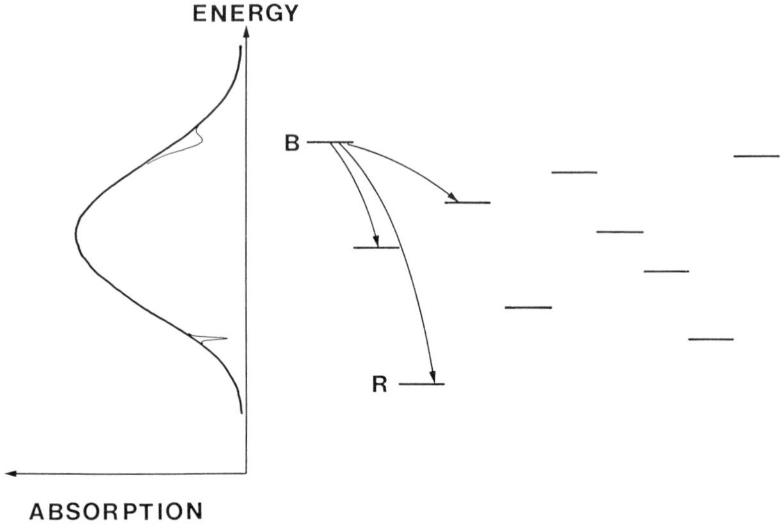

FIG. 13.3. Energy transfer between sites in the case of an inhomogeneously broadened absorption band. A donor molecule in the blue wing of the energy distribution easily finds energy acceptors to transfer efficiently. A molecule in the red wing finds no acceptor. Therefore, holes must be broader in the blue wing of the absorption band than in the red wing.

much broader in the center of the band than in the red wing (Fig. 13.4). This interpretation was confirmed by a recent study of the same system where the porphyrin concentration was varied.[42]

One question still unanswered concerns the nature of the transfer (nonresonant via phonon bands, or quasiresonant via the zero-phonon line). Comparison to energy transfer in nonresonant donor-acceptor systems should help to clear up this point. The study of such systems would also extend earlier studies of energy transfer in LB films by Kuhn and colleagues[3] to shorter transfer times, when transfer is much faster than fluorescence.

13.3.2. Low-Temperature Dynamics of LB Films

The structure and dynamics of amorphous systems (or glasses) present unique features compared to crystals. These features become particularly conspicuous at low temperatures. Having no single equilibrium configuration, glasses are frustrated and wander even at helium temperatures in a complex potential energy landscape with many local minima. Several dynamic properties of glasses (like specific heat or ultrasound attenuation)

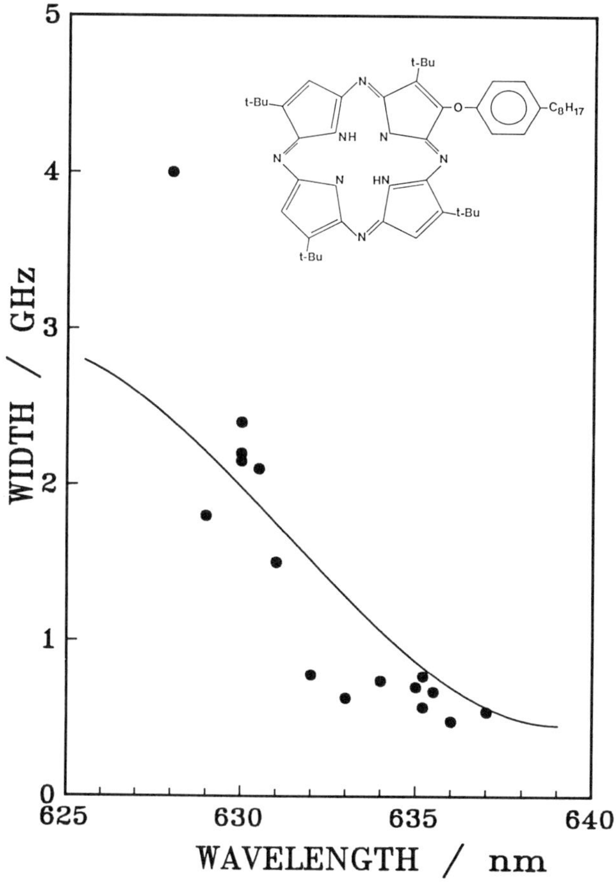

FIG. 13.4. Wavelength dependence of the width (●) of a spectral hole burned in the excitation spectrum of an LB film doped at fairly high concentration with porphyrins (1 dye molecule for 100 chains). The width increases suddenly below 632 nm, because energy transfer to lower lying acceptors starts to take place. The solid line is the absorption band profile.

have been described successfully[43,44] by modeling the transitions from one local minimum to another by a simple two-level system (TLS). Insight into the nature and properties of TLSs is desirable in its own right, but is also needed for the understanding of the dynamics of glasses and polymers below the temperature of their glassy transition. We started to investigate LB films as model systems for the low-temperature dynamics of glasses. As our results show, they present specific dynamic properties that have no equivalent in bulk systems.

Experiment and Results

The most striking of our findings is the large difference in hole width between a monolayer and a multilayer at helium temperature. This observation was made for several different systems (resorufin,[34] porphyrins,[35] cyanine dye[45,46]) and must therefore be regarded as a property of the LB matrix itself. Figure 13.5 presents two holes burned in the spectra of a multilayer and a monolayer of Cd-arachidate doped with porphin (the simplest of porphyrins, see the structure on Fig. 13.6). The ratio of widths

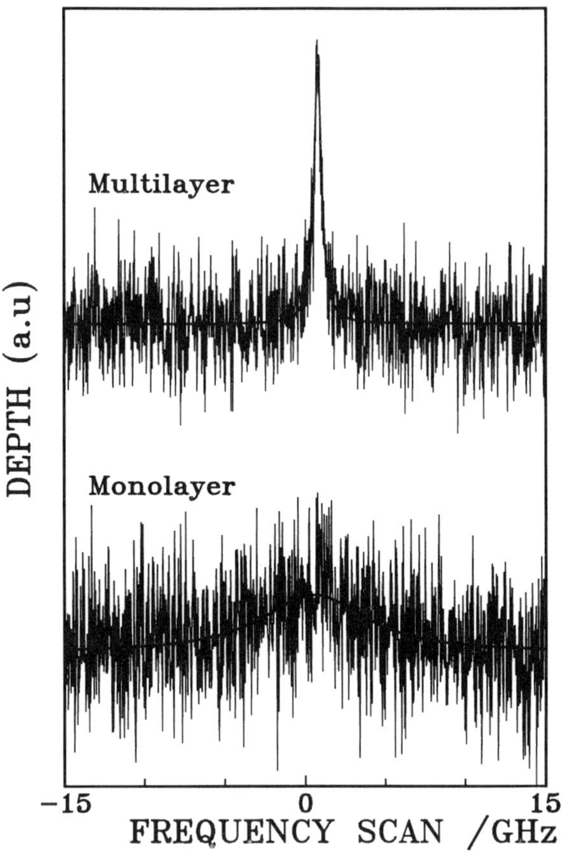

FIG. 13.5. Comparison of hole profiles for porphin (see structure on Fig. 13.6) imbedded at low concentration either in a monolayer LB film or in a multilayer film of Cd-arachidate at 2 K. The hole in the monolayer is about 20 times broader than in the multilayer. Similar observations were made for different dyes (although with somewhat less difference between the mono- and multilayers) and indicate that the broadening mechanism is linked to the LB matrix.

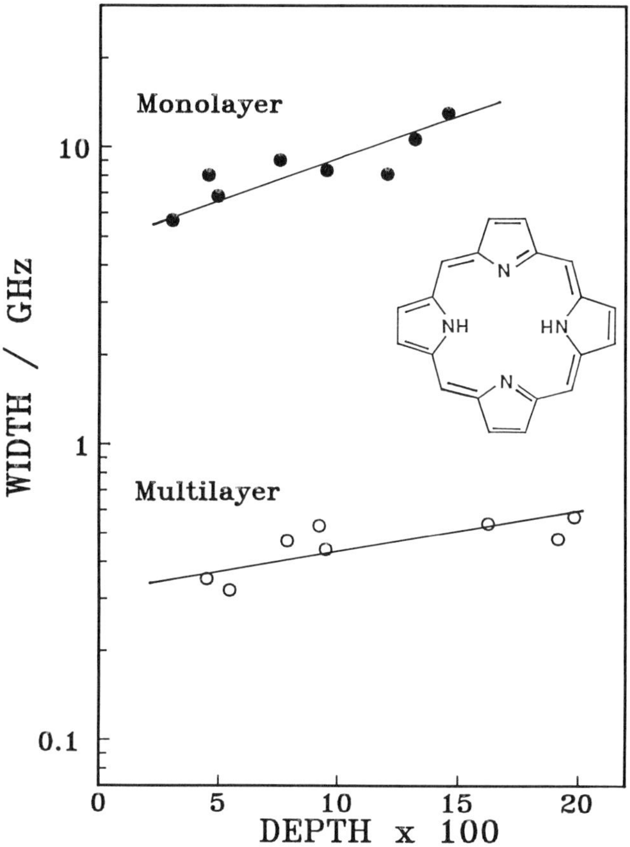

FIG. 13.6. The holes are broadened by saturation at high power and a high irradiation time. To obtain an intrinsic width, holes are burned at the lowest power possible and the hole width is plotted as a function of irradiation time, fluence, or hole depth (as here). Extrapolating to zero depth gives the intrinsic hole width free from saturation effects and determined only by the homogeneous processes: dephasing and spectral diffusion. These plots show immediately that the difference between mono- and multilayers is not a saturation effect.

is about 20. The difference is not an effect of saturation, as the study of the width as a function of hole depth testifies (Fig. 13.6).

We studied extensively the hole width (extrapolated to low-burning fluences) as a function of temperature for different positions of a dye-doped monolayer (S20 in Cd-arachidate) in a Cd-arachidate multilayer assembly of 11 monolayers on a glass substrate. In this way, the dynamics at a particular point in the assembly could be probed by S20 molecules.[46] The results are summarized in Fig. 13.7. We number the position of the

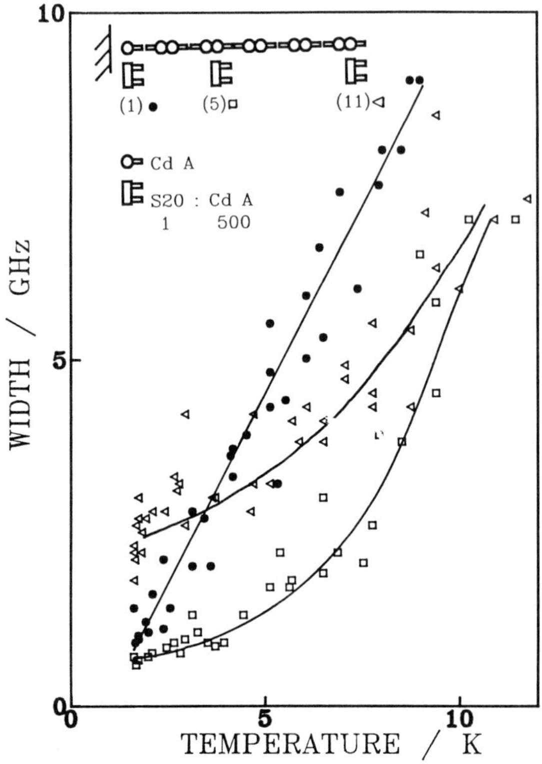

FIG. 13.7. Dependence of the extrapolated hole width (see Fig. 13.6) on temperature for the amphiphilic cyanine dye S20 (see dye structure on Fig. 13.9). The 11 Cd arachidate layers were deposited on a glass substrate. To prepare sample number *n*, the *n*'th Cd-arachidate layer was replaced by a Cd-arachidate layer doped with S20 at low concentration (1 dye molecule for 500 chains). The temperature dependences of the hole width in samples 1, 5, and 11 are markedly different, with the broadest holes at low temperature for sample 11 (as was observed for the monolayer sample of Fig. 13.5).

dye-doped monolayer from 1 (against the glass) to 11 (the last surface layer). The narrowest holes are obtained when the dye is included in the bulk of the multilayer, for example, in position 5; an identical dependence was measured for positions 7 and 9 (not shown in Fig. 13.7).

When the dye is against the glass backing, a steep temperature dependence is observed, but the holes are narrow at low temperature, as in most three-dimensional (3-D) samples. This result indicates that the glass/monolayer interface is more disordered than the hyrophilic interface between identical layers in a multilayer, a conclusion reached independently in infrared absorption studies.[47,48] Finally, broad holes and a slow temperature dependence are obtained when the dye is in the last surface monolayer. Since the last monolayer differs from the others only by its position, a lifetime effect is excluded: The observed broadening must arise from specific dynamics of the surface layer.

We attempted complementary experiments to examine the surface dynamics. The hole width did not change between the timescale of the hole-burning experiment (a few minutes) and several hours. This means that the broadening is either due to dephasing (motion faster than 10 ps) or to spectral diffusion on a timescale shorter than a few minutes. We varied the fatty acid in the last layer, from Cd-palmitate (16 C atoms) to Cd-lignocerate (24 C atoms), but differences in hole widths were within our experimental noise and no clear-cut conclusion could be reached.

Discussion and Interpretation

The main result of our study of the hole width as a function of dye position in a multilayer was the different temperature dependence of the hole width in the surface monolayer. Since this observation was made for different dyes, and for the same dye in different positions, the additional broadening of spectral holes in the surface monolayer can only arise from a specific motion of the molecules in this layer, which has no equivalent in the bulk of the assembly (surface dynamics). The temperature dependence of the hole width shows some saturation of the broadening effect between 2 and 5 K, in contrast to the usual width dependences in bulk systems, which are often power laws of the type T^{α}, with $\alpha \approx 1.3$ for a distribution of TLS and $a \approx 2$ for harmonic motion at high temperatures. Such a saturation is the expected behavior when the temperature is higher than the activation energy of all TLSs to which the dye molecule is coupled. Therefore, we think that the surface motion responsible for the observed broadening is due to TLSs with activation temperatures lower than 2 K. The timescale of the motion cannot be determined accurately from our experiments: It is probably slower than 10^{-11} s and certainly faster than 10^{2} s. The holes burned for position 10 (next to the last layer) show a somewhat lower width than for position 11, although the chromophore is at the same distance (27 Å) from the surface. This indicates that the molecule's hydrophobic chains are involved in the surface motion, because the chains' directions are the only difference between the two positions. However, the motion is not restricted to the chains of S20; otherwise the hole widths would be identical for positions 9 and 10.

We interpret our observations by means of the following picture: We assume that the surface motion involves the long saturated chains of the last (surface) monolayer. A faulty packing of the chains could lead to two-dimensional (2-D) dynamics of double-well configurations (TLSs) with low energy differences between the wells. Because of the partial interdigitation of the chains' ends between two adjacent monolayers, such a motion is blocked in the bulk of the multilayer, where the dynamics is 3-D.

Several techniques have been applied to the characterization of the surface layer of LB assemblies.[49-51] Infrared absorption spectra[46] clearly show the less ordered structure of the surface layer, but slow electron diffraction[49] and atomic force microscopy[51] point to a good positional order within the layer. The low-temperature dynamics that we observe could therefore be assigned to collective motions of several chains, involving both translational and rotational degrees of freedom. The large mass of the chains and the weak van der Waals interaction between them can explain the low activation energies observed.

13.4. The Hole as a Spectral Label

One of the interests of hole-burning spectroscopy is the high sensitivity of holes to external perturbations. These perturbations can be applied homogeneously (electric or magnetic fields, pressure etc.) to bulk materials. In that case the sample can be an LB multilayer. The external perturbation can also be applied to the surface of the sample. LB monolayers are especially adapted to study the effect of a surface perturbation, because all molecules are at the same distance from the surface.

13.4.1. Effect of the Immersion in Superfluid Helium

When a dye is included in the hydrophilic part of the surface layer, it lies some 23 to 30 Å away from the surface and is therefore sensitive even to small changes occurring across the interface. In the present experiment, a hole was burned and recorded, the sample being surrounded by a low-pressure helium atmosphere. The sample was later immersed in superfluid helium and the hole was recorded again. Finally, the hole was recorded a third time, after the liquid helium was evacuated. All three hole spectra are shown in Fig. 13.8. The hole of spectrum 2, recorded in the liquid, is shifted and broadened relative to the other holes. Strictly speaking, spectra 1, 2, and 3 should not be compared directly because they were recorded at different temperatures (1 and 3 at 2.6 K; 2 at 1.8 K). However, we determined that the temperature shift was negligible. The reverse cycle (burning the hole with an immersed sample and measuring in the gas atmosphere) leads to the opposite shift, but to a larger broadening, because the broadening effect of the immersion is partly compensated in

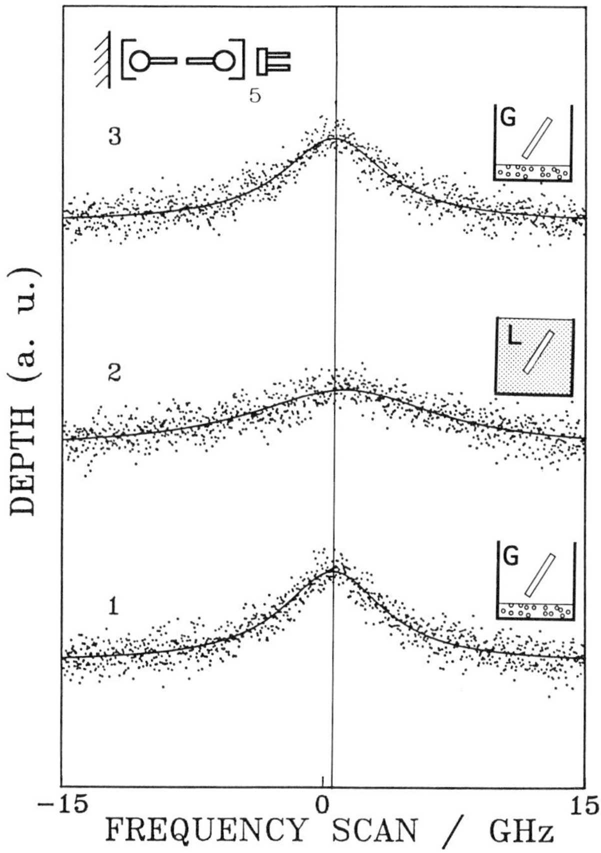

FIG. 13.8. Change of shape of a spectral hole burned in a surface monolayer under a gas helium atmosphere (bottom) upon immersion in superfluid helium (middle). The red shift and broadening of the hole are reversible (top). These spectral changes occur because the chromophores feel the change of environment across the saturated chains, 27 Å away. The effect of the immersion can be modeled by a change in the magnitude of the (static and dynamic) image charges in the interface between the LB film and the outer atmosphere.

Fig. 13.8 by a small narrowing due to cooling. Very similar observations were made using the dye resorufin.[34]

Several effects can contribute to the observed spectral change. The van der Waals interaction between chromophore and helium atoms changes when the molecules passes from the ground to the excited state; therefore, we expect a change in the absorption frequency when the sample is immersed in liquid helium. This effect can be conveniently modeled by replacing the liquid helium with a dielectric half-space and considering the interaction of the transition dipole moment of the molecule with its image in the half-space (this is the dynamic effect responsible for van der Waals forces). However, there is also a static image effect from the static charges in the neighborhood of the dye molecule. If we assume that the field from these charges is not correlated with the chromophore's axes, the static effect will lead only to a broadening of the hole upon immersion, while the dynamic effect always leads to a red shift (a molecule's transition is stabilized when polarizable units are added in its neighborhood). The dynamic image effect can be calculated easily by classical electrodynamics. We find for the frequency shift the result:

$$\Delta\nu = -\frac{3}{4\pi}\gamma_r\left(\frac{\lambda}{4\pi L}\right)^3\frac{\Delta\varepsilon_2}{(\varepsilon_1+\varepsilon_2)^2} \qquad (13.1)$$

where γ_r is the purely radiative width; λ the wavelength of the transition; ε_1, ε_2, and $\varepsilon_2 + \Delta\varepsilon_2$ the permittivities of the LB film, of the vacuum, and of liquid helium, respectively; and L is the distance from the chromophore to the surface. This expression differs from the erroneous result in Ref. 34 by a factor α due to the fact that in Ref. 34 only the 0–0 transition had been considered. The order of magnitude of the observed shift (630 $\pm$ 100 MHz) agrees with the estimate of Eq. (13.1), but a quantitative comparison is impossible because the unknown static shift must be taken into account.

13.4.2. Stark Effect

Instead of image fields, the molecules can be subject to a uniform external field (Stark effect), which will affect the spacing of electronic levels and the transition frequency. For noncentrosymmetric molecules, the shift of the transition frequency is given by the product $-\overrightarrow{\Delta\mu}\cdot\overrightarrow{E_{loc}}$ of the (local) applied field by the change in dipole moment of the molecule and its solvent shell upon excitation: The Stark effect is linear. For molecules that would be centrosymmetric in their isolated state, a linear Stark effect is nonetheless observed in disordered matrices,[28,52] due to the breaking of molecular symmetry by short-range forces from the solvent. With this latter effect, it would be possible to probe the local forces on centrosymmetric molecules in complex systems like LB films.

In the work discussed here, we used the noncentrosymmetric cyanine S20 in a multilayer assembly to measure the Stark effect.[38,39] A diagram of the multilayer assembly and of the structure of S20 is shown in Fig. 13.9. To study the orientation of the chromophores in the assembly, we prepared an alternating multilayer in which all dye molecules have the same orientation with respect to the electrodes. Figure 13.10 shows the change

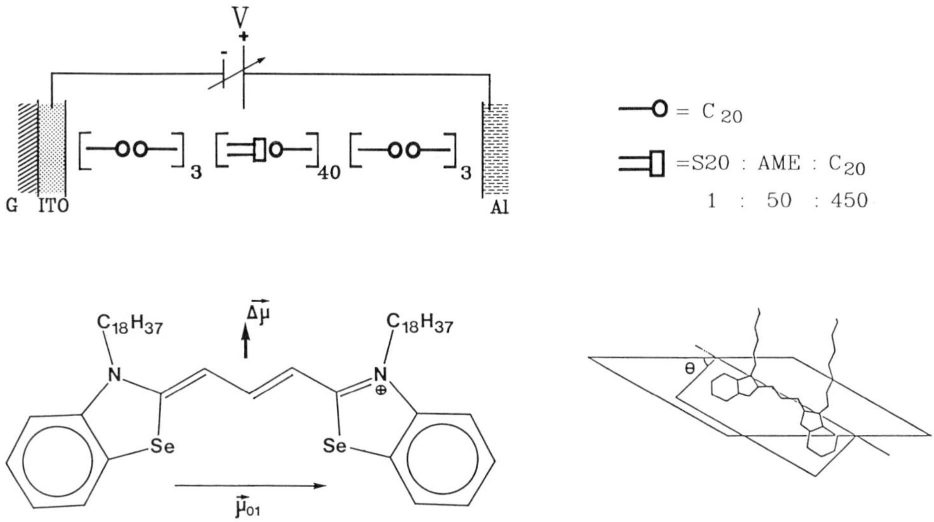

FIG. 13.9. Schematic diagram of the cell used for Stark effect measurements. The alternate LB assembly is of the type $(AB)_n$ with B being a pure Cd-arachidate layer and A a mixture of Cd-arachidate and arachidic methyl ester doped with the cyanine S20. All dye molecules in the sample have the same orientation with respect to the semitransparent ITO electrode and to the aluminum electrode. The shift of a molecule's resonance under an applied field is related to the orientation of its dipole moment change $\overrightarrow{\Delta\mu}$ with respect to the field.

of shape under electric field of a hole burned in the excitation spectrum of the LB film. The solid lines are Lorentzian fits to the hole profile drawn for measuring shift and broadening of the hole. These quantities are shown in Fig. 13.11 as functions of the applied field. The broadening is an even function of the field, which behaves linearly for high field (linear Stark effect), while the shift due to the asymmetry of the sample is an odd function of the field. The shift is much smaller than the broadening, indicating that the distribution of $\overrightarrow{\Delta\mu}$ directions is nearly symmetrical with respect to the layer plane.

The long chromophore axis carries the transition dipole moment and is known from linear dichroism measurements to be parallel to the surface within a few degrees.[37] The average orientation of the short chromophore axis, which carried $\overrightarrow{\Delta\mu}$, is obtained by comparing the hole shift and broadening in the LB film to the much larger hole broadening in bulk Cd-arachidate. We infer an average absolute angle $\langle|\theta|\rangle$ of 18 deg between $\overrightarrow{\Delta\mu}$ and the layer plane, whereas the average bias angle related to the hole shift is much smaller, about 2 deg.

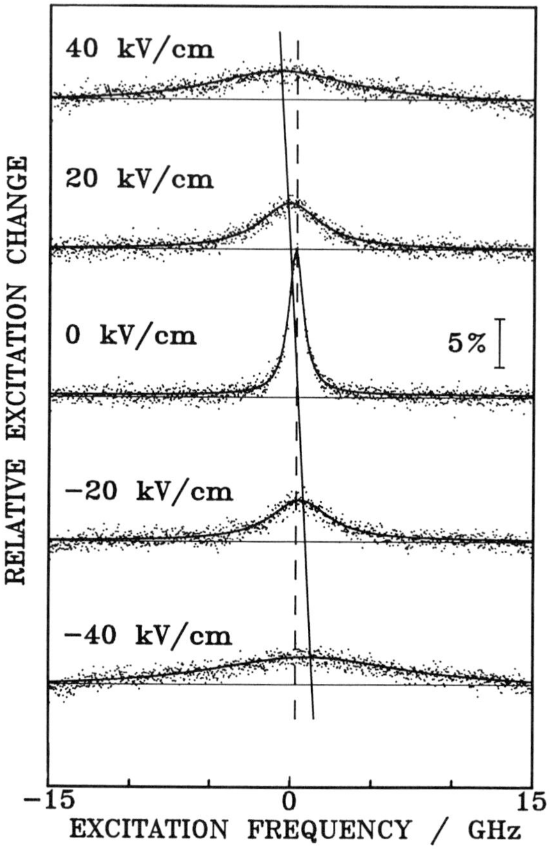

FIG. 13.10. Shape of a hole burned in the excitation spectrum of S20 in the alternate multilayer of Fig. 13.9 for different applied field strengths. The hole broadens considerably, but the shift is much smaller, indicating that the molecules' $\overrightarrow{\Delta\mu}$ is nearly perpendicular to the film surface. The solid lines are Lorentzian fits to the hole profile, drawn to evaluate width and shift of the holes.

The cyanine chromophore is thus lying flat on the LB film surface, with larger disorder for its short axis than for its long axis. This result agrees with a simple picture of the multilayer, where the ionic chromophores tend to lie flat on the surface but can be tilted by Cd ions or residual water molecules. The same disorder leads to a larger tilt angle for the short axis than for the long one. The smallness of the bias angle $\langle \theta \rangle$ as compared to the width of the angular distribution was quite unexpected. It means that the direction of the chains of the amphiphilic dye has little influence over

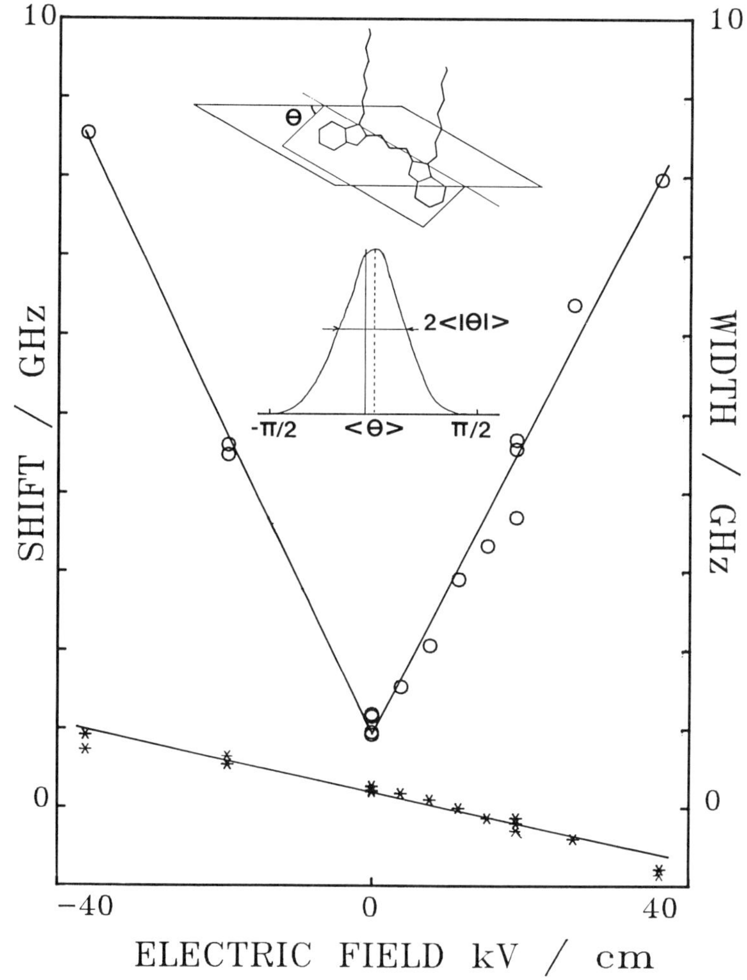

FIG. 13.11. Dependence on electric field of the hole width (open circles) and shift (asterisks). The broadening of the hole is related to the width of the distribution of angles θ between the dipole moment change $\overrightarrow{\Delta\mu}$ and the surface plane. The hole shift is related to the bias in the distribution, or average angle $\langle \theta \rangle$. The smallness of the shift indicates that the distribution of θ is nearly centered around $\theta = 0$ deg, and the comparison of the hole broadening with that in a bulk sample gives a width of the distribution of 30 to 40 deg.

the orientation of the short chromophore axis, and presumably on the general structure of the hydrophilic region. The chains are long and flexible and interact weakly with neighboring matrix chains. They must accommodate the hard structure build by strong polar and ionic forces in the hydrophilic interface between LB layers.

13.5. Conclusion and Prospects

The aim of this chapter was to show the power and interest of the high-resolution spectroscopy of LB films. We have seen that the width of a spectral hole provides information about the relaxation processes affecting the excited state lifetime, and about the dynamics of the matrix surrounding the molecule. Spectral holes may also be used as sensitive probes of external perturbations. In particular, the Stark effect on holes allows more reliable measurements of the dipole moment changes than the conventional electrochromic method,[53,54] in which the change of broadbands under an alternating field is measured. The electrochromic spectra, being derivatives of the absorption spectra, are difficult to interpret when several broadbands overlap. Most of the results described here would have been very difficult to obtain by other methods; they are interesting for the characterization of LB films, and therefore for their potential applications, but LB films are interesting systems in which to study such fundamental problems as the dynamics and structure of amorphous solids, or energy transfer in complex molecular systems.

We can propose some possible directions to extend this study:

1. Energy transfer can proceed toward nonresonant or quasiresonant acceptors. Because they allow control of the concentration and distance of acceptors, LB films are unique systems for energy transfer studies. The hole-burning method has the potential to test energy transfer even when it is much faster than fluorescence.
2. The linear Stark effect on centrosymmetric molecule is connected to the way the molecule's symmetry is broken by the matrix. A current question is: Can this symmetry breaking be described by a microscopic field from the matrix? Hole-burning studies could answer this question and complement the measurements by nonlinear optical methods.
3. The dynamics we have discovered in the last surface layer has no equivalent in other systems. It deserves closer examination as a model for the dynamics of amorphous systems. The activation temperature and the timescale of the motion must be determined in order to test the proposed model of a collective chain motion.
4. LB films could have applications in the optical processing and storage of information. The possibility of bringing about different molecules to organize and cooperate like the pieces of a machine in complex multilayers could have implications for the photon gating or electric field gating of the hole-burning process, a prerequisite to practical uses of hole-burning memories. Finally, fast electro-optical modula-

tion of light[55,56] could be achieved with low applied voltages (less than 1 V) in LB films with deep spectral holes.

Acknowledgments

We are glad to thank Dr. D. Möbius for his help in devising, preparing, and characterizing the LB films, and Prof. R. I. Personov for common work on energy transfer and the Stark effect.

References

1. Langmuir, I. (1920). *Trans. Faraday Soc.* **15,** 62.
2. Blodgett, K. B. (1935). *J. Am. Chem. Soc.* **57,** 1007.
3. Kuhn, H., Möbius, D., and Bücher, H. (1972). In: *Physical Methods of Chemistry, Vol. 1, Part IIIB* (A. Weissberger, and A. Rossiter, eds.). Wiley, New York, Ch. VII.
4. Grüniger, H., Möbius, D., and Meyer, H. (1983). *J. Chem. Phys.* **79,** 3701.
5. Möbius, D., and Kuhn, H. (1979). *Israel J. Chem.* **18,** 375 and references therein.
6. Ulman, A. (1991). *An Introduction to Ultrathin Organic Films*. Academic Press, New York, p. 339.
7. Sluch, M. I., Vitukhnowsky, A. G., Warren, J. C., and Petty, M. C. (1992). *Thin Solid Films* **210 / 211,** 211.
8. Tran-Thi, T. H., Lipskier, J.-F., Simoes, M., and Palacin, S. (1992). *Thin Solid Films* **210 / 211,** 150.
9. Nagamura, T., Toyozawa, K., Kamata, S., and Ogawa, T. (1992). *Thin Solid Films* **210 / 211,** 332.
10. Lukas, B., Lovett, D. R., and Silver, J. (1992). *Thin Solid Films* **210 / 211,** 213.
11. Bosshard, C., Küpfer, M., Flörsheimer, M., Borer, T., Günter, P., Tang, Q., and Zahir, S. (1992). *Thin Solid Films* **210 / 211,** 198.
12. Nagamura, T., Sakaguchi, H., and Matsuo, T. (1992). *Thin Solid Films* **210 / 211,** 160.
13. Verbiest, T., Samyn, C., and Persoons, A. (1992). *Thin Solid Films* **210 / 211,** 188.
14. Dentan, V., Blanchard-Desce, M., Palacin, S., Ledoux, I., Barraud, A., Lehn, J.-M., and Zyss, J. (1992). *Thin Solid Films* **210 / 211,** 221.
15. Kajikawa, K., Anzai, T., Shirota, K., Takazoe, H., and Fukuda, A. (1992). *Thin Solid Films* **210 / 211,** 699.
16. Kajzar, F., Girling, I. R., and Peterson, I. R. (1988). *Thin Solid Films* **160,** 209.
17. Friedrich, J., and Haarer, D. (1984). *Angew. Chem.* **96,** 96; (1984). *Int. Ed. Engl.* **23,** 113.
18. Orrit, M., Bernard, J., Zumbusch, A., and Personov, R. I. (1992). *Chem. Phys. Lett.* **196,** 595 and references therein.
19. Moerner, W. E., ed. (1988). *Persistent Spectral Hole-Burning: Science and Applications.* Springer Verlag, Berlin.
20. Johnson, S. G., and Small, G. J. (1989). *Chem. Phys. Lett.* **155,** 371.
21. van der Laan, H., Schmidt, T., Visschers, R. W., Visscher, K. J., van Grondelle, R., and Völker, S. (1990). *Chem. Phys. Lett.* **170,** 231.
22. Johnson, S. G., Tang, D., Jankowiak, R., Hayes, J. M., Small, G. J., and Tiede, D. M. (1990). *J. Phys. Chem.* **94,** 5849.

23. Silbey, R., and Kassner, K. (1987). *J. Luminescence* **36,** 283.
24. Littau, K. A., and Fayer, M. D. (1991). *Chem. Phys. Lett.* **176,** 551.
25. Müller, K. P., and Haarer, D. (1991). *Phys. Rev. Lett.* **66,** 2344.
26. Samoilenko, V. D., Rasumova, N. V., and Personov, R. I. (1982). *Opt. Spectrosc.* (*USSR*) **52**(4), 346.
27. Meixner, A. J., Renn, A., Bucher, S. E., and Wild, U. P. (1986). *J. Phys. Chem.* **90,** 6777.
28. Kador, L., Haarer, D., and Personov, R. I. (1987). *J. Chem. Phys.* **86,** 5300.
29. van den Berg, R., van der Laan, H., and Völker, S. (1987). *Chem. Phys. Lett.* **142,** 535.
30. Sesselmann, T., Richter, W., Haarer, D., and Morawitz, H. (1987). *Phys. Rev. B* **36,** 7601.
31. Bernet, S., Kohler, B., Rebane, A., Renn, A., and Wild, U. P. (1992). *J. Opt. Soc. Am. B* **9,** 987.
32. Wild, U. P., Renn, A., de Caro, C., and Bernet, S. (1990). *Appl. Opt.* **29,** 4329.
33. Bogner, U., Röska, G., and Graf, F. (1983). *Thin Solid Films* **99,** 257.
34. Orrit, M., Bernard, J., and Möbius, D. (1989). *Chem. Phys. Lett.* **156,** 233.
35. Bernard, J., Orrit, M., Personov, R. I., and Samoilenko, A. D. (1989). *Chem. Phys. Lett.* **164,** 377.
36. Talon, H., Orrit, M., and Bernard, J. (1990). *Chem. Phys.* **140,** 177.
37. Orrit, M., Möbius, D., Lehmann, U., and Meyer, H. (1986). *J. Chem. Phys.* **85,** 4966.
38. Orrit, M., Bernard, J., Mouhsen, A., Talon, H., Möbius, D. and Personov, R. I. (1991). *Chem. Phys. Lett.* **179,** 232.
39. Bernard, J., Talon, H., Orrit, M., Möbius, D., and Personov, R. I. (1992). *Thin Solid Films* **217,** 178.
40. Shu, L., and Small, G. J. (1992). *J. Opt. Soc. Am. B* **9,** 724.
41. Kokai, F., Tanaka, H., Brauman, J. I., and Fayer, M. D. (1988). *Chem. Phys. Lett.* **143,** 1.
42. Romanovskii, Y., Personov, R. I., Samoilenko, A. D., Holliday, K., and Wild, U. P. (1992). *Chem. Phys. Lett.* **197,** 373.
43. Anderson, P. W., Halperin, B. I., and Varma, C. M. (1972). *Phil. Mag.* **25,** 1.
44. Philips, W. A. (1972). *J. Low Temp. Phys.* **7,** 351.
45. Bernard, J., and Orrit, M. (1990). *J. Luminescence* **45,** 70.
46. Orrit, M., Bernard, J., Talon, H., and Mouhsen, A. (1992). *Thin Solid Films* **210 / 211,** 141.
47. Arndt, T., and Bubeck, C. (1988). *Thin Solid Films* **159,** 443.
48. Blaudez, D., Buffeteau, T., Desbat, B., Orrit, M., and Turlet, J.-M. (1992). *Thin Solid Films* **210 / 211,** 648.
49. Vogel, V., and Wöll, C. (1986). *J. Chem. Phys.* **84,** 5200.
50. Böhm, C., Seitz, R., and Riegler, H. (1989). *Thin Solid Films* **178,** 511.
51. Meyer, E., Howald, L., Overney, R. M., Heinzelmann, H., Frommer, J., Güntherodt, H.-J., Wagner, T., Schier, H., and Roth, S. (1991). *Nature* **349,** 398.
52. Bogner, U., Schätz, P., Seel, R., and Maier, M. (1983). *Chem. Phys. Lett.* **102,** 267.
53. Blinov, L. M., Ivashchenko, A. V., Palto, S. P., and Yudin, S. G. (1988). *Thin Solid Films* **160,** 271.
54. Grewer, G., and Lösche, M. (1992). *Thin Solid Films* **210 / 211,** 670.
55. Kanaan, Y., Attenberger, T., Bogner, U., and Maier, M. (1990). *Appl. Phys. B* **51,** 336.
56. Hartmannsgruber, N., Bogner, U., and Maier, M. (1991). *Opt. Quant. Electron.* **23,** 361.

Dynamic Processes

Evaluation of a Transfer Process for Langmuir-Blodgett Films by Means of a Quartz-Crystal Microbalance

YOSHIO OKAHATA, KATSUHIKO ARIGA, and KENTARO TANAKA

Tokyo Institute of Technology, Department of Biomolecular Engineering, Nagatsuda, Midori-ku, Yokohama 227, Japan

14.1. Introduction

Langmuir-Blodgett (LB) films have been widely studied as a highly oriented organic thin film. The characterization of monolayers on a subphase and LB films after deposition on a substrate has been carried out by various methods such as FTIR or fluorescence spectroscopy, x-ray diffraction, ellipsometry, and an x-ray photoelectron process.[1-4] However, the *in situ* evaluation of LB films during a dipping process has not been fully studied because of the absence of suitable techniques to analyze transfer processes such as a transfer ratio for monolayers. Recently, we observed that a transfer mass on a substrate can be determined directly and quantitatively from frequency changes on a nanogram level when a quartz-crystal microbalance (QCM) is used as a substrate for the vertical dipping or horizontal lifting methods of LB films.[5-9] QCMs are known to be very sensitive mass measuring devices because of the resonance frequency changes upon the deposition of a given mass on the QCM plate.[10]

Two methods for deposition of LB films on a substrate are known: the conventional vertical dipping method and the horizontal lifting method. The former has been commonly used since the pioneering work of Blodgett[11] and a transfer ratio of monolayers can be estimated from a moving area of a barrier on a trough. The latter begun as an X-type deposition method: A substrate is touched horizontally to monolayers surrounded by barriers, then lifted into the air, and one layer is deposited for each cycle.[12-15] Recently, a new Y-type horizontal lifting method has

ISBN 0-12-523485-6

been reported[16–19] in which a substrate is lowered and lifted horizontally from the usual monolayer without being surrounded by barriers. Therefore, two monolayers are deposited for each cycle. Horizontal lifting methods are reported[16–19] to be applicable for transferring lipid monolayers, and the structures of the obtained LB films are very well oriented compared to those obtained by the conventional vertical dipping method. However, the transfer mechanism or the transfer ratio of the horizontal lifting method has not been fully studied because of difficulties with the *in situ* measurements of horizontal transfer processes.[16–19]

In this chapter, we describe the evaluation of the transfer processes of LB films (vertical dipping and horizontal lifting processes) by using a QCM as a substrate. In the conventional vertical dipping method, water is incorporated with lipid monolayers from a water phase during the "rising-up" process of a substrate and the incorporated water gradually evaporates through the layers when exposed to air.[5] In the horizontal lifting process, water was hardly incorporated into the layers of LB films during the transfer process, and transfer ratios of various LB films on the substrate could be determined in a nanogram level from frequency changes. The transfer mechanisms of the Y-type and X-type horizontal lifting methods are also discussed from the transfer ratio.

14.2. Quartz-Crystal Microbalances

An AT-cut, 9-MHz quartz-crystal plate (0.567 cm^2), on each side of which a 0.159-cm^2 Ag electrode has been deposited, and a homemade oscillator designed to drive the quartz at its resonance frequency were employed for frequency measurements.[5–8] The QCM was driven at 5-V DC and the frequency of the vibrating quartz was measured with an Iwatsu frequency counter (Model SC 7201) attached to the microcomputer system (NEC, Model PC 8801). The following equation has been obtained for the AT-cut shear mode QCM[10,20]

$$\Delta F = \frac{-2F_0^2}{A\sqrt{\rho_q \mu_q}} \Delta m \qquad (14.1)$$

where ΔF is the measured frequency shift (Hz), F_0 the parent frequency of the QCM (9×10^6 Hz), Δm the mass change (g), A the electrode area (0.159 cm^2), ρ_q the density of quartz (2.65 g cm^{-3}), and μ_q the shear modulus (2.95×10^{11} dyn cm^{-2}). Calibration of the QCM used in our experiments showed that a frequency change of 1 Hz corresponded to a mass increase of 0.80 ± 0.01 ng on the electrode of the QCM[5–9]:

$$\Delta m = (0.80 \pm 0.01) \times 10^{-9}\, \Delta F \qquad (14.2)$$

14.3. Vertical Dipping Depositions of LB Films on a QCM Plate

A benzene solution of stearic acid was spread on Milli Q water containing 0.1 mM $CdCl_2$ (aqueous phase, pH 6.5) in a Teflon-coated trough. The QCM plate, whose surface is relatively hydrophilic (the contact angle against water is 60 deg), was lowered and raised vertically through the monolayer on a water surface, depositing two layers on each side of the QCM with each successive cycle. The QCM's frequency was measured each time the QCM was raised out of the water phase.

Typical time courses for the frequency change of the QCM in air are shown in Fig. 14.1. The QCM plate was lowered at point A and lifted out of the $(C_{17}H_{35}COO^-)_2Cd^{2+}$ monolayer-covered surface (surface pressure 20 mN/m) at point B four times, then lowered (point A′) and lifted out of the aqueous phase without deposition of a monolayer (point B′) at a rate of 100 mm/min. The frequency of the QCM gradually increased with time and reached a constant value after 15 min in air each time the QCM was lifted through the monolayer-covered subphase. The increased mass on the

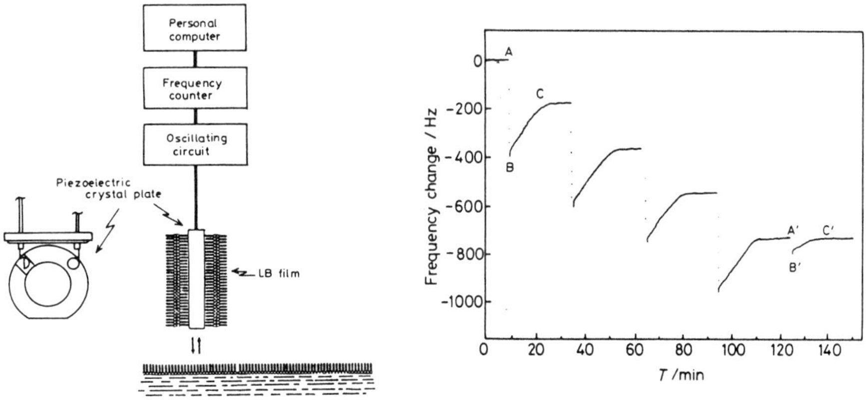

FIG. 14.1. Frequency changes of a QCM in air. The QCM was lowered (A) and raised (B) four times through the cadmium stearate monolayer on the water surface at 20 mN/m, then lowered (A′) and raised (B′) through the water phase with no surface monolayer, at a rate of 100 mm/min, at 20°C.

QCM with each cycle (M_{LB}) was calculated from the decrease of the frequency of 183 ± 3 Hz from points A to C, to be 232 ± 3 ng. This value was consistent with the theoretical mass of four dry monolayers of cadmium stearate (225 ng) calculated from the surface area per molecule in the monolayer (0.237 nm^2 from a π-A curve). The increase of the frequency between points B and C in Fig. 14.1 (the decrease of the mass) can be explained by evaporation of water from interlayers of the LB films. The amount of incorporated water M_w and its evaporation speed v_w were calculated from the frequency change and the initial slope of the curves between points B and C, respectively. Those values are summarized in Table 14.1 (run 1), together with the mass of the four dry deposited monolayers (M_{LB}) and the transfer ratio.

When a QCM with eight monolayers on each side was lowered at point A′ and lifted out of the aqueous solution with no monolayer on its surface, the frequency reverted almost to the starting value after the slow evaporation of a small amount of water (points B′ to C′ in Fig. 14.1; run 2, Table 14.1). This shows that the LB film is not removed from the QCM plate by dipping in water. The evaporated water (30 ng) was probably incorporated by swelling of the LB films,[8] and it represented only one-seventh of the amount incorporated when the QCM was dipped in a monolayer-covered water surface (209 ng). This shows that most of the water in LB films is from the water surface with monolayer.

When monolayers were deposited on the QCM at a low surface pressure (5 or 10 mN/m), both the amounts of lifted water and its evaporation

TABLE 14.1

CHARACTERIZATION OF LB FILMS DEPOSITED ON A QCM PLATE WITH A VERTICAL DIPPING PROCESS

Run	Surface Pressure mN/m	Dipping Speed (mm/min)	M_{LB}[a] (ng)	Transfer Ratio	M_w[b] (ng)	v_w[c] (ng/min)
1[d]	20	100	232	1.03	209	15.7
2[e]	—	100	7	—	30	4.20
3[d]	10	100	177	0.86	361	26.7
4[d]	5	100	150	0.78	399	36.2
5[d]	20	60	232	1.03	212	16.0
6[d]	20	40	231	1.03	262	7.70

[a] M_{LB} = dry mass of deposited LB monolayers.
[b] M_w = mass of incorporated water.
[c] v_w = evaporation rate of incorporated water.
[d] The QCM was lowered and raised through the cadmium stearate monolayers on the water phase.
[e] The QCM was lowered and lifted out of the subphase with no monolayer.

speed were increased, but the mass deposited LB film was less than that lifted at the high surface pressure of 20 mN/m (runs 3 and 4, Table 14.1). Thus, a relatively large amount of water is lifted with the disordered monolayer, and this incorporated water evaporates easily through the disordered monolayers. When the monolayer was lifted at slower speeds (runs 5 and 6), the evaporation rate of incorporated water decreased, suggesting that a better oriented monolayer was deposited on the substrate compared to that deposited at the higher speed (100 mm/min). When monolayers were deposited before complete evaporation of water from the proceeding layers, the transfer ratios decreased and much water was lifted with the disordered monolayers.

McCaffrey and coworkers reported on deposition of calcium stearate monolayers on a QCM by the LB technique and showed that the change in frequency was a linear function of the number of layers deposited.[21] However, the weight associated with each layer is 25% larger than the calculated mass of the LB layers, probably because they weighed the deposited LB films with the water lifted still incorporated in them.

14.4. Horizontal Lifting Depositions of LB films on a QCM

In the horizontal lifting method, the QCM plate is touched horizontally but slightly slants to a monolayer surface and is lifted to the air phase. LB films were deposited, two layers for each successive cycle, on one side of the QCM plate (a Y-type deposition). When monolayers were surrounded by barriers on a subphase, LB films were deposited one layer for each cycle (an X-type deposition). After drying in air at each cycle, the deposited amount (a transfer ratio) of LB films was confirmed from the frequency decrease (mass increase) of the QCM according to Eq. (14.2).

14.4.1. Transfer Ratios for Y-Type Horizontal Lifting Methods

Typical time courses of the frequency change of the QCM in air are shown in Fig. 14.2, when monolayers of cadmium stearate were deposited on the QCM at the surface pressure of 25 mN/m by a Y-type horizontal lifting method. The QCM was touched to the monolayer surface at point A and lifted into the air at point B at a speed of 5 mm/min. The frequency of the QCM reached to a constant value in a few minutes and decreased 120 ± 4 Hz each cycle from points A to B. In the horizontal lifting process,

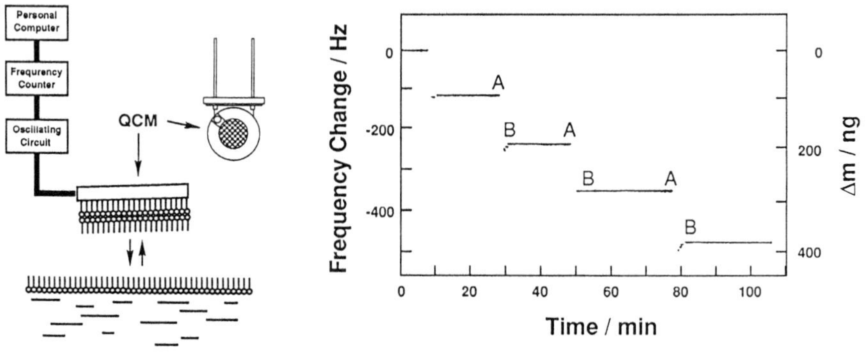

FIG. 14.2. Typical frequency changes of a QCM as a Y-type horizontal lifting plate for monolayers of cadmium stearate. At point A the QCM plate was touched to the monolayer and lifted horizontally at point B. Surface pressure of the monolayer, 25 mN/m; subphase, pH 7.1, 0.4 mM $KHCO_2$, 0.1 mM $CdCl_2$; lowered and raised speed of the QCM, 5 mm/min.

water was hardly incorporated into LB layers when the substrate was raised up from the subphase, compared with that in the vertical dipping process.

The increased mass on the QCM was calculated from the frequency changes using Eq. (14.2) to be 97 $\pm$ 5 ng. This value was consistent exactly with the theoretical mass of two dry monolayers of cadmium stearate (97 ng) calculated from the area per molecule in the monolayer (0.185 nm^2 at 25 mN/m from the *p-A* curve) and the area of the QCM electrode (0.159 cm^2). Thus cadmium stearate monolayers could be deposited on the QCM plate by using the Y-type horizontal lifting method with a transfer ratio of 2.00 $\pm$ 0.05. The transfer amount of Y-type horizontal depositions was not affected by the lowering and raising speed of a QCM plate in the range of 1 to 100 mm/min.

Figure 14.3 shows the correlation between transferred weight on the QCM plate and dipping cycles for dipalmitoylphosphatidylcholine (DPPC) monolayers at a surface pressure of 45 mN/m, when Y-type horizontal and vertical dipping methods were carried out. In the case of the Y-type horizontal lifting method, a good linear correlation was obtained, and dipping cycles and two DPPC monolayers (92 $\pm$ 5 ng) could be calculated to be transferred per cycle from the slope. Such a Y-type deposition of DPPC LB films could continue for more than 100 layers. On the contrary, in the vertical dipping process, DPPC could be transferred for only two layers (98 ng per one side of a QCM plate) at the first cycle and could not be obtained as LB films. In the vertical dipping method, the outer monolayer of DPPC films on the QCM is easily reverted to the water

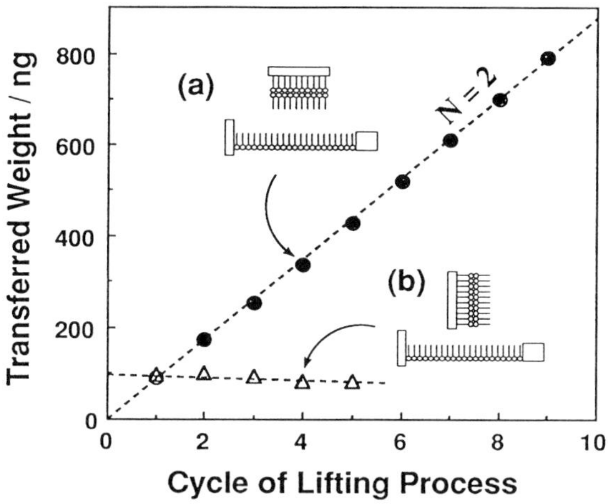

FIG. 14.3. Correlation between transferred weight on the QCM plate and dipping cycles in (a) a horizontal lifting method and (b) a vertical dipping method for DPPC monolayers (subphase: pure water, pH 5.8, 20°C). The QCM plate was moved horizontally or vertically at a rate of 5 mm/min.

surface when the plate is lowered through the air/water interface.[6] Therefore, the vertical dipping method is not suitable to transfer the relatively hydrophilic DPPC monolayers.

We applied the Y-type horizontal lifting methods using the QCM plate as a substrate for various monolayers, and transfer ratios are summarized in Table 14.2 in comparison with those of the vertical dipping method obtained by the QCM methods. Transfer ratios were obtained from slopes of the correlation between the transferred weight of LB films and dipping scales, as shown in Fig. 14.3. Phosphatidylethanolamine (DPPE), DPPC monolayers, and lipid-coated enzymes (glucose oxidase)[22] could not be transferred sufficiently in the conventional vertical dipping method. On the contrary, most of the monolayers except for oleic acid monolayers, could be transferred with a transfer ratio of 2 by using the Y-type horizontal lifting method. These results indicate that the horizontal lifting method is widely applicable for various monolayers compared with the conventional vertical dipping method.

Figure 14.4 shows x-ray diffraction patterns of cadmium stearate LB films (20 layers with a transfer ratio of 2.0) obtained by the Y-type horizontal lifting method. A series of peaks can be assigned[18] to well-oriented Y-type LB films with a long spacing of 5.0 nm. These x-ray diffraction patterns were very consistent with those obtained by the

TABLE 14.2

TRANSFER RATIOS OF VARIOUS MONOLAYERS WITH Y-TYPE HORIZONTAL LIFTING AND VERTICAL DIPPING METHODS OBTAINED BY QCM TECHNIQUES

Monolayers 1	Surface Pressures (mN/m)	Transfer Ratios[a] Horizontal Lifting Method	Vertical Dipping Method
Cadmium stearate	25	○ 1.98	○ 2.08
Stearyl alcohol	25	○ 2.07	○ 2.14
Ethyl stearate	25	○ 2.08	○ 2.00
Choresterol	45	○ 1.90	○ 1.98
DPPE	45	○ 2.12	▲ 1.05
DPPC	45	○ 2.01	× 0
Dialkylammonium amphiphile[b]	40	○ 2.01	× 0
Enzyme-lipid complex[c]	40	○ 2.01	× 0.16
Oleic acid	20	× 0.59	× 0.04

[a] Obtained from a slope of correlations between the transferred amount on the QCM and the dipping cycles.

[b]

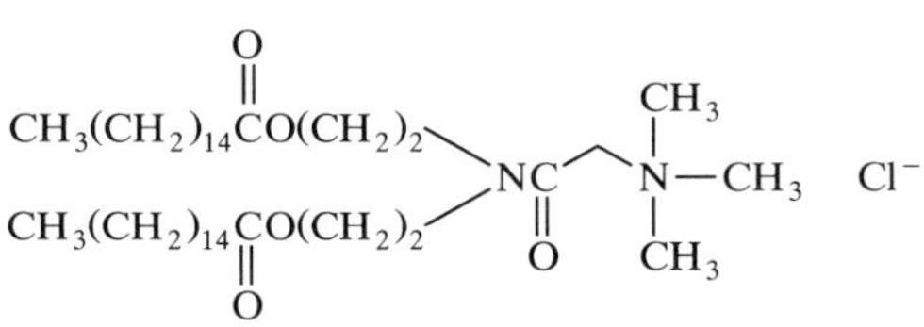

$2C_{16}$-glu-N^+

[c] Glucose oxidase coated with a monolayer of dihexadecyl N-(trimethylammoniomethylcarbonyl)glutamate ($2C_{16}$-glu-N^+). Glucose oxidaze content is 24% (see Ref. 20).

vertical dipping method. Similar well-oriented Y-type structures of LB films were also obtained for some LB films deposited by the horizontal lifting methods.[16–18]

In Figure 14.5, we compared transfer weights obtained by the frequency changes of the QCM with those obtained by a moving area of a barrier on a trough (a barrier method). When ethyl stearate monolayers were deposited with a Y-type horizontal lifting method at the high surface pressure of 25 mN/m, transfer weights obtained by both the QCM and the conventional barrier methods were consistent and showed two layers of deposition for each cycle. At the lower surface pressure of 5 mN/m, the QCM method indicates that one monolayer of ethyl stearate could be deposited for one cycle; however, the conventional barrier method shows that monolayers are hardly transferred on the substrate. The QCM method expresses the true transferred weight of monolayers on the substrate. On

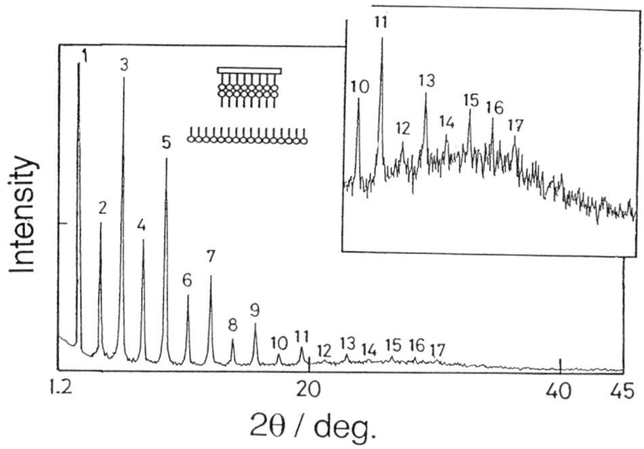

FIG. 14.4. x-ray (Cu K_α) diffraction patterns of cadmium stearate LB films (20 layers) deposited on a glass plate (3 × 5 cm) by the Y-type horizontal lifting method (two layers one cycle). Transfer conditions are the same as those in Fig. 14.1.

the contrary, the conventional barrier method just indicates the decreasing area of the monolayer obtained from a barrier moving. The reason one layer is deposited by the Y-type horizontal deposition at the lower surface pressure is described later (see Fig. 14.7 in Section 14.4.2).

When polystyrene monolayers and polyion-complex monolayers of dioctadecyldimethylammonium poly(styrene sulfonate) were deposited with a horizontal lifting method, the QCM method showed the real transfer ratio on the substrate, but the conventional barrier method did not. Similar behaviors were observed when the relatively rigid (fragile) monolayers of lanthanum stearate $[CH_3(CH_2)_{16}COO^-)_3\ La^{3+}]$ were deposited with a

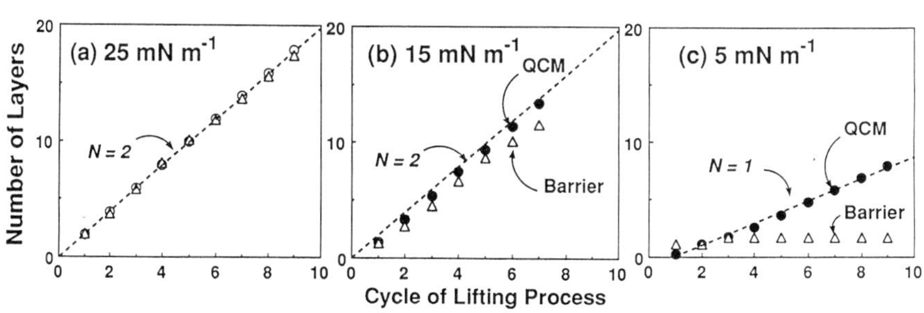

FIG. 14.5. Comparison of transfer ratios calculated from (○) the frequency changes of the QCM plate and (△) the moving area of a barrier on a trough, in the Y-type horizontal lifting method of ethyl stearate monolayers at surface pressures of (a) 25, (b) 15, and (c) 5 mN/m.

horizontal lifting method. These results indicate that the QCM method gives a real transfer ratio of LB films on the substrate compared with the conventional method, especially when the monolayers at the lower surface pressure, the polymeric, and the rigid monolayers are employed.

14.4.2. TRANSFER MECHANISMS OF HORIZONTAL LIFTING METHODS

Figure 14.6 shows schematic illustrations of the Y-type and X-type horizontal lifting processes. In the Y-type deposition, when the first layer is lifted with a substrate, the second layer is turned down and then two layers are transferred for each cycle. In the X-type deposition,[13–15] the QCM plate was lowered on a monolayer that was surrounded by a barrier on the subphase and then only the monolayer contacted with the substrate was lifted horizontally. Thus, in the X-type horizontal lifting method, one layer is deposited for each cycle.

Figure 14.7 shows the effect of the surface pressures of ethyl stearate monolayers on the transfer ratio. In the case of Y-type deposition, two layers could be deposited at the surface pressure above 15 mN/m. The transfer ratio decreased with a decrease in the surface pressure and reached nearly to the transfer ratio of 1.0 around the surface pressure of 3 to 5 mN/m. On the other hand, the X-type horizontal deposition could transfer a monolayer with a transfer ratio of 1 under a wide surface pressure range of 3 to 25 mN/m. This could be explained as follows. The first monolayer contacted to a substrate could be transferred even in the expanded or phase-separated structures at the lower surface pressure. However, these monolayers could hardly be turned over when the sub-

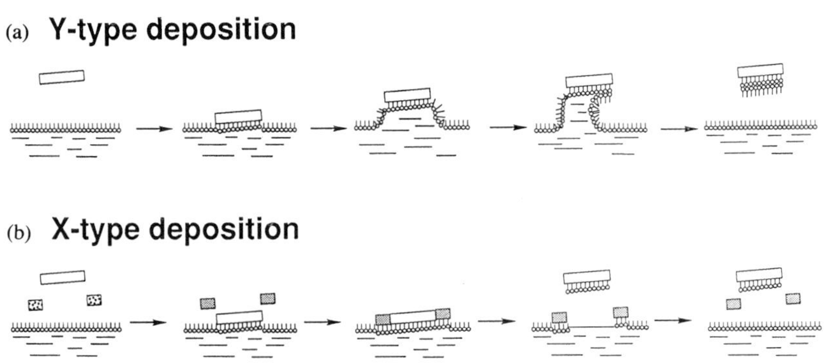

FIG. 14.6. Schematic illustrations of (a) the Y-type and (b) the X-type horizontal deposition processes.

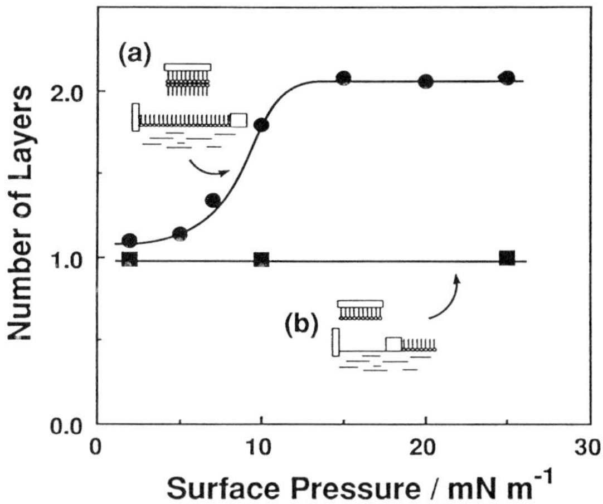

FIG. 14.7. (a) Y-type and (b) X-type horizontal depositions of ethyl stearate monolayers at various surface pressures at 20°C. The QCM plate was lifted at a rate of 5 mm/min.

strate was lifted. Therefore, the transfer ratio of the Y-type horizontal deposition decreases to 1 at the lower surface pressure [see also Fig. 14.5(b)].

Figure 14.8 shows the effect of surface pressures for the horizontal depositions of DPPC monolayers. At high surface pressures (45 to 30 mN/m), DPPC monolayers could be deposited two layers for each cycle

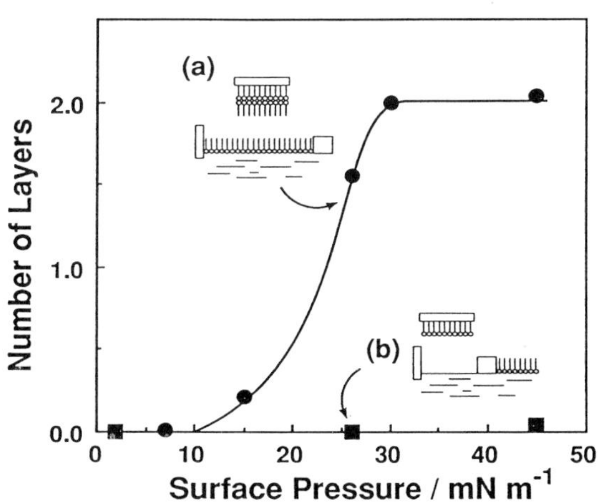

FIG. 14.8. (a) Y-type and (b) X-type horizontal depositions of DPPC monolayers at various surface pressures at 20°C. The QCM plate was lifted at a rate of 5 mm/min.

with the Y-type deposition. However, the transfer ratio decreased with a decrease in the pressure and reached to nearly zero below 10 mN/m. In the X-type horizontal lifting method, DPPC could not be deposited on a substrate at all. Since the hydrophilic surface of DPPC monolayers faces the hydrophobic air phase in X-type deposition, DPPC monolayers could not be transferred on a substrate because the hydrophilic surface of DPPC is more stable on the water subphase than in the air phase. At the lower surface pressure in Y-type deposition, the expanded or phase-separated monolayer could not be turned over as a second layer, and the first layer is not transferred on the lifted substrate to avoid exposing the hydrophilic surface of the first layer to air.

In the case of ethyl stearate monolayers, X-type deposition could have occurred due to the hydrophobic ester headgroup of monolayers. Thus, to achieve the Y-type horizontal deposition of monolayers having a hydrophilic headgroup, the second layer should be turned over, which should cover the hydrophilic surface of the first layer.

14.5. Conclusions

QCMs are useful for measuring a real transfer weight (a transfer ratio) of LB films deposited both with vertical dipping and horizontal lifting methods. In the vertical dipping method, we have observed the incorporation of water into LB films and have correlated the amount of incorporated water and its evaporation rate with the molecular orientation of the monolayers. In the horizontal lifting method, we could estimate transfer mechanisms of Y-type horizontal lifting processes in comparison with X-type horizontal lifting processes from the transfer ratios. The Y-type horizontal lifting method is widely applicable for various monolayers, when monolayers cannot be transferred by the conventional vertical dipping method. The obtained LB films are as well oriented as those obtained by the vertical dipping method. The horizontal lifting method also has the advantage of allowing LB films to be prepared by semi-dry processes, because the substrate does not directly touch the water subphase.

The QCM methods are also effective for observing other characterizations of LB films. We have observed from the frequency change that the outer monolayer of LB films partly flakes when the LB film-deposited QCM passes through the air/water interface.[6] We have also observed the swelling behavior of LB films in the aqueous solution decrease (mass increase) when the LB film-deposited QCM is soaked in the solution.[8] The QCM is also a useful tool for detecting the fluidity change from the solid to the liquid crystalline state of LB films on the QCM in the water phase.

The frequency of the LB film-deposited QCM increased abruptly due to the fluidity change at the phase transition temperature of LB films when the temperature was raised gradually in a water phase.[7]

References

1. Ellis, J. W., and Pauley, J. L. (1964). *J. Colloid Interface Sci.* **19,** 755.
2. Kimura, F., Umemura, J., and Takenaka, T. (1986). *Langmuir* **2,** 96.
3. Saperstein, J. (1986). *J. Phys. Chem.* **90,** 1408.
4. Shimomura, M., and Kunitake, T. (1985). *Thin Solid Films* **132,** 243.
5. Okahata, Y., and Ariga, K. (1987). *J. Chem. Soc. Chem. Commun.* 1535; Ariga, K., and Okahata, Y. (1994). *Langmuir* **10,** 3255.
6. Okahata, Y., and Ariga, K. (1989). *Langmuir* **5,** 1261; Ariga, K., and Okahata, Y. (1994). *J. Colloid Interface Sci.* **167,** 275.
7. Okahata, Y., Kimura, K., and Ariga, K. (1989). *J. Am. Chem. Soc.* **11,** 9190.
8. Okahata, Y., and Ariga, K. (1989). *Thin Solid Films* **178,** 465; Ariga, K., and Okahata, Y. (1994). *Langmuir* **10,** 2272.
9. Okahata, Y., Ariga, K., and Tanaka, K. (1992). *Thin Solid Films* **210 / 211,** 702.
10. Sauerbrey, G. (1959). *Z. Physik* **155,** 206.
11. Blodgett, K. B. (1935). *J. Am. Chem. Soc.* **56,** 495; (1935). *J. Am. Chem. Soc.* **57,** 1007.
12. Langmuir, I., and Schaefer, V. J. (1938). *J. Am. Chem. Soc.* **60,** 1351.
13. Fukuda, K., Nakahara, H., and Kato, T. (1976). *J. Colloid Interface Sci.* **54,** 430.
14. Nakahara, H., and Fukuda, K. (1979). *J. Colloid Interface Sci.* **69,** 24.
15. Kato, T. (1986). *Jpn. J. Appl. Phys.* **27,** L1358, L2128.
16. Iwahashi, M., Naito, F., Watanabe, N., and Seimiya, T. (1985). *Chem. Lett.* 187.
17. Iwahashi, M., Itoh, K., Ohshima, K., Seimiya, T., Yoshida, A., Kobayashi, K., and Tanaka, K. (1988). *Bull. Chem. Soc. Jpn.* **61,** 3061.
18. Kamata, T., Umemura, J., and Takenaka, T. (1988). *Chem. Lett.* 1231.
19. Kawai, T., Umemura, J., and Takenaka, T. (1990). *Langmuir* **6,** 672.
20. Ebersole, R., and Ward, D. M. (1988). *J. Am. Chem. Soc.* **110,** 8623.
21. McCaffrey, R. R., Bruckenstein, S., and Prasad, P. N. (1986). *Langmuir* **2,** 228.
22. Okahata, Y., Tsuruta, T., Ijiro, K., and Ariga, K. (1988). *Langmuir* **4,** 1373. Okahata, Y., Tsuruta, T., Ijiro, K., and Ariga, K. (1989). *Thin Solid Films* **180,** 65.

Translational Diffusion and Electron Hopping in Monolayers at the Air/Water Interface

MARCIN MAJDA

Department of Chemistry, University of California at Berkeley, Berkeley, California 94220

15.1. Electrochemistry on the Water Surface

This chapter highlights recent results of our electrochemical investigations of the dynamics of lateral processes in monolayer assemblies at the air/water interface. Lateral diffusion of phospholipids and proteins in monolayer and bilayer assemblies has been a subject of considerable interest among biophysicists as a part of broad research on the structure and functioning of cell membranes.[1-10] Extensive investigations of molecular mobility in these types of systems were carried out, in most cases, by fluorescence microphotolysis known also as FRAP techniques (fluorescence recovery after photobleaching).[1-3] Our electrochemical approach is substantially different although fundamentally analogous.[11,12] In both cases, an initial perturbation generates a concentration gradient of probe species in the monolayer, which then leads to a temporal evolution of a measured signal interpretable in terms of a lateral diffusion coefficient.

In our approach [see the experimental setup in Fig. 15.1(a)], application of a potential to the electrode contacting a monolayer assembly containing electrochemically active amphiphiles, as shown schematically in Fig. 15.1(b), sets in a concentration gradient of redox species in the monolayer. Lateral diffusion of those species to the electrode results in current that is proportional to the rate of their arrival and can, therefore, be interpreted in terms of their lateral molecular mobility or, speaking more generally, lateral charge transport in the monolayer.[12] Indeed, since electrochemical techniques involve, fundamentally, measurements of the rate of electron flow, a unique feature of these experiments is a capability to follow not

ISBN 0-12-523485-6

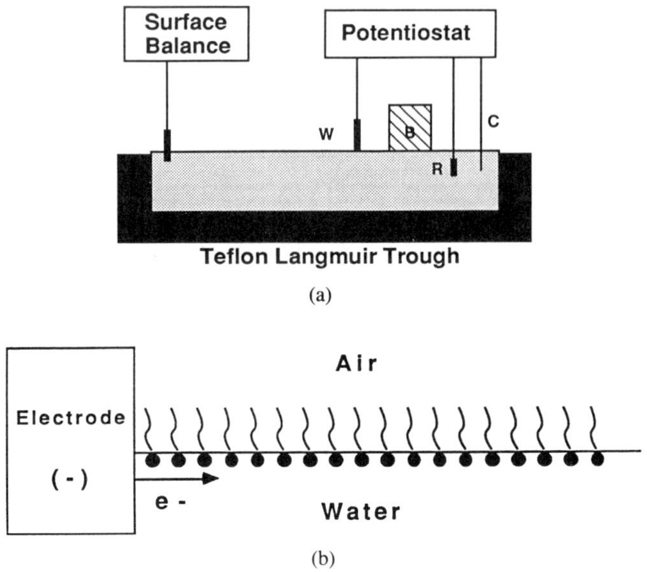

FIG. 15.1. (a) Experimental setup used in the electrochemical measurements of lateral processes in Langmuir monolayers: W, working microelectrode; B, movable barrier of the Langmuir trough; R, reference electrode; C, counter electrode. (b) A schematic diagram of a microelectrode and its perpendicular orientation with respect to a Langmuir monolayer at the air/water interface. This type of orientation of an electrode at the water surface is essential in electrochemical measurements of the dynamics of lateral diffusion processes. A detailed presentation of the design of microelectrodes used in such measurements is shown in Fig. 15.2.

only translational diffusion of molecules at the air/water interface but also lateral electron hopping in a two-dimensional (2-D) network of redox sites in monolayer films at that interface. We present here results dealing with both types of phenomena as an illustration of the capabilities of the electrochemical techniques in the studies of the dynamics of lateral processes in Langmuir monolayers. Electrochemical characterization of the lateral processes mentioned is possible as a result of the development of microelectrodes that can be positioned in the plane of the air/water interface and that contact the monolayer assembly spread at that interface as shown in Fig. 15.1. They key features of the microelectrode design and its essential fabrication details are reproduced in Fig. 15.2.

To address a monolayer assembly, a 2-D system, we designed and fabricated an electrode functioning as a one-dimensional (1-D) "line" electrode in the plane of the air/water interface.[12] We accomplished this goal by creating a sharp gradient of wettability along a fracture line of a 500-Å-thick gold substrate vapor deposited on a glass slide. All electrochemical experiments were carried out in a Langmuir trough under controlled surface-pressure conditions. The subphase is a solution of electrolyte; the other two electrodes of the usual three-electrode potentiostatic circuit are placed in the subphase on the other side of the movable barrier.

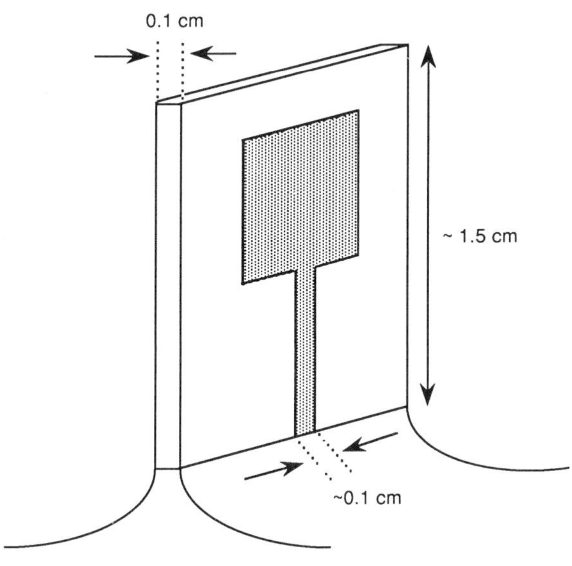

(a)

FIG. 15.2. Microelectrode and its positioning at the air/water interface. (a) A glass slide with the vapor-deposited gold film (marked by the shaded area) is held above the water surface. Water forms a positive meniscus by wetting the bottom edge of the glass slide. The electrical contact (not shown) is formed in the upper section of the gold pattern. The one-dimensional electrode is the 0.1-cm-long line in the plane of the water surface, defined by the difference in wettability of the clean gold edge and of the octadecylthiol-coated gold surface above the water, as shown in part (b). (b) A schematic (out of scale) cross-sectional side view of the microelectrode at the air/water interface. The bottom edge of the glass slide (including the gold film) in contact with water was created by breaking in half a gold-coated glass slide, which initially carried a twin-pair electrode pattern.[12] All gold and glass surfaces of the glass slide shown in part (a) are nonwettable, since they are coated by a self-assembled monolayer of octadecyl mercaptan (OM) and octadecyltrichlorosilane (OTS). For simplicity, the OTS + OM monolayer is shown on the gold surface only. The amphiphilic molecules on the water surface represent C_{18}Fc. (Reprinted with permission from Charych, D. H. *et al.*, *J. Am. Chem. Soc.* **113,** © 1991, American Chemical Society.)

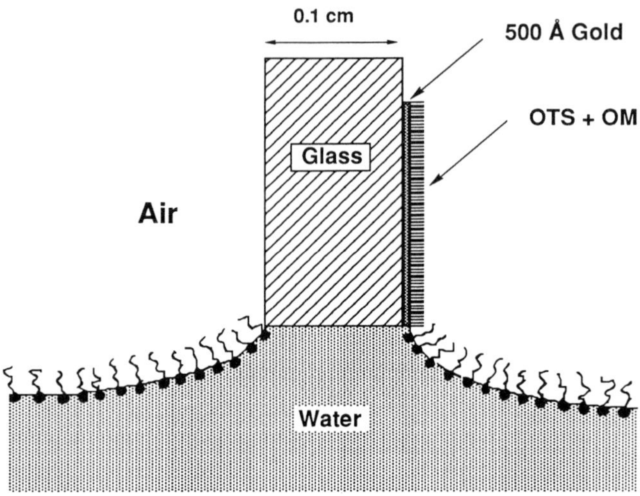

(b)

FIG. 2. (*Continued*)

15.2. Translational Diffusion in Octadecyl Ferrocene Monolayers

We look first at molecular mobility of N-octadecylferrocenecarboxamide ($C_{18}Fc$, see the structure in Fig. 15.3), which forms fluid monolayers on a 50 mM $HClO_4$ subphase. Figure 15.3 shows a pressure-area diagram and the molecular structure of this amphiphile. The experiments described below were done at 28°C unless specified otherwise. When a 0.1-cm-long "line" microelectrode is positioned at the water surface covered with a $C_{18}Fc$ monolayer at any surface concentration in the range between approximately 53 and 250 Å^2/molecule, a well-defined cyclic voltammogram can be recorded (at a moderate scan rate of 0.1 V/s), which fulfills the usual criteria of a reversible redox couple.[12]

In other words, the reduction of dimensionality (1-D electrode contacting a 2-D system) does not affect the shape of the current-voltage curves, which are determined by the geometry of the diffusion processes controlling the observed current. Specifically, under the conditions mentioned, linear diffusion prevails, generating a diffusion zone in the $C_{18}Fc$ monolayer that is narrow in comparison to the electrode length. We showed that a radial (hemicircular) diffusion zone develops when these experiments are done with much shorter electrodes (5 to 100 μm in length), leading to the characteristic sigmoidal shape of the voltammograms.[13]

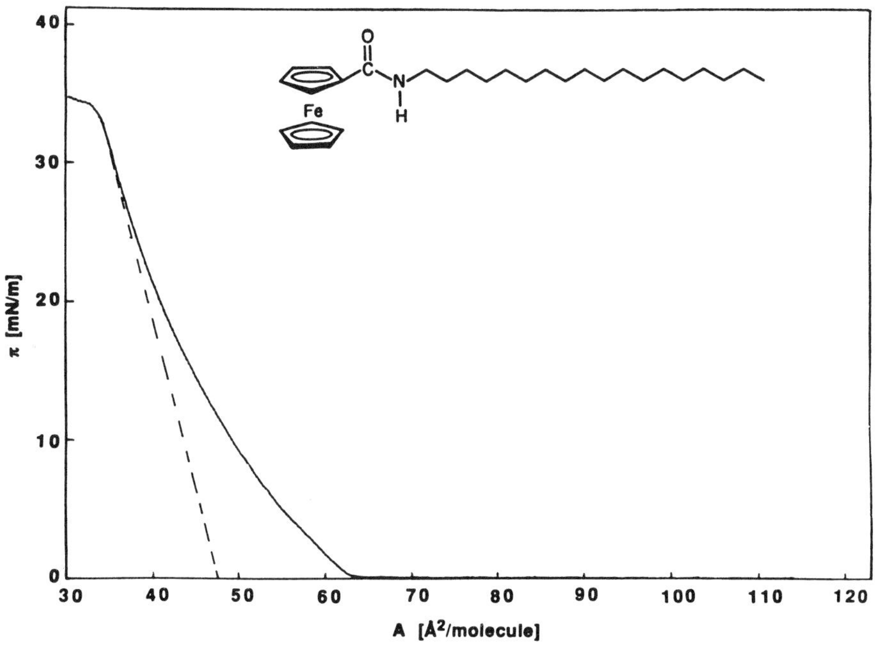

FIG. 15.3. The structural formula and a π-A isotherm of C_{18}Fc recorded on 0.05 M $HClO_4$ subphase at 28°C. The average area per molecule obtained by extrapolation of the steepest part of the isotherm to $\pi = 0$ is 48 ± 2 Å/molecule.(Reprinted with permission from Charych, D. H. *et al., J. Am. Chem. Soc.* **113,** © 1991, American Chemical Society.)

In the concentration range mentioned above (corresponding to a surface pressure of 0 to 8 mN/m), interpretation of the magnitude of the voltammetric peak current produced a plot of the lateral diffusion coefficient versus surface concentration shown in Fig. 15.4(a). As expected, a decreasing mobility of C_{18}Fc in monolayers of increasing surface concentration reveals decreasing fluidity of the system. We found that this behavior is consistent with a free-volume model of diffusion in the hard-sphere fluids of Turnbull and Cohen[14] adapted for a 2-D system of hard disks. In their theory,[15] the diffusion coefficient D is expressed by the following equation:

$$D = gu \int_{v^*}^{\infty} p(v) f(v) a(v) \, \mathrm{d}v \qquad (15.1)$$

where g is a geometric factor; u is a gas-phase velocity; $p(v)$ is the probability of finding free volume between v and $v + \mathrm{d}v$; $a(v)$ is the average displacement assumed to be a linear function of free volume:

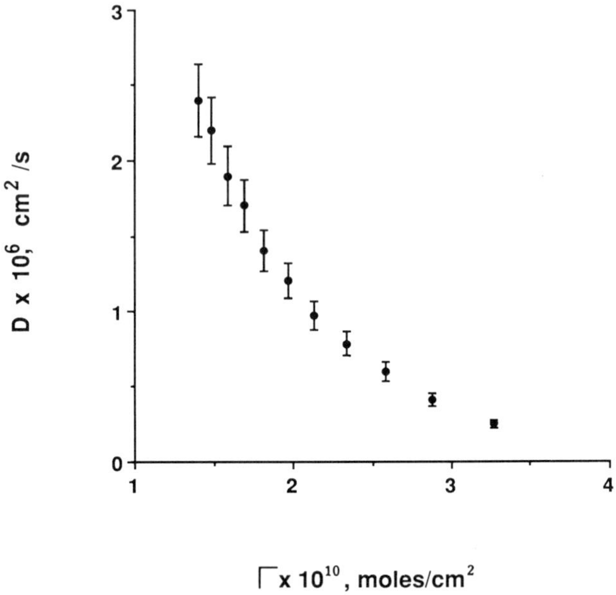

(a)

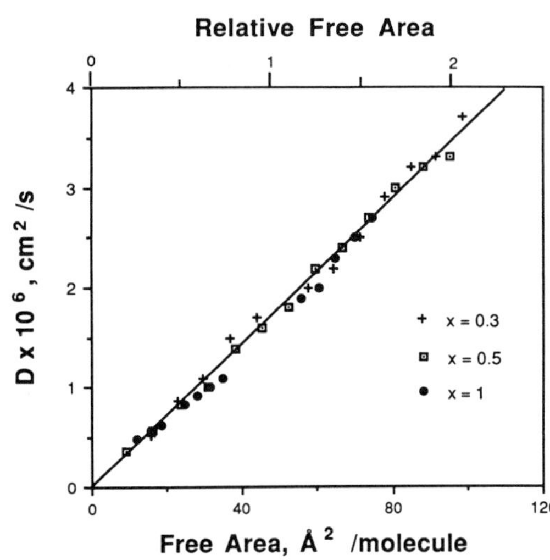

(b)

FIG. 15.4. (a) Lateral diffusion coefficient of $C_{18}Fc$ at the air/water interface plotted versus $C_{18}Fc$ surface concentration and (b) as a function of the average free area per molecule and relative free area per molecule. In the latter case, also included are data from two additional series of experiments in which $C_{18}Fc$ was diluted with octadecanol. Mole fractions (x) of $C_{18}Fc$ are given in the figure. (Reprinted with permission from Charych, D. H. *et al.*, *J. Am. Chem. Soc.*, **113,** © 1991, American Chemical Society.)

$a(v) = av$; and $f(v)$ is the displacement correlation factor expressing whether the system is fluid or "solid-like." The latter is approximated by a step function jumping from 0 when $v < v^*$ to 1 when $v > v^*$. The probability distribution function $p(v)$ is given by

$$p(v) = (\gamma/v_f)\exp(-\gamma v/v_f) \tag{15.2}$$

where v_f is the average free volume (average volume per molecule less molecular volume) and γ is a numerical factor between 0.5 and 1 and is introduced to account for an overlap of free volume. Integration of Eq. (15.1) yields

$$D = gu\alpha(v^* + v_f/\gamma)\exp(-\gamma v^*/v_f) \tag{15.3}$$

which can be reduced to

$$D = gu\alpha v_f/\gamma \tag{15.4}$$

when critical volume $v^* \ll v_f/\gamma$. Physically, the latter limit assumes that no minimum void volume is necessary for diffusion to take place and that the system is "fluid" at all values of void volume $v > 0$. We showed[12] that the same result [Eq. (15.4)] can be obtained by changing the limits of integration in Eq. (15.1) to 0 and ∞.

That our results can indeed be described by Eq. (15.4) is demonstrated in Fig. 15.4(b), where the data from Fig. 15.4(a) are replotted versus free area (A_f) and relative free area per molecule. Note that the linearity of D versus A_f holds equally well for pure $C_{18}Fc$ monolayers as well as mixed monolayers of $C_{18}Fc$ and octadecanol, a nonelectroactive amphiphile that simply limits the free area available to $C_{18}Fc$. Overall, this analysis of our results indicates that the $C_{18}Fc$ monolayer remains fluid (no phase transitions) in a substantial range of surface pressures and concentrations.

It is also interesting and useful to relate the data in Fig. 15.4(b) to the kinetic theory of diffusion in "dense gases" of hard spheres.[16] We refer here to the Enskog theory[17] and later modifications of Alder based on his molecular dynamics calculations.[18–20] As we show later, molecular dynamic calculations predict a linear dependence of the diffusion coefficient on the relative free area per molecule in agreement with our experimental findings. Interpretation of our results in view of these theories yields an *apparent molecular weight* of $C_{18}Fc$, which in turn reveals some elements of intermolecular interactions between $C_{18}Fc$, molecules in the monolayer at the water surface.

The Enskog diffusion coefficient is described by

$$D_E = \frac{3}{8n\sigma^2}\left(\frac{kT}{m\pi}\right)^{1/2}\frac{1}{g(\sigma)} \tag{15.5}$$

where n is number density, σ and m are the molecular diameter and mass, respectively, and $g(\sigma)$ is the radial distribution function when the hard spheres are in contact.[16] The latter is related to the compressibility factor $Z = PV/RT$, while $(Z - 1)^{-1}$ is a function of relative free volume v_f/v_0:

$$g(\sigma) = \frac{3(Z - 1)}{2\pi n \sigma^3} \tag{15.6}$$

Combining Eqs. (15.5) and (15.6), one obtains:

$$D_E = \frac{\sigma}{4}\left(\frac{\pi kT}{m}\right)^{1/2}(Z - 1)^{-1} \tag{15.7}$$

Alder and coworkers carried out extensive molecular dynamics studies in order to take into account many-body correlations, which become important at higher densities.[16,18–20] A major discovery resulting from these studies was the divergence of the velocity-velocity correlation function in the hard-disk system, which makes the 2-D diffusion coefficient an ill-defined quantity.[19] Since our monolayer system is not, naturally, a strictly 2-D system, the values of the diffusion coefficients obtained in our experiments are not subject to this limitation. Another result of the calculations done by Alder and coworkers[20] is the tabulated correction to the Enskog diffusion coefficient, D/D_E in the higher range of densities, v/v_0. Plotting their data reveals, indeed, a linear dependence of the corrected diffusion coefficient D on the relative free volume. The slope calculated with the parameters of our system (m = 425 g/mol, T = 293 K) is 8.9 ± 10^{-5} cm^2/s.

Comparison of the theoretically predicted slope and our results in Fig. 15.4(b) [the experimentally measured slope in Fig. 15.4(b) is 1.7×10^{-6} cm^2/s] allows us to calculate [see Eq. (15.7)] an apparent molecular weight of C_{18}Fc. Its value is 1.2×10^6 g/mol, which is 2.7×10^3 times larger than the actual molecular weight of C_{18}Fc. We postulate that this perhaps surprising result reflects interactions of C_{18}Fc molecules in the hydrocarbon region, which thus far were neglected in our 2-D hard-disk model of the system. It is likely that entanglement of the octadecyl chains leads to drastically lower diffusion coefficients in comparison with the theoretical predictions presented earlier. It is also important to notice that the linearity of the plot in Fig. 15.4(b) extends to a region where the relative free area per molecule is significantly lower than 0.5. The latter value marks a point of freezing of hard-sphere fluids,[20] whereas our system does not undergo a phase transition but remains in a gel-like supercooled state. Hence, it appears that the entanglement of the hydrocarbon chains mentioned earlier also prevents film crystallization.

To confirm the hypothesis involving hydrocarbon chain entanglement, we measured[21] lateral diffusion coefficients in monolayers of di-octadecyl ferrocene (Di-C_{18}Fc, m = 720 g/mol). In this system, the diffusion coefficient also varies linearly with the free area per molecule. However, the slope of that dependence is 7.1 times smaller than that reported already for the C_{18}Fc system. On the basis of Eq. (15.7), one would expect a decrease in the slope by a factor of 1.3 only (a square root of the ratio of the molecular weights of Di-C_{18}Fc and C_{18}Fc). Therefore addition of the second hydrocarbon chain to C_{18}Fc increases, predictably, the extent of intermolecular interactions and entanglement between the molecules in the monolayer at the air/water interface. We intend to carry out additional experiments with ferrocene derivatives with shorter hydrocarbon chains expecting an opposite trend in the variation of the slope of the diffusion coefficients versus free area per molecule.

15.3. Electron Hopping in Langmuir Monolayers

In the previous section, we dealt with a monolayer system in which translational diffusion of amphiphiles constituted the sole mode of transport of charge to the microelectrode contacting the surface assembly. In principle, however, electron hopping initiated at the electrode and involving a series of electron transfer steps between adjacent redox centers can also contribute to the observed current.[22] Charge transport via electron hopping is analogous to the translational diffusion and likewise leads to a diffusive propagation of the redox conversion front in a bucket-brigade fashion as illustrated in Scheme 15.1. We did not observe this effect in C_{18}Fc monolayers because the electron exchange rate constant is too small relative to the translational diffusion coefficient of the ferrocene amphiphile.[12]

In general, electron hopping proceeds independently of translational diffusion although it may operate in conjunction with the latter when the two diffusive fluxes combine to yield current. We present here a summary

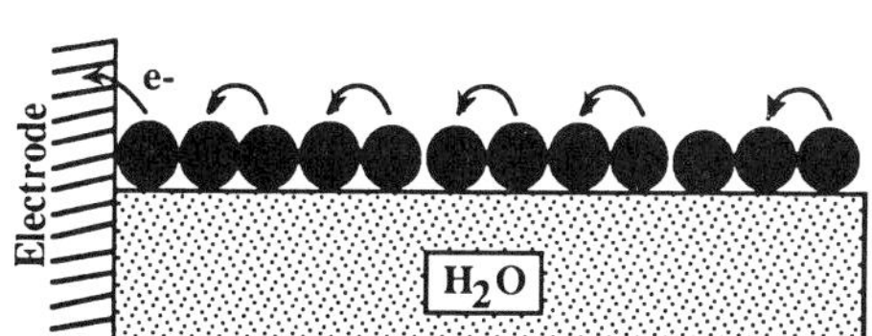

SCHEME 15.1

of recent results describing one such case. As shown later, electrochemical measurements in this case can be interpreted in terms of the kinetics of electron transfer. This opens new possibilities in the studies of electron transfer processes. The system being investigated consists of the perchlorate salt of osmium tris-diphenylphenanthroline, $Os(DPP)_3$ (see structure in Fig. 15.5).[21] In spite of their nonamphiphilic character, $Os(DPP)_3$ molecules can be compressed on a 0.05 M $HClO_4$ subphase as shown in Fig. 15.5.

Brewster-angle microscopic images[23] of these monolayers taken at various stages of compression and reproduced in Fig. 15.6 clearly indicate that the $Os(DPP)_3$ forms solid-like amorphous 2-D islands, which upon compression coalesce to produce a rather homogeneous film.[24] Consistent with the solid-state character of $Os(DPP)_3$ monolayers, we observed[21] no current in our 2-D electrochemical experiments until the surface pressure reached approximately 7 mN/m. At this point further compression resulted in a rapid increase of the voltammetric signal completely opposite to the trend observed in $C_{18}Fc$ monolayers.[12]

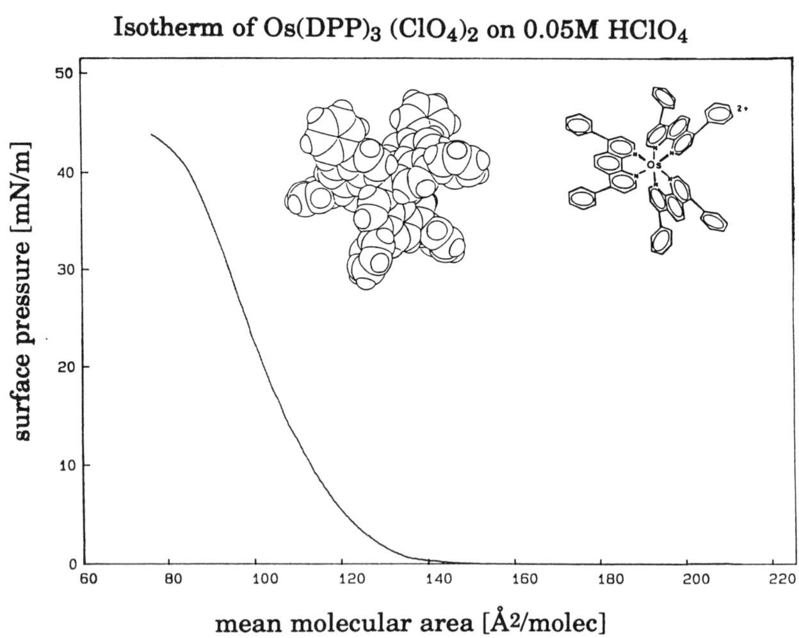

FIG. 15.5. Structural formulas and a π-A isotherm of $Os(DPP)_3$ recorded on 0.05 M $HClO_4$ subphase at 23°C. (Reprinted with permission from Charych, D. H. *et al.*, *Thin Solid Films*, © 1994, Elsevier Sequioa.)

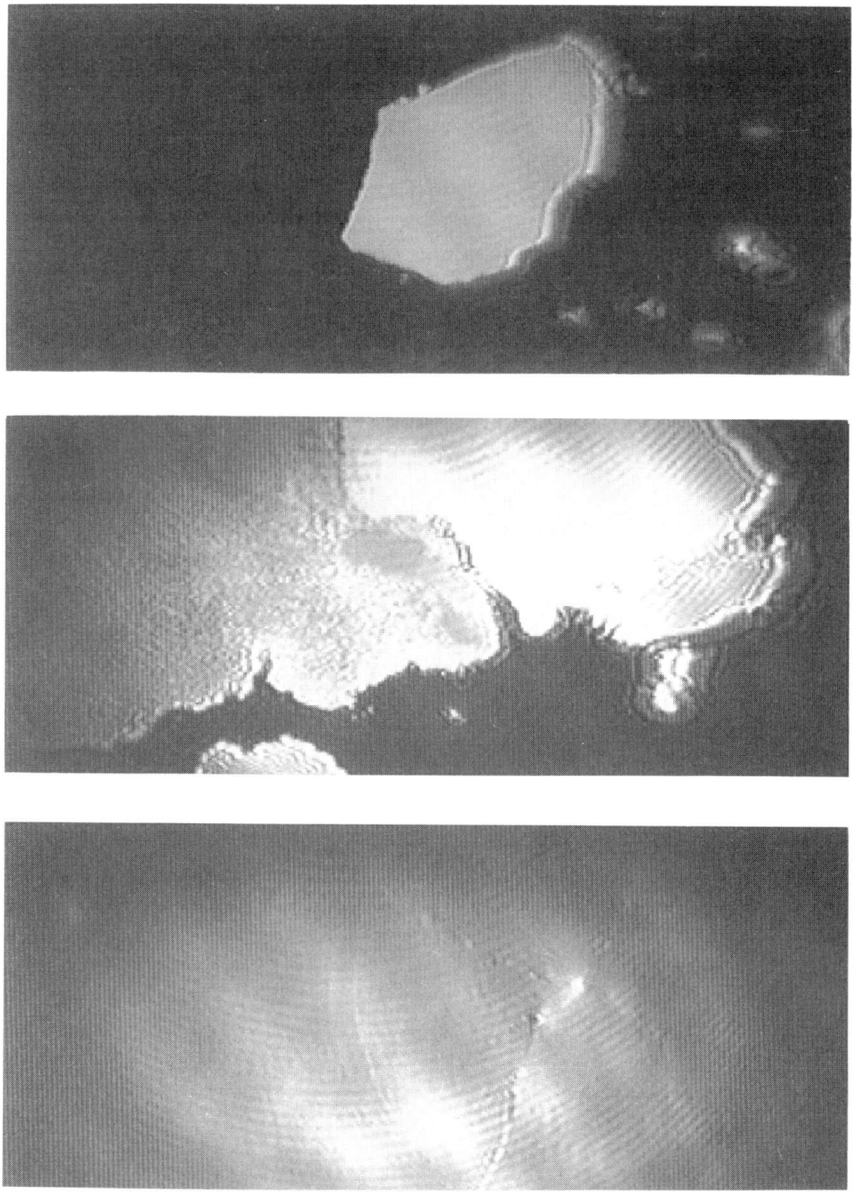

FIG. 15.6. Brewster-angle micrographs of $Os(DPP)_3$ monolayer at the air/water interphase (0.05 M $HClO_4$) spread from a 0.198-mM chloroform solution. Average areas per molecules: top, 360; middle, 150; and bottom, 100 Å^2/molecule. The width of each image corresponds to 730 μm. The differences in brightness as well as the interference fringes are the instrumental artifacts.

This behavior and the Brewster-angle microscopic images give a clear indication that electron hopping is the sole mechanism responsible for the observed current. The increase in the observed current that parallels the increase of surface pressure, as shown in Fig. 15.7, reflects, most likely, an increase in connectivity between individual islands and a concurrent increase of the extent of electroactivity of the system.

Thus the highest value of the diffusion coefficient D_{max} obtained from the current-voltage curves can be associated with full electroactivity of the $Os(DPP)_3$ monolayer. Further compression leads to an irreversible monolayer collapse at the line of the microelectrode and a decrease of current. In view of this understanding of the system's behavior, we can interpret D_{max} in terms of the unimolecular rate constant of the $Os(DPP)_3(III+/II+)$ electron exchange (k_1) knowing the distance (r_{NN}) between the neighboring sites from the corresponding Langmuir isotherm:

$$D_{max} = \tfrac{1}{4} k_1 r_{NN}^2 \tag{15.8}$$

Thus, the data in Fig. 15.7 yield $k_1 = 2.9 \pm 0.3 \times 10^8\ s^{-1}$, which corresponds to a bimolecular rate constant of electron exchange of $8.7 \times 10^7\ M^{-1}\ s^{-1}$. The latter can be obtained in the framework of the precursor

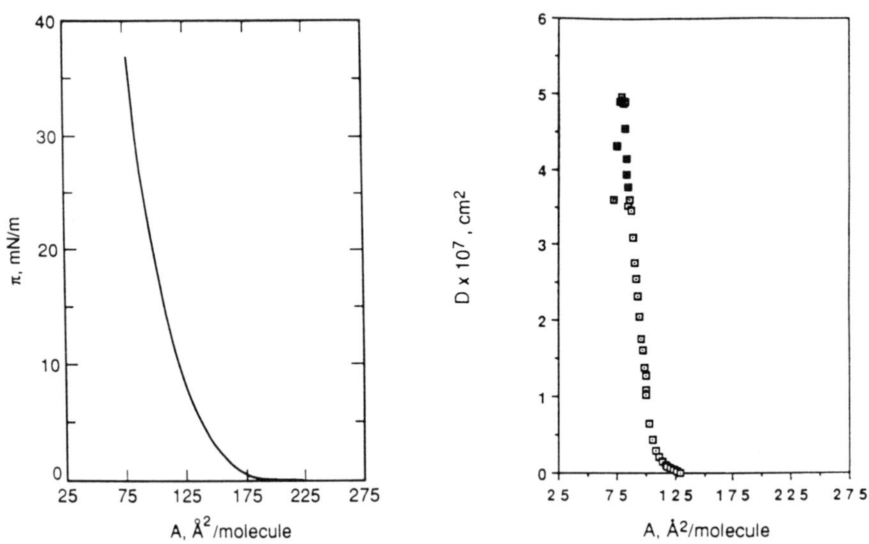

(a) (b)

FIG. 15.7. (a) π-A isotherm of $Os(DPP)_3$ at 0.05 M $HClO_4$. (b) Plot of the apparent diffusion coefficient of electron hopping in $Os(DPP)_3$ monolayers obtained from the 2-D cyclic voltammetric experiments concurrently with the data in part (a).

model of electron transfer.[12,25,26] This value is of the same order of magnitude as the electron exchange rate constant of the osmium couple measured in acetonitrile.[27]

This agreement is perhaps fortuitous considering how different are the environments of the $Os(DPP)_3$ species in both cases. It is particularly interesting to ask a question about the extent of solvation of $Os(DPP)_3$ centers at the air/water interface compared to bulk solvent conditions and the role solvent dynamics might play in electron hopping along the air/water interface. It has long been postulated that electrostatic coupling between polar solvents and the precursor complex during electron transfer involves *dielectric friction*, a phenomenon that significantly affects the rates of adiabatic electron transfer depending on the nature of the solvent.[28–30] To this end, we discovered that injection of small volumes of different solvents to the 0.05 M $HClO_4$ subphase results in significant changes of the electrochemically measured rate constant k_1 in $Os(DPP)_3$ monolayers.[31] Apparently, solvent partitioning to the surface monolayer affects solvation of the $Os(DPP)_3$ sites. We examined five different solvents, as listed in Fig. 15.8, and measured k_1 values in each case under plateau conditions where an additional aliquot of a given solvent injected into the subphase no longer resulted in a change of the rate constant, suggesting that a complete change of the solvation environment of the $Os(DPP)_3$ monolayer has been accomplished.[31]

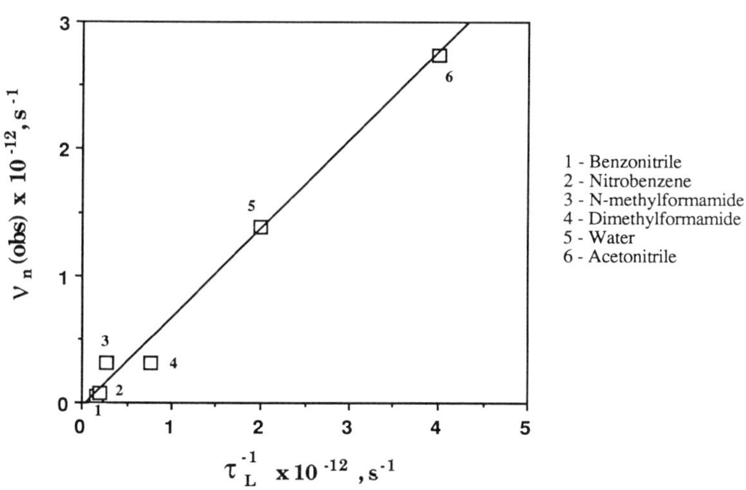

FIG. 15.8. Correlation of the experimentally obtained values of nuclear frequency factor [Eqs. (15.9) and (15.10)] and the inverse longitudinal relaxation time for the solvents listed in the figure.

Our data analysis was based on the Marcus theory.[25,32] To test whether the observed changes in the rate constant are due to the dynamic solvent effects, we correlated the nuclear frequency factor ν_n in the Marcusian expression of the electron transfer rate constant measured in various solvents with the inverse longitudinal relaxation time (τ_L) of these solvents. The nuclear frequency factors were extracted from the rate expression:

$$k_1 = \kappa_{el} \nu_n \exp(-\Delta G^*/RT) \tag{15.9}$$

In the case of the outer sphere, adiabatic electron transfer κ_{el}, the electronic transmission coefficient equals unity. Since the activation energy ΔG^* expresses in this case only the energy of solvent reorganization (the outer shell terms), its magnitude can be calculated in view of the solvent dielectric continuum model:

$$\Delta G^* = \frac{1}{4} e^2 \left(\frac{1}{2r_1} + \frac{1}{2r_2} - \frac{1}{2r_{1,2}} \right) \left(\frac{1}{D_{op}} - \frac{1}{D_s} \right) \tag{15.10}$$

Thus, the observed value of the frequency factor can be calculated from the measured rate constant [Eq. (15.9)]. The correlation of the measured frequency factor and the literature values of the inverse longitudinal relaxation times for the solvents used in our studies is shown in Fig. 15.8. A rather good correlation of the experimental and literature frequency factors indicates that our solvent injection method indeed results in changes of the solvation of $Os(DPP)_3$ at the water/air interface. One of the most intriguing questions remaining to be answered concerns the extent of solvation of the $Os(DPP)_3$ centers on the water surface relative to the bulk solvent conditions. The linearity of the plot in Fig. 15.8 suggests further a strong electronic coupling of the osmium centers in the surface monolayer. A weak coupling leading to $\kappa_{el} < 1$ would result in a negative deviation of the ν_n versus τ_L^{-1} plot for solvents with the highest frequencies of dielectric relaxation.

Finally, we describe briefly our initial attempts to study long-range electron transfer kinetics. The ideally 2-D character of monolayers at the air/water interface and the control of the surface film composition allow us to design experiments where the lateral electron hopping is investigated in two-component monolayers consisting of $Os(DPP)_3$ and a nonelectroactive diluent species. Under conditions of complete miscibility of the two components, we can obtain the distance dependence of the electron transfer kinetics. This is illustrated in Scheme 15.2 and Eq. (15.11):

$$k_1 = k_0 \exp[-\beta(\bar{r}_{NN} - r_0)] \tag{15.11}$$

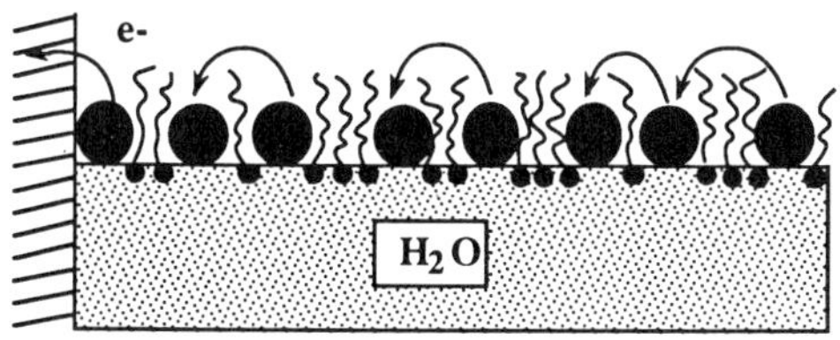

SCHEME 15.2

Of particular interest in such experiments would be the dependence of the decay parameter β on the chemical character of the diluent species.

Our initial measurement[31] involved $Os(DPP)_3$ and octadecanol ($C_{18}OH$). A variation of the electrochemically obtained diffusion coefficient of electron hopping with the mole fraction of $OS(DPP)_3$ in the two-component monolayers is shown in Fig. 15.9. These data are compared there with the predictions of the percolation theory[33,34] in absence of long-range electron transfer. It is readily apparent that the experimental values of D decay faster than expected. This finding strongly indicates self-segregation of the two components. The lack of miscibility of $Os(DPP)_3$ and $C_{18}OH$ does not allow us to interpret these results in terms of the electron transfer kinetics.

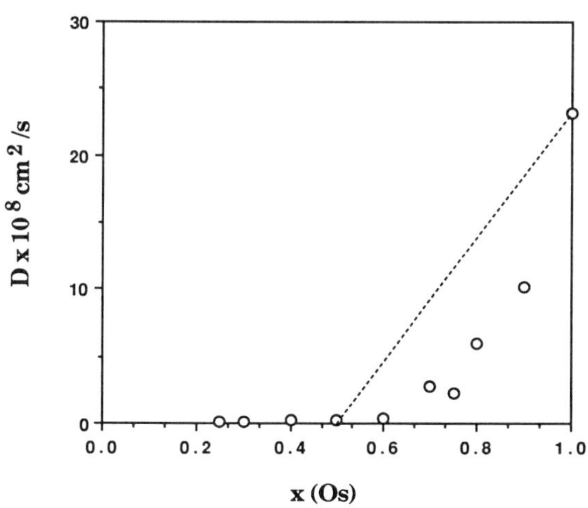

FIG. 15.9. Plot of the diffusion coefficient of the lateral electron hopping in $Os(DPP)_3$/octadecyl alcohol mixed monolayers at the water surface versus $Os(DPP)_3$ mole fraction. The dashed line is an approximate decay curve of the diffusion coefficient expected on the basis of the percolation theory (not involving long-range electron hopping) for a two-component system of disk on the triangular lattice.

Since our previous results have already demonstrated the solid-like character of $Os(DPP)_3$ monolayers, it becomes important to investigate strategies designed to solubilize $Os(DPP)_3$ in Langmuir monolayers before we can return to the measurements of the long-range electron transfer kinetics.

15.4. Conclusions

In summary, the results presented in this chapter demonstrate that electrochemical techniques can be used in the studies of monolayers on the water surface. In these studies a combination of the low current measurements offered by the electrochemical methods with the capabilities of the Langmuir experiments, which provide an exquisite control of the structure, composition, and environment of monolayer assemblies, presents a rather broad range of promising capabilities in the investigations of the dynamics of lateral processes as well as the kinetics of electron transfer phenomena.

Acknowledgments

I would like to thank Dr. B. J. Alder for a number of most interesting and illuminating discussions and for sharing with us the results of his molecular dynamics calculations. Research presented here was funded by the National Science Foundation under grants CHE-8807846 and CHE-9108378. We are grateful for this support.

References

1. Cherry, R. J. (1979). *Biochim. Biophys. Acta* **559,** 289.
2. Peters, R. (1981). *Cell Biol. Int. Rep.* **5,** 733.
3. Cadenhead, D. A. (1985). In: *Structure and Properties of Cell Membranes* (G. Benga, ed.). Vol. 3, CRC Press, Boca Raton, Florida, p. 21.
4. Edidin, M. (1981). In: *Membrane Structure* (J. B. Finean, and R. H. Michell, eds.). Elsevier, New York, Chap. 2, p. 37.
5. Vaz, W. L. C., Derzko, Z. I., and Jacobson, K. A. (1982). *Cell Surface Rev.* **8,** 83.
6. Wade, C. G. (1985). In: *Structure and Properties of Cell Membranes* (G. Benga, ed.). Vol. 1, CRC Press, Boca Raton, Florida, p. 51.
7. Peters, R., and Beck, K. (1983). *Proc. Natl. Acad. Sci. USA* **80,** 7183.
8. Vaz, W. L. C., Goodsaid-Zalduondo, F., and Jacobson, K. A. (1984). *FEBS Lett.* **174,** 199.
9. Vaz, W. L. C., Clegg, R. M., and Hallmann, D. (1985). *Biochem.* **24,** 781.
10. Almeida, P. F. F., Vaz, W. L. C., and Thompson, T. E. (1992). *Biochem.* **31,** 6739.
11. Widrig, C. A., Miller, C. J., and Majda, M. (1988). *J. Am. Chem. Soc.* **110,** 2009.

12. Charych, D. H., Landau, E. M., and Majda, M. (1991). *J. Am. Chem. Soc.* **113,** 3340.
13. Charych, D. H., Goss, C. A., and Majda, M. (1992). *J. Electroanal. Chem.* **323,** 339.
14. Turnbull, D., and Cohen, M. H. (1970). *J. Chem. Phys.* **52,** 3038.
15. Clegg, R. M., and Vaz, W. L. C. (1985). In: *Progress in Protein-Lipid Interactions* (A. Watts, and J. J. H. H. M. DePont, eds.). Elsevier, New York, Chap. 5, p. 173.
16. Tyrrell, H. J. V., and Harris, K. R. (1984). *Diffusion in Liquids, A Theoretical and Experimental Study*. Butterworths, London, Chap. 6.2, p. 266.
17. Chapman, S., and Cowling, T. G. (1970). *The Mathematical Theory of Non-Uniform Gases*. Cambridge University Press, Cambridge, Massachusetts.
18. Alder, B. J., and Wainwright, T. E. (1967). *Phys. Rev. Lett.* **18,** 988.
19. Alder, B. J., and Wainwright, T. E. (1970). *Phys. Rev. A1* 18.
20. Alder, B. J., Gass, D. M., and Wainwright, T. E. (1970). *J. Chem. Phys.* **53,** 3813.
21. Charych, D. H., and Majda, M. (1992). *Thin Solid Films* **210 / 211,** 348.
22. Majda, M. (1992). In: *Molecular Design of Electrode Surfaces* (R. W. Murray, ed.). *Techniques of Chemistry Series, Vol. XXII*, John Wiley, New York, Chap. IV, p. 159.
23. Honig, E., and Mobius, D. (1991). *J. Phys. Chem.* **95,** 4590.
24. Anvar, D., and Majda, M. (1992). Unpublished results.
25. Sutin, N. (1983). *Prog. Inorg. Chem.* **30,** 441.
26. Hupp, J. T., and Weaver, M. J. (1983). *J. Electroanal. Chem.* **152,** 1.
27. Chan, M.-S., and Wahl, A. C. (1978). *J. Phys. Chem.* **82,** 2542.
28. Weaver, M. J. (1992). *Chem. Rev.* **92,** 463.
29. Marconelli, M., MacInnis, J., and Fleming, G. R. (1989). *Science* **243,** 1674.
30. Barbara, P. F., and Jarzeba, W. (1988). *Acc. Chem. Res.* **21,** 195.
31. Charych, D. H., Anvar, D. J., and Majda, M. (1994). *Thin Solid Films* **242,** 1.
32. Marcus, R. A., and Sutin, N. (1985). *Biochim. Biophys. Acta* **811,** 265.
33. Zallen, R. (1983). *The Physics of Amorphous Solids*. John Wiley, New York, Chap. 4.
34. Blauch, D. N., and Savéant, J.-M. (1992). *J. Am. Chem. Soc.* **114,** 3323.

On-Line Structure Control of Langmuir-Blodgett Films

HANS RIEGLER AND KARL SPRATTE

Institut für Physikalische Chemie, Universität Mainz, Welder-Weg 11, D-55099 Mainz

16.1. Introduction

The Langmuir-Blodgett (LB) process is defined as the sequential transfer of ultrathin (typically monomolecular) organic layers onto a solid support to generate a lamellar film architecture on the solid substrate. The deposition of the monolayers is achieved by the consecutive submerging (withdrawal) of the substrate into (from) a liquid subphase with the monolayer floating on this subphase.[1-3] Through these dipping cycles, a layer-by-layer transfer is achieved. This qualifies the technique for the build-up of complex lamellar architectures consisting of stacks of ultrathin organic monolayers, called *LB multilayers*. The fabrication of LB films is an elaborate procedure and the final product is the result of the interplay of the various intralayer and interlayer interactions, of drainage hydrodynamics, and of dilational and shear stresses. The scientific interest in this field pertains to those interactions and forces whereas applications concentrate on the final product, the LB multilayer. Of course, both, science and application, are linked together. LB multilayer design with specific sample properties requires the knowledge and control of the physics and chemistry of the preparation of LB films. The process starts with the preparation of the precursor Langmuir monolayer and the substrate surface, followed by the contacting of both surfaces (the LB deposition), to result finally in the LB multilayer.

The field of Langmuir monolayer research has developed extensively in recent years, independent from its relation to the LB technique. LB multilayers also have been investigated for a long time. The essential part of the LB technique, the transfer itself, has not yet been the focus of detailed research although it is of considerable scientific interest and, in

ISBN 0-12-523485-6

view of applications, opens manifold opportunities for layer control and modification.

During the transfer the system is in a specific configuration, the *LB transfer* or *Langmuir wetting configuration* (see Figs. 16.3 and 16.5 later in this chapter). In this formation the three bulk phases—solid, liquid, and gas—contact at the three-phase line, and the quasi two-dimensional (2-D) Langmuir monolayer forms the interface between the gas and the solid or liquid (hydrophilic configuration), or the liquid and the solid or gas (hydrophobic configuration).

This Langmuir wetting configuration offers unique possibilities in interface science to investigate adhesion, local phase transition, and wetting phenomena. The contacting of the Langmuir monolayer with the substrate surface during the LB transfer represents, on a molecular scale, a one-dimensional (1-D) physisorption. The monolayer molecules are deposited consecutively at the advancing (receding) three-phase line, well defined in space and time. Compared to other preparation techniques for ultrathin organic layers (e.g., SA, MBE) this additionally allows for the local lateral structure and composition control during the preparation. In this chapter, we outline the scope of the structural variety of LB multilayers, analyze the LB technique with respect to structure control and sample design, and finally present several examples of structure control and modification performed during the monolayer deposition.

16.2. Structural and Compositional Variety of Organic Multilayers

The most general LB film is a multilayer consisting of different monolayers of locally varying structure and composition (Fig. 16.1). The individual monolayer is characterized by the local distribution of its molecular packing and composition, i.e., its phase (type and size of the unit cell) for monomeric substances, or the packing density and molecular orientation for polymers. The local distribution of the various packings and/or compositions specifies the domain morphology and/or density distribution. Areas of different molecular packing are separated by transition regions, which can be described by various types of defects. The multilayer (superlattice) structure is determined by the type, sequence, and spatial orientation of its constituting monolayers: the spacing between individual layers, the layer orientation normal to the layer plane (e.g., head/head or head/chain), the layer orientation parallel to the plane (e.g., the tilt orientation of molecules in each monolayer), the lateral shift between consecutive layers, and the local angle between two layers (buckling). Further, lateral structuring of the superlattice can occur on a larger scale (e.g., patterns or grooves).

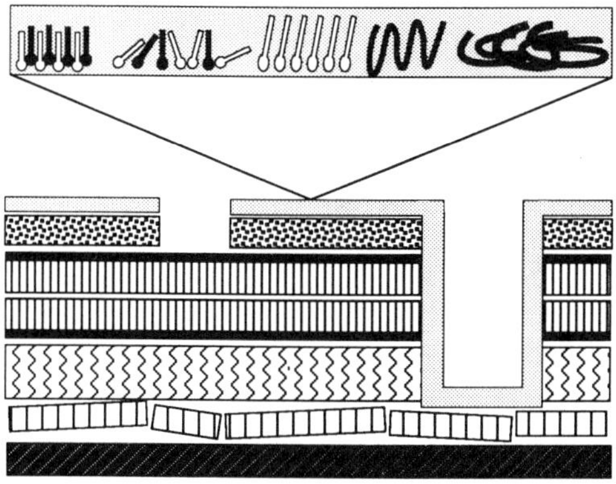

FIG. 16.1. Sketch of the structural variety of LB multilayers. The structure can vary on different levels: Within each monolayer the molecular packing and composition may be locally different. On the superlattice level the structure is defined by the sequence, composition, and relative positioning of the individual layers. On the microscopic scale, there can be grooves or missing sections, etc.

16.3. Schematic of the Preparation of Langmuir-Blodgett Films

A scheme of the various steps taken in the preparation of LB films is presented in Fig. 16.2. For each step several examples of preparational modifications (e.g., for Langmuir monolayers, its surface pressure, subphase ionic conditions) and the corresponding affected system properties (e.g., phase, defect densities, etc.) are presented. The process commences with the separate preparation of substrate surface and Langmuir monolayer. This aspect is often overlooked because the substrate surface preparation is regarded as less important compared to that of the Langmuir monolayer. This originates in two widely spread premises: (1) The carrier surface is not affected by the LB deposition, and (2) the deposition does not affect the structure of the Langmuir monolayer. Accordingly, LB film preparation is regarded as multiple layering of sections of the same monolayer (the one prepared at the air/water interface).

Under these propositions, the structure of the multilayer depends only on the preparation of the precursor Langmuir monolayer; substrate surface properties are irrelevant. However, in reality most monolayers are altered by their deposition (especially the first few layers on the solid substrate) so that in fact two different surfaces are contacting during deposition. In the deposition step both systems are combined, resulting in the layer/substrate system. From the view of sample control and modifica-

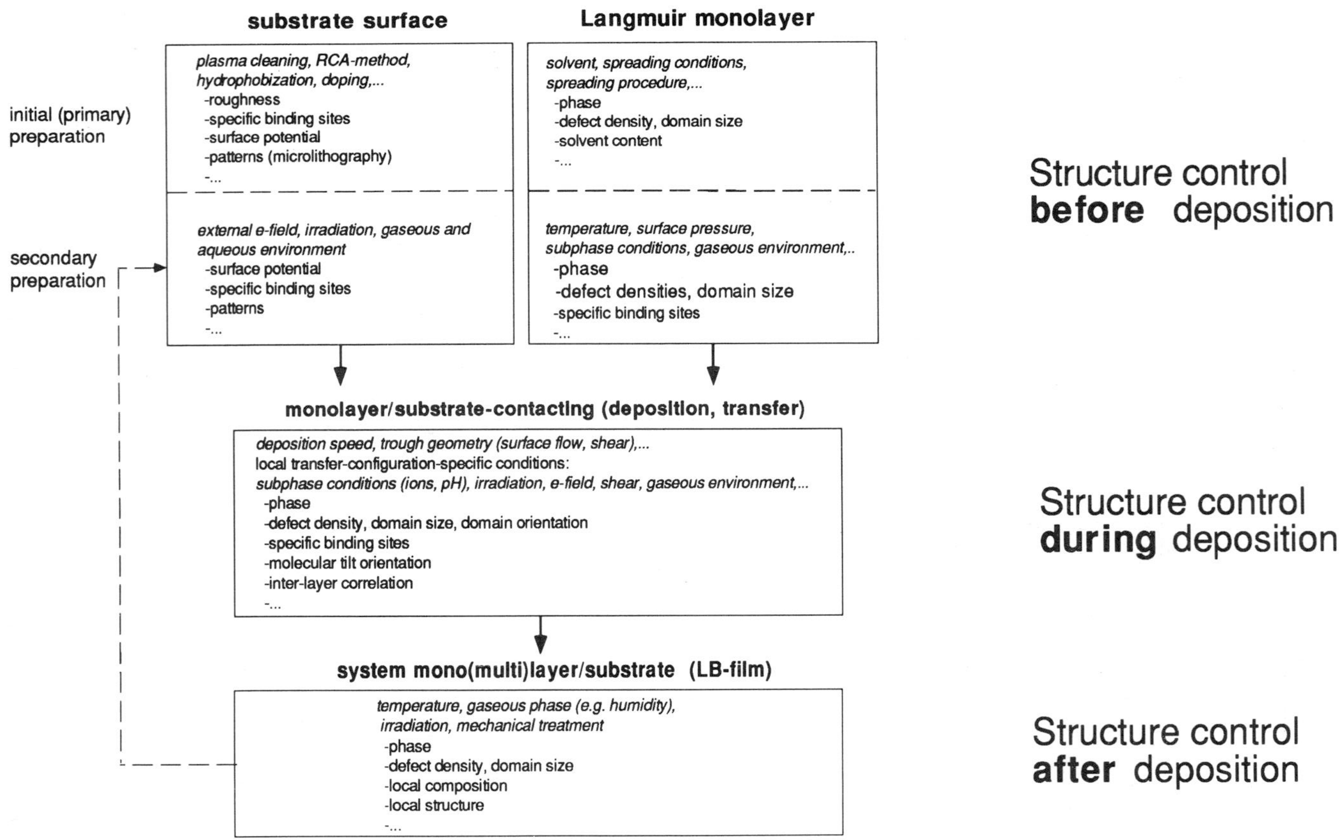

FIG. 16.2. Schematic of the preparation of LB films with emphasis on the possibilities of structure control. At each preparation step, examples of specific experimental conditions and corresponding, affected system properties are listed.

tion, it is useful to split the preparation into sections before the deposition, during the deposition, and after the deposition. All these stages offer possibilities of system control and allow modifications of the structure of the final LB film.

16.3.1. Structure Control Before Deposition

Before monolayer deposition, the solid surface and the monolayer are separate units. One has to distinguish between the preparation of the substrate surface and that of the Langmuir monolayer. An additional distinction may also be useful between the initial (primary) preparation and the secondary preparation. Per definition the primary treatment concerns those parameters which cannot be modified later during the (repeated) deposition of monolayers. Typical primary operations are the preparation of the substrate surface (e.g., hydrophilization or hydrophobization) or the use of specific spreading solvents or techniques. Secondary monolayer treatments typically adjust the monolayer phase (lateral pressure, temperature, surface pH, etc.). More complex structural control can also be performed at this stage (compare numerous experiments on monolayer polymerization,[3] local thermal treatment,[4,5] and other types of defect annealing).[6]

16.3.2. Structure Control During Deposition

The question may arise as to why structural control during deposition should be treated as an individual processing stage, different from structure control before and after the deposition. This is explained as follows: Generally the system preparation and structure control before deposition concern separately the Langmuir monolayer (on the plane liquid surface) and the substrate. Both are not in a typical LB deposition configuration or, in case they are, influences due to this configuration are neglected because at any given time most of the monolayer or substrate surface is not affected by the specific conditions of the transfer configuration. However, in the course of the transfer, sequentially every section of the layer and substrate are brought into contact (Fig. 16.3). As a consequence, every section of the monolayer and the substrate will eventually be affected by the local transfer-specific conditions.

Several preparational parameters are only defined in the context of the transfer configuration (e.g., shear-induced alterations due to water drainage and subphase flow[7–9]). Per definition, all the local differences of relevant parameters (pH, temperature, gaseous composition, electrical field, etc.) between their global values (at the plane water or substrate surface) and

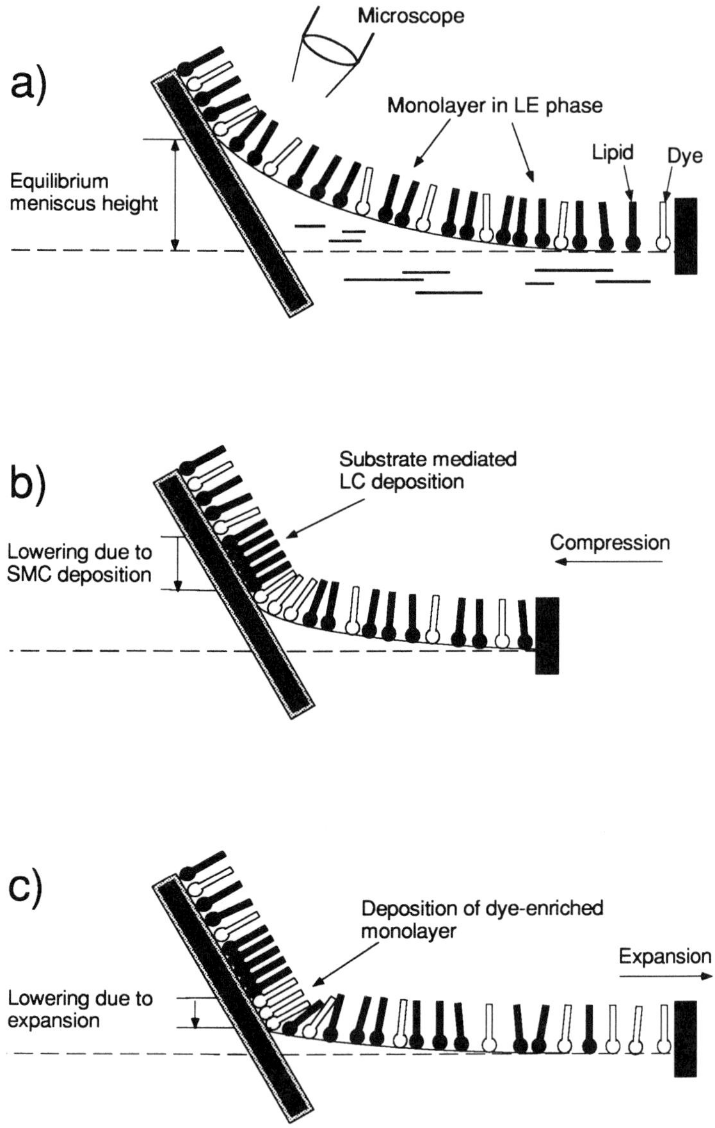

FIG. 16.3. Preparation of substrate-mediated dye-depleted stripes (SMC deposition) and dye-enriched stripes.

their values in the vicinity of the meniscus are transfer-related parameters. It is further useful to distinguish two types: (1) alterations caused by the specific monolayer and substrate environment in the meniscus area (e.g., the local pH,[10] flow-induced shear, local contamination pile-up,[11] and (2) alterations due to monolayer/substrate interactions in the moment of contact, a process that can be described as 1-D physisorption of the monolayer molecules onto the substrate.

16.3.3. Structure Control After Deposition

The modification of LB films has been extensively researched in recent years. Numerous groups have investigated the patterning, polymerization, and thermal treatment of LB films.[3] In the context of this chapter, it is important to note that all films manipulated after deposition can of course serve as substrates for further layer deposition.

16.4. Structure Control During the Transfer Due to Substrate /Monolayer Interactions

The deposition of a monolayer from the aqueous subphase onto the solid substrate surface requires a significant change of the monolayer environment and will induce its structural reorganization. Structural and morphological alterations in monolayers can often be visualized by fluorescence microscopy.[12–14] For this purpose the monolayer is doped with minor amounts (~ 1 mol%) of a fluorescence dye, which has different solubilities in different monolayer phases. Especially suited for this technique are monolayers of phospholipids, which show within a certain pressure/temperature range the coexistence of a higher ordered liquid-condensed (LC) phase and a lower ordered liquid-expanded (LE) phase. Typically, the dye is less soluble in the LC phase, so that domains of this phase appear as probe-depleted, nonfluorescing areas in a fluorescing LE matrix. The addition of fluorescence dye has two implications: (1) It is possible to visualize and localize different monolayer phases, and (2) the monolayer behaves like a mixture of a solvent (the phospholipid) and a solute (the dye).

In the following we present some experiments performed with monolayers of L-α-dimyristoylphosphatidylethanolamine (DMPE) doped with a few mole percent of fluorescence dye [L-α-phosphatidylcholine-β-(NBD-aminohexanoyl)-γ-palmitoyl (P 3412, SIGMA)]. The monolayers were deposited on silicon wafers with a thermally grown SiO_2 layer of several 100

Å thickness.[11] The on-line observations were performed with the transfer microscope.[15]

16.4.1. Deposition of Dark, Probe-Depleted Stripes Via Pressure-Induced Substrate-Mediated Condensation

Figure 16.3 shows a sketch of the modified LB procedure, which can be used to prepare samples with patterns of defined geometry and composition. The method utilizes the effect of substrate-mediated condensation (SMC) in combination with fractional condensation.[11,16,17] According to the schematic, the monolayer initially covers the aqueous subphase and the substrate [Fig. 16.3(a)]. The monolayer is in its LE phase at low pressure (≈ 1 mN/m) and the meniscus height has relaxed to its equilibrium value. When the monolayer is now compressed slowly (< 0.1 Å^2 mol^{-1} s^{-1}) and reaches a certain threshold pressure, the deposition of a probe-depleted monolayer stripe is observed [Figs. 16.3(b) and 16.4]. The threshold pressure is several mN/m below the coexistence pressure of the LE and LC phases on the plane water surface.

The stripe deposition is accompanied by the lowering of the meniscus height. According to this lowering of the meniscus, one can postulate that the interface energy of the substrate/monolayer/air interface is lower for the condensed probe-depleted stripe than for the originally deposited LE phase mixed with the dye [Fig. 16.3(a)].[11] The condensation is accompanied

FIG. 16.4. SMC deposition of a DMPE monolayer. The dark stripe (width ≈ 6 μm) is being created according to the process of Fig. 16.3(b). The micrograph has been taken on-line with the transfer microscope, showing on the right side the bright, floating monolayer. The monolayer deposited before SMC deposition is on the left and separated from the SMC stripe by a narrow dye-enriched stripe produced according to Fig. 16.3(c).

by the segregation of dye into the floating LE phase. Therefore, the deposited stripe appears dark. Such a dye pileup in front of the three-phase line is also observed during the regular slow dynamic Langmuir wetting (i.e., substrate-upstroke at constant pressure). In the case of continuous transfer, phenomena analogous to those occurring during the solidification of three-dimensional (3-D) alloys are observed.[18,19] The SMC deposition can be terminated by stopping the slow compression and waiting for the equilibration of the three-phase line.

16.4.2. Deposition of Dye-Enriched Bright Stripes

Geometrically defined stripes containing high concentrations of fluorescence dye are deposited when the monolayer is reexpanded after SMC deposition [Fig. 16.3(c)]. This deposition upon expansion is observed when the floating monolayer adjacent to the three-phase line contains an increased amount of dye due to the previous segregation in the course of SMC deposition. The reason for this deposition is not yet well understood. It is probably caused by the increase of the interface energy at the air/water interface during expansion, whereas the other interface energies (substrate/monolayer/air, substrate/water) remain about constant. The dye concentration of the deposited stripe can be controlled by various parameters: It depends on the time that has elapsed after the end of the SMC deposition step. This waiting time determines the diffusion-controlled decrease of the dye concentration profile adjacent to the three-phase line. The width of the precursing SMC deposited stripe also affects the following deposition through the meniscus height and the degree of local dye enrichment.

Some other parameters are the compression and expansion speeds, the pressures applied during the various steps, and the inweight probe concentration. The alternating deposition of dye-depleted SMC stripes and stripes of increased dye concentration can be repeated infinitely by inserting suitable upstrokes of the substrate during or after the various compression, expansion, and relaxation steps. A simple procedure to obtain alternating stripes is, for example, the deposition of SMC stripes upon monolayer compression, followed by the deposition of dye-enriched stripes. During both processes the substrate is kept stationary (static Langmuir wetting). Then the original meniscus height is reestablished before SMC deposition through a substrate upstroke (under constant low pressure). Then the sequence is repeated.

The deposited stripes span the entire breadth of the substrate (typical dimensions: centimeters). Their widths typically range from submicrons to several tens of microns. The width of the SMC deposition can, in principle, be extended to infinity as can be shown through ultraslow upstrokes (< 1 $\mu m/s$) of the substrate while the monolayer is held above the threshold pressure. In this way, SMC stripes of more than 1 mm width have been fabricated in our laboratory. During the very slow upstroke and concomittant SMC deposition process time is sufficient for the segregated dye to dilute into the plane floating monolayer. It is also possible to enforce the dye dilution and transport by induced convection. Thus the process resembles the preparation of ultrapure substances (zone refinement) in three dimensions, where the solution is continuously stirred for contaminant removal.

16.4.3. Influence of an Inhomogeneous External Electrical Field at the Three-Phase Line

The structure control described in the previous paragraphs can be augmented by the application of an external electrical field. A suitable field configuration is sketched in Fig. 16.5. An electrical potential difference is applied between the silicon substrate and a gold wire submerged into the aqueous subphase. The SiO_2 layer on the substrate surface is electrically insulating. Thus only a negligible current flows and the silicon carrier and the aqueous subphase serve as two differently charged poles. In this case local monolayer modifications are caused by electrostatic interactions with the electrostatic field. Chemical modifications due to electrolysis[20,21] can be neglected.

The effect of electrical fields on the morphology of plane monolayers has been described before,[22] with the water surface as plane equipotential surface and a metal tip electrode some 10 μm above the monolayer. In this case the monolayer interacts with the electrical field and field gradients at the plane water surface. The energy contribution of the electrical field has been shown to be sufficient to move molecule aggregates although its share per molecule is less than kT. A direct influence on molecular ordering (e.g., melting or crystallization) can therefore be neglected. The situation presented here is significantly different. The aqueous subphase represents a wedge-like electrode with the three-phase line at its tip. The counter electrode is the plane substrate body covered with an insulating SiO_2 layer only several 100 Å thick. The monolayer covers this wedge electrode and the SiO_2 surface and is directly exposed to the field at the three-phase line at the tip of the wedge. Due to the tip geometry the local field strength can be high and energetic contributions of more than kT for individual molecules cannot be excluded. Accordingly, a direct influence of the electrical field on the monolayer structure might be feasible.

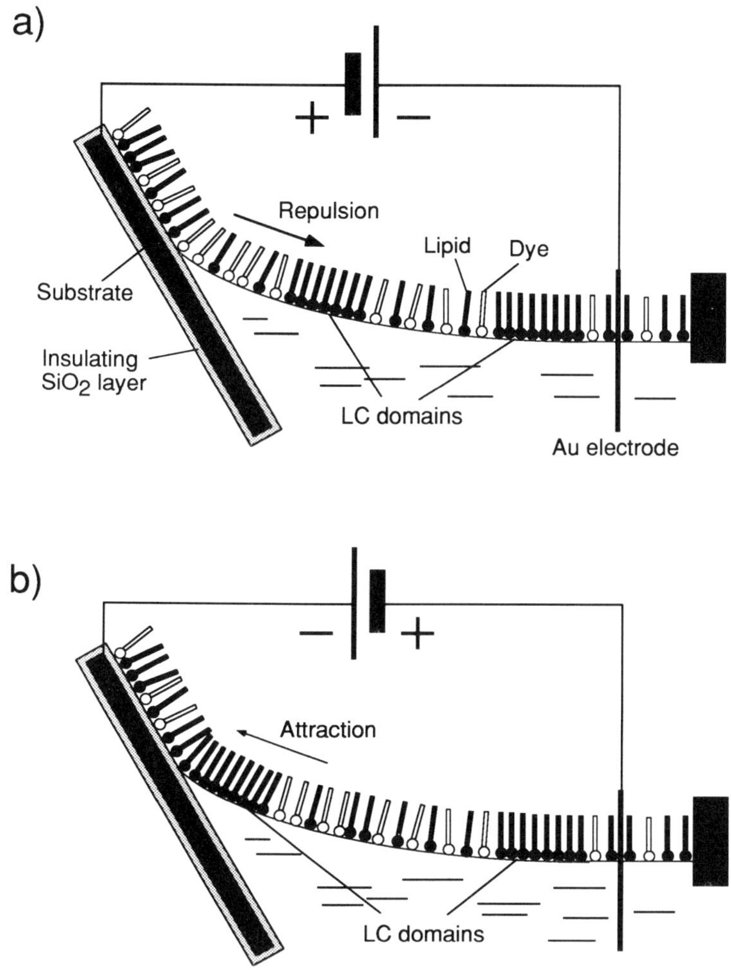

FIG. 16.5. Schematic of the electrical field configuration and its effect on LC domains.

Manipulations of the domain arrangements caused by the electrical field can readily be observed. The influences are very similar to those observed with a tip electrode above the water surface. The schematics of Fig. 16.5 show the field configurations under which the micrographs of Fig. 16.6 were obtained. The floating DMPE monolayer of Fig. 16.6 is in its LC/LE phase coexistence. The three-phase line bisects the frames into the deposited monolayer part (left) and the floating section (right). Without an electrical field [Fig. 16.6(b)] the domains are evenly distributed in front of

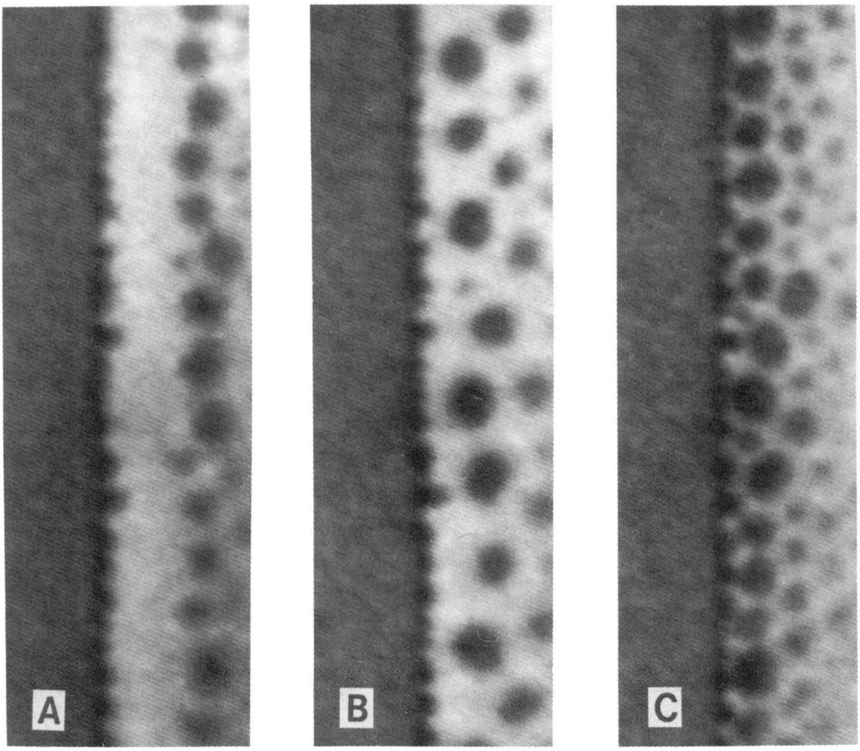

FIG. 16.6. Domain repulsion and attraction in an electrical field as sketched in Fig. 16.5. The three-phase line runs vertically in the middle of the frames with some fingerlike, floating LC domains attached to it on the right. On the left the monolayer is deposited, on the right it is floating. Frame A shows the situation with the substrate charged positively (+300 V); frame B is without the electrical field. In the case of frame C, the substrate is charged negatively (−300 V). (Each micrograph: 30 μm × 100 μm.)

the three-phase line. If the substrate is positively charged versus the subphase (+300 V), the domains are repelled from the three-phase line (Fig. 16.6(a)]. With the substrate surface charged negatively (−300 V), domain attraction is observed [Fig. 16.6(c)]. This electrical field effect can be explained analogously to experiments performed on the plane water surface: Due to the different dipole density of the LC phase compared to the LE phase, the inhomogeneous field component parallel to the water surface attracts or repulses the domains.

In all three frames one can also observe fingerlike, floating LC domains pinned to the three-phase line. This specific domain growth has been described elsewhere and is caused by the three-phase line serving as a 1-D nucleation site. The attraction and repulsion of the floating domains as it is shown in Figs. 16.6(a) and (c) is completely reversible, whereas the fingerlike domains are fixed to the three-phase line, albeit floating. The effect of the electrical field on the nucleation properties of the three-phase line is still under investigation.

The micrographs of Fig. 16.6 demonstrate that an external electrical field can be used to modify the morphology and composition of the monolayer selectively in the vicinity of the three-phase line. This can be combined with the patterning capabilities provided by the alterations between SMC deposition, deposition upon expansion, and conventional LB deposition with a substrate upstroke.

Figure 16.7 shows lines of various fluorescence intensities and widths (1 to 7), which are separated by dark, low-fluorescence monolayer sections (A to H). The photograph has been taken with a conventional microscope from a deposited monolayer. The fluorescence profiles of the stripes are plotted below the micrograph. The stripes have widths of typically 5 to 20 μm and span the whole breadth of the substrate. Remarkable are the sharp and well-defined edges between the dye-enriched and dye-depleted stripes. Particularly for the transition B-2 and 2-C the edges are as sharp as 1 μm. The stripes have been produced under various different conditions (compression speed, relaxation time, electrical fields of various strength and polarity).

It turns out that the correlation between stripe widths, composition, and fabrication parameters is not perceived easily and is still under investigation. Nevertheless, in agreement with the observations of the floating domains (Fig. 16.6) the width of the dye-free SMC-deposited stripes is affected by the applied field and polarity. Stripes E, F, and G have been deposited with the substrate charged negatively, neutrally, and positively. Analogous to the attraction of the LC domains for negatively charged substrates, the SMC stripe is broader for the negatively charged substrate. In agreement with this, the concentration of fluorescence dye in the bright stripes increases with positive polarity.

16.5. Conclusions

The preparation of LB films is a complex procedure in which monolayer molecules preorganized at the air/water interface are deposited onto the air/substrate interface. It is a "soft" process performed around room temperature and thus especially suited for the organization of fragile molecules (e.g., biomolecules) into complex stratified layer structures. The various preparational steps offer manifold possibilities to study the

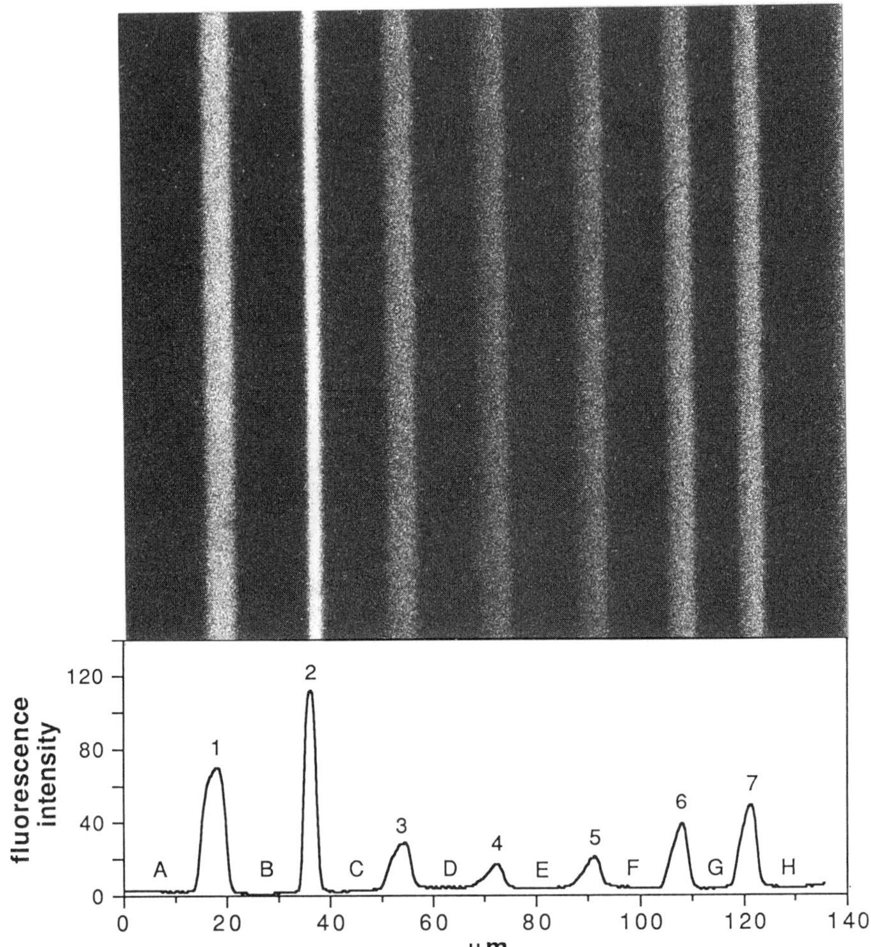

FIG. 16.7. Deposited stripes of alternating high and low fluorescence dye concentration. The fluorescence intensity is plotted below the micrograph. The alternating stripes have been created in cycles of SMC deposition, expansion, and upstroke to restore the original equilibrium mensicus height (compare Fig. 16.3). During the process, electrical fields of varying strength and polarity were applied.

behavior of 2-D organic molecule arrangements and, concomitant, their purposeful structure control. The pivot of the LB technique is the system in its LB transfer configuration.

The contacting of monolayer and substrate surface occurs at the three-phase line and the deposition of the molecules can be described as a 1-D physisorption. In the course of this physisorption the molecular arrangement will change. In the case of mixed monolayers, local alterations are induced analogous to those occurring during the solidification of 3-D alloys,[23] where they are used for processes such as directional solidification, zone melting, or zone refinement.

Potential applications of deposited ultrathin organic films are often based on the functionalization of the monolayer by incorporation of active molecules (e.g., proteins) into a passive carrier matrix. In this case it is important to control and modify the local distribution of the active components. The few examples presented in this chapter show that this can be accomplished.

It is further demonstrated that the application of an external electrical field can be used to manipulate film structure and composition. Not too much imagination is necessary to think of complicated conducting patterns (e.g., insulated, crossed conducting lines) on the substrate that allow the selective local charging of any point on the substrate (Fig. 16.8). Thus the local 2-D addressing and modification of any monolayer location would be possible during deposition and it is possible to generate purposeful, complicated 3-D lamellar structures with custom-designed properties. The modifications presented are based on selective electrostatic interactions between different monolayer compositions or structures. This is not a harsh restriction because significant differences in dipole density or charge often exist or can be induced by tailor-made molecules.

Apart from these more technological considerations, we must emphasize that the monolayer in its LB transfer configuration (Langmuir wetting

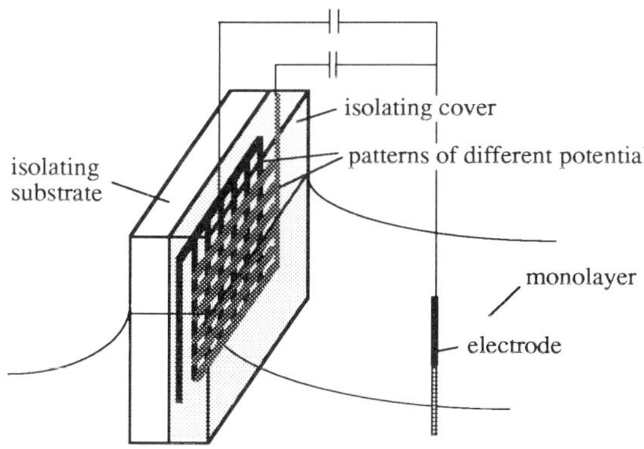

FIG. 16.8. Schematic of conducting pattern for a locally varying electrical field. The potential variation can be used to replicate a similar structural and/or compositional pattern in the LB film. Time-dependent fields and more complicated conducting structures allow the purposeful build-up of much more complex 3-D layered structures.

configuration) and the interpretation of the transfer as a 1-D physisorption offer unique possibilities for fundamental research in the field of 2-D phase transitions, adsorption, and wetting.

Acknowledgments

We thank the Deutsche Forschungsgemeinschaft for financial support (Projekt Ri 529) and Wacker AG, Burghausen, for providing the silicon wafers. This work would not have been possible without fruitful discussions with many of our colleagues, including Helmuth Möhwald in whose group this work has been done.

References

1. Swalen, J. D., Allara, D. L., Andrade, J. D., Chandross, E. A., Garoff, S., Israelachvili, J., McCarthy, T. J., Murray, R., Pease, R. F., Rabolt, J. F., Wynne, K. J., and Yu, H. (1987). *Langmuir* **3,** 932.
2. Ulman, A. (1991). *An Introduction to Ultrathin Organic Films from Langmuir-Blodgett to Self-Assembly*. Academic Press, New York.
3. Barraud, A., and Palacin, S., eds. (1992). "Langmuir-Blodgett Films 5," *Thin Solid Films* **210 / 211.**
4. Kasuga T., Kumehara, H., Watanabe, T., and Miyata, S. (1989). *Thin Solid Films* **178,** 183.
5. Shimomura, M. (1992). *Thin Solid Films* **210 / 211,** 98.
6. Bibo, A. M., and Peterson, I. R. (1992). *Thin Solid Films* **210 / 211,** 515.
7. Tabe, Y., Sugi, M., Ikegami, K., Kuroda, S., Saito, K., and Saito, M. (1992). *Thin Solid Films* **210 / 211,** 32.
8. Decher, G., and Klinkhammer, F. (1989). *Thin Solid Films* **178,** 445.
9. Duda, G., Schouten, A. J., Arndt, T., Lieser, G., Schmidt, G. F., Bubeck, C., and Wegner, G. (1988). *Thin Solid Films* **159,** 221.
10. Riegler, H. E., and LeGrange, J. D. (1988). *Phys. Rev. Lett.* **61**(21), 2492.
11. Riegler, H., and Spratte, K. (1992). *Thin Solid Films* **210 / 211,** 9.
12. McConnell, H., Tamm, L. K., and Weis, M. (1984). *Proc. Natl. Acad. Sci.* **81,** 3249.
13. Lösche, M., and Möhwald, H. (1984). *Rev. Sci. Instrum.* **55,** 1968.
14. Riegler, H., and Möhwald, H. (1992). In: *Fluorescence Spectroscopy* (O. S. Wolfbeis, ed.). Springer Verlag, Berlin, p. 111.
15. Riegler, J. E. (1988). *Rev. Sci. Instrum.* **59**(10), 2220.
16. Riegler, H., and Spratte, K. (1990). In: *Dynamical Processes in Condensed Molecular Systems* (A. Blumen, J. Klafter, and D. Haarer, eds.). World Scientific, Singapore, p. 105.
17. Spratte, K., and Riegler, H. (1991). *Makromol. Chem. Macromol. Symp.* **46,** 113.
18. Spratte, K., Chi, L. F., and Riegler, H. (1994). *Europhys. Lett.* **25,** 211.
19. Spratte, K., and Riegler, H. (1994). *Langmuir* **10,** 3161.
20. Charych, D. H., and Majda, M. (1992). *Thin Solid Films* **210 / 211,** 348.
21. Tieke, B., Wegmann, A., Fischer, W., Hilti, B., Mayer, C. W., and Pfeiffer, J. (1989). *Thin Solid Films* **179,** 233.
22. Heckl, W. M., Miller, A., and Möhwald, H. (1988). *Thin Solid Films* **159,** 125.
23. Kurz, W., and Fisher, D. J. (1989). *Fundamentals of Solidification.* Trans Tech Publications Ltd., Switzerland.

Structure

Phase Diagrams and Chain Order in Monolayers of Aliphatic Chains

I. R. PETERSON

Nima Technology, The Science Park
Coventry CV4 7EZ, England

17.1. Phases of Amphiphilic Monolayers

The idea that water-surface monolayers of amphiphilic molecules can have an internal structure corresponding to that in the gas, liquid, and solid states of bulk matter can be traced back to Devaux.[1] Following the work of Adam,[2,3] mainstream opinion accepted in addition the existence of a fourth state, the liquid condensed phase, whose structure was subject to debate. However, the existence of four phases apparently spoiled the simplicity of the initial concept, and even quite recently,[4,5] serious attempts have been made to reduce the accepted number of surface phases to three. In both of these, experimental evidence was interpreted in terms of specific models to mean that the liquid condensed and the liquid expanded phases are one and the same. Anticipating a demonstration of this sort, many authors have assumed implicitly that there are only three phases. Isolated reports of a rich polymorphism in monolayers of well-known amphiphiles[6-9] have been largely ignored.

However, the latter minority opinion has been recently and unambiguously confirmed by synchrotron x-ray diffraction studies from several independent groups.[10-12] The original phase diagrams and phase assignments of Stenhagen and Lundquist have stood up very well to the latest investigations, with only minor modifications to the positions of phase boundaries, and new recognition of some phase distinctions that they had overlooked.[13-15] Figure 17.1 shows the latest phase diagram for docosanoic acid and schematics of the local molecular packing.

It turns out that the phases originally identified as gaseous and liquid were in fact correctly classified, and that their molecular organization corresponds to that in their higher dimensional analogue. The solid phase was, however, wrongly identified. At least one crystalline solid phase does

ISBN 0-12-523485-6

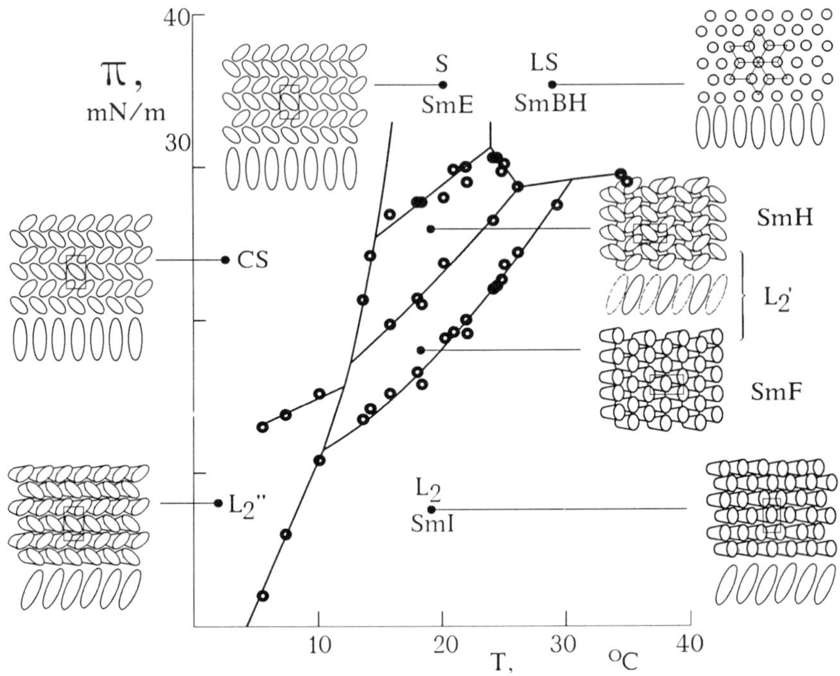

FIG. 17.1. π-T phase diagram for a monolayer of docosanoic acid on water, with schematics of the local molecular packing, both top and side views, showing the two-dimensional (2-D) unit cell seen in diffraction. The Harkins-Stenhagen-Lundquist (HSL) nomenclature is given in each case, and the miscibility category is shown for those phases demonstrated to be mesophases.

exist, but only at temperatures much lower than the range of stability of the monolayer phase usually called "solid." The intermediate phases have some similarities to liquids, because they flow in response to surface pressure gradients, with threshold stresses for plastic flow usually zero, and in the remaining cases much lower than the MPa typical for organic solids.[16] In other respects, they resemble solids, for example, they show x-ray diffraction peaks. Because of their intermediate behavior they are neither true liquids nor true solids. However, just like the other surface phases they have analogues in three dimensions, which were known even at the time of Devaux. These are the mesophases. In hindsight, this appears unsurprising, because amphiphiles have long been known to form bulk mesophases, not just lyotropic but also thermotropic.[17,18]

It is of interest to assign the 2-D phases to known categories of bulk mesophases. For example, a phase that is isotropic in the plane could be called a 2-D liquid, and a material with just one orientation related to the molecular axes could be called a 2-D nematic. Unfortunately, this obvious assignment cannot be extended to the phases shown in Fig. 17.1, which have directors associated with both the lattice and the molecular axes. An unambiguous assignment can be made by restricting the bulk mesophases to lamellar phases, i.e., smectics. A monolayer phase can be identified with the smectic category which has the same molecular arrangement within each of its smectic planes. This nomenclature has advantages related to two experimental techniques. In the first, it is possible to prepare free-standing films of smectics as little as two layers thick. Conversely, using the Langmuir-Blodgett (LB) technique, it is possible to build up macroscopically thick films from amphiphilic monolayers. When using the proposed system, the phase nomenclature does not have to change in the thin layer limit. In addition to the Harkins-Stenhagen symbol, Fig. 17.1 gives the smectic nomenclature defined in this way for each of the mesophases.

The evidence for the phase diagram of Fig. 17.1 was obtained mainly from x-ray diffraction[10] and isotherm studies[14] of docosanoic acid monolayers. However, two of the details shown are only inferred. The distinction between the SmH and SmF subphases of the L'_2 region of long- and short-range correlation, respectively, of the axial order parameter, was deduced[14] from a consideration of areas per molecule[10] together with the second-order nature of the L'_2-S transition.[6,10,14] The structure of the L''_2 phase was inferred by analogy from an x-ray diffraction study of the corresponding phase in heneicosanoic acid[11] together with the second-order nature of the L''_2-CS transition.[9]

If many lamellar phases are mesophases, it follows that to understand them, it is necessary to have some insight into the structure and behavior of mesophases. In the light of present knowledge, a mesophase can be defined to be one showing long-range orientational order, but with incomplete translational order. The orientational order may be represented mathematically[19] by a director field **n**, which can correspond physically to the average orientation of the molecular axis or to the direction from one molecule to its nearest neighbor. The lack of long-range translational order means that it is much easier to cause fluctuations of **n** than it is in a crystalline material. For example, in the "off" state of a twisted-nematic display, the director describes a helix from one wall of the cell to the other, while the application of an electric field in the "on" state results in a new continuous distribution representing an equilibrium between the forces due to the field and those due to surface alignment.

There is a widespread belief that all materials consist of domains of constant orientation. Although this can only be an approximation, because the application of a bending moment will cause a domain of any material to take on different orientations at different points, it is an experimental fact that the idea holds to good accuracy for polycrystalline materials. However, in light of the above, it is a very poor approximation to the behavior of mesophases. Unfortunately, this idea is taking a long time to die out, specifically in connection with aliphatic chain monolayers.

There is, in fact, no mystery about the observed continuous variations of orientation within a mesophase domain. The splay, bend, and twist behavior of smectics was treated by Oséen and Zöcher in the early years of this century, and the theory was put in its most general form by Frank.[20] The predictions of this theory have been verified, and the elastic constants measured, for a vast number of liquid crystals. The theory shows that even the apparently well-defined "domain walls" which occur in the more highly ordered mesophases are not always molecularly sharp.

Variations of orientation are thus characteristic for mesophases and distinguish them from true crystals. Nevertheless, the fully aligned state has a special significance. In the absence of perturbations from the surface, and if the mesophase is not chiral, its thermodynamic ground state has **n** parallel everywhere. When fully aligned, such a mesophase—even a nematic—gives rise to diffuse but discrete diffraction spots,[19] indicating the presence of local translational order both parallel and normal to **n**. Figure 17.2 shows a cross section through a region of splay, together with a deformed rectangle related to the molecular order. Sides AB and CD of the rectangle are parallel to the director everywhere, while sides BC and DA are perpendicular everywhere. Clearly, as a result of the splay, sides BC and DA are of different length and so are crossed by different numbers of molecular rows. By construction, AB and CD are not crossed by any rows. As an inescapable conclusion, some of the rows crossing side BC must terminate within the rectangle. Terminating rows are of course well known in crystallography and are called *dislocations*.

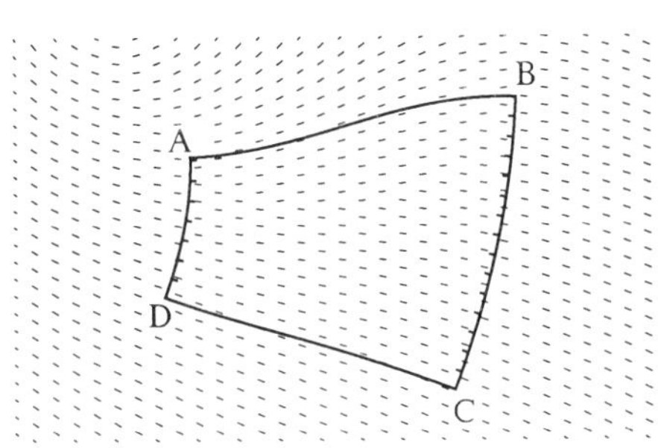

FIG. 17.2. The sides of distorted rectangle ABCD are everywhere either parallel or perpendicular to the local director **n** of a mesophase subject to splay. Side BC is longer than side AD, so that some of the molecular rows crossing BC terminate within ABCD.

For a 2-D mesophase, **n** is fully specified by the polar angle ϑ. It is possible to analyze the diagram of Fig. 17.2 mathematically, while allowing the rectangle to shrink to zero. It is readily shown that:

$$\nabla\vartheta = \boldsymbol{\beta} \tag{17.1}$$

where $\boldsymbol{\beta}$ is the areal density of dislocation Burgers vectors **b**. Hence, it can be seen that the relationship between dislocations and splay is mutual: Continuous variations of orientation are indicative of a distribution of dislocations throughout the material, while an (unbalanced) distribution of dislocations, of whatever origin, will cause the director orientation to change. In polycrystalline materials, orientational change occurs only at grain boundaries, indicating that dislocations are actively expelled from the interiors of domains. Conversely, it follows that dislocations must be present in high densities and at thermodynamic equilibrium in all mesophases.

A theory of the thermodynamics of dislocated lattices has been put forward, and it has been shown that mesophases do arise in a natural way in such systems.[21,22] The calculations are particularly simple for a smectic BH phase, which has isotropic mechanical behavior. In this case it is possible to analyze the profiles of the diffraction spots, which are distinctly different from those of a crystalline phase. It has been shown that the transmission electron diffraction profiles from monolayers of two different types of amphiphiles in this phase (on thin polymeric supports rather than on the water surface) correspond to those expected from the Nelson and Halperin theory.[23]

17.2. The Principle of Corresponding States

There is no necessary correspondence between the phase diagrams of two different materials. However, new isotherm studies of a number of long-chain amphiphiles, justified by the synchrotron data, have revealed interesting regularities. Perhaps not surprisingly, the phase diagrams plotted against surface pressure π and temperature T for the homologous series of the straight-chain fatty acids[13] are very similar to one another, just as reported by Lundquist for the long-chain ethyl and acetate esters.[9] Figure 17.3 gives a composite phase diagram for a series of different acids, showing that phase regions of similar shape occur in the same positions relative to one another, but with systematic shifts of position as the chain length changes. More interestingly, the condensed phases of the ethyl esters and fatty acids show the same close correspondence,[7,13,14] as can be seen by comparing the two phase diagrams of Fig. 17.4. Although the

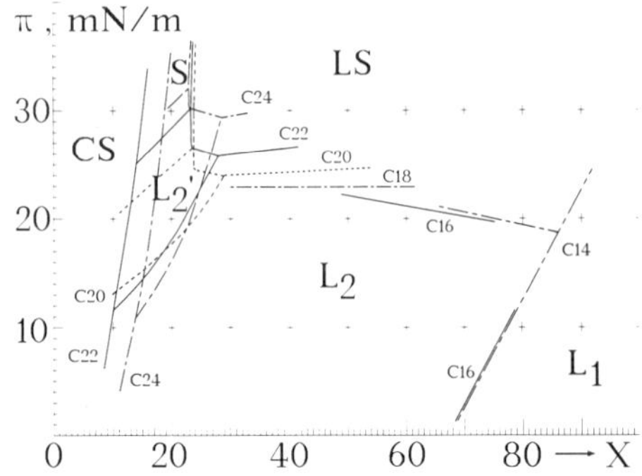

FIG. 17.3. A composite phase diagram of the even-chain fatty acids from tetradecanoic (C14) to tetracosanoic (C24), compiled from data given in Refs. 7 and 13. The abscissa x represents temperature, with one division representing 1 K, but with a different offset for each fatty acid. For chain length N, $x = T\,[\mathrm{K}] + (16 - 166N + 3N^2)/8$. The phase regions are labeled with their Harkins-Stenhagen symbol, and only those phase transitions recognized by Stenhagen are shown.

long-chain alcohols have not been dealt with in the same detail, there is considerable evidence that they also fit this pattern.[14,24] Based on these similarities, the suggestion appears plausible that the one generic phase diagram can be made applicable to very many materials by merely shifting and perhaps distorting the surface pressure and temperature scales.

From Fig. 17.3, it might appear that the major effect on the fatty acid phase diagram of changing chain length is to change the temperatures of corresponding triple points. However, the relative changes of absolute

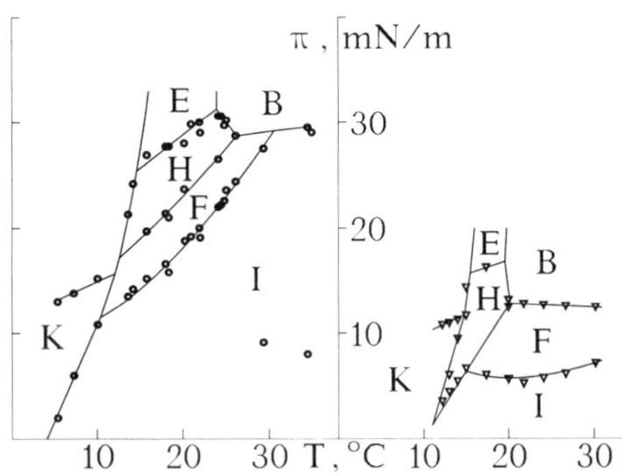

FIG. 17.4. Phase diagrams showing the smectic miscibility categories of each phase for water-surface monolayers of two substances, docosanoic acid and ethyl eicosanoate. These have the same chain length when close packed in an oriented monolayer, but different headgroups.

temperature are quite modest, typically 2% per additional carbon atom, i.e., proportionately less than the change of chain length. Close inspection shows that the characteristic surface pressures also change systematically. It is clear that there is a strong linear dependence of characteristic surface pressures on chain length. The intercept at zero chain length is not far from zero. It is well known that surface pressure in a monolayer can be considered to be due to a bulk pressure P within the monolayer acting over the monolayer molecular length λ, so that

$$\pi = \lambda P \tag{17.2}$$

Hence, it appears possible that corresponding monolayer phase transitions of fatty acids with differing chain lengths occur when the bulk pressure P within the monolayer reaches a certain critical value, of the order of magnitude 10 MPa, dependent on the temperature and the particular phase transition, but essentially independent of chain length.

A closely related conjecture was made quite a few years ago by a number of groups working in the field of biomembranes.[25,26] Although the biological lipids in cell membranes exist in the form of bilayers, it is common to justify the investigation of their properties in monolayer form by the claim that the structures in the two cases are connected. According to the principle of corresponding states, the molecular packing in a monolayer and in a bilayer may be the same, but the different environments of the individual monolayers in the two cases will lead to different tensions at their boundaries. Figure 17.5 shows the systems of forces acting in the two different cases. In the case of an unconstrained bilayer lipid membrane, the tension on any boundary must be zero. Albrecht, Gruler, and Sackmann[26] argued that a monolayer in exactly the same state but on the water surface would exert a surface pressure of 12.5 mN/m. Hence, it is possible on this scheme to talk about the effective surface pressure in a monolayer, differing from the actual surface pressure by an amount equal to the difference $\Delta\varphi$ between the total interfacial tension and that occurring in a water-surface monolayer.

Unfortunately, the experimental systems used by these investigators had only one accessible phase transition, and there was no way of changing the effective surface pressure in the bilayers. Hence, the only experimental confirmation they could offer for the conjecture was the approximate agreement of just two measured transition temperatures. Moreover, uncertainties exist about the effects of interaction between monolayers in a bilayer membrane. Not many were convinced, and there appears to be at least as much support in the literature for the incompatible conjecture that the structure of a monolayer might be independent of its environment.[27]

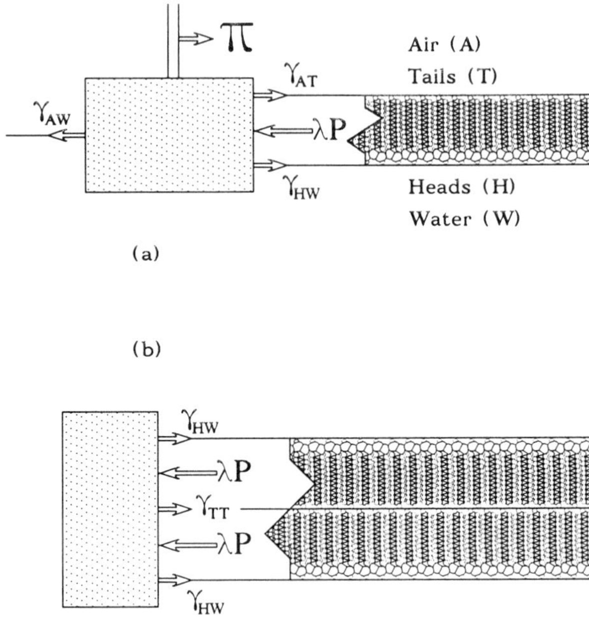

FIG. 17.5. Schematic consistent with the principle of corresponding states, showing the forces presumed to act (a) on the pontoon of a Langmuir balance when measuring the surface pressure of a monolayer-covered water surface (b) on the boundary of a lipid bilayer.

The recent vindication of the monolayer phase diagrams of Stenhagen[6,7] and Lundquist[9] for a range of substances, together with more recent work, has provided the opportunity for a much more thorough test of the conjecture. Because there is still no quantitation in the literature of the interaction between monolayers in a bilayer membrane, monolayer-bilayer comparisons such as those made by Albrecht, Gruler, and Sackmann[26] must remain unconvincing for the time being. However, even ruling these out, there are many conceivable verifications of the scheme of Fig. 17.5.

The most powerful of these verifications involves monolayers on a liquid surface, because in this case it is possible to vary the surface tension by adjusting the available area, so that rather than just comparing individual values of phase transition temperature, it is possible to compare whole sequences of values. Thanks to the richness of the polymorphism, it is possible to compare not just 1 phase transition, but up to 17, including as many as 10 triple points.

The first of these verifications involves the comparison of monolayers with the same headgroup and the same subphase but with different chain lengths, as was done in Fig. 17.3. The triple-point surface pressures should vary linearly with chain length. This is indeed found to be the case.

Secondly, it is possible to compare substances with the same subphase and chain length, but with different headgroups, as is done in Fig. 17.4. The scheme of Fig. 17.5 suggests that the temperatures of triple points should remain unchanged, but that their surface pressures should be systematically different by an amount equal to the difference $\Delta\varphi$ between the tensions at the headgroup/water interface. The ester group, showing weaker interaction with water, would be expected to give the headgroup interface of higher energy, so that the surface pressure of triple points should be lower than that of the corresponding acid. Comparing the phase diagrams of Fig. 17.4, it can be seen that these expectations are largely fulfilled. There are slight discrepancies. The following values of $\Delta\varphi$ can be determined from triple points: B/F/H ~ 17 mN/m; E/H/B ~ 14 mN/m; CS/K/H ~ 4 mN/m; and I/F/K ~ 5 mN/m. However, the variations could perhaps be explained by the slightly different molecular shape of acid and ester: Like is not really being compared with like.

In such comparisons it is unfortunate that it is as yet impossible to derive the difference $\Delta\varphi$ independently. Recent studies of surface interactions have indicated the need for at least three parameters to quantify each substance, characterizing the strength of its van der Waals, electrophilic, and nucleophilic interactions, respectively.[28-32] As yet, these parameters are only available for a very small number of surfaces, and there is far from universal agreement even about the number of parameters required.

17.3. Unambiguous Tests of the Principle

It is possible to test the conjecture so as to avoid completely the problem of changes of molecular shape. Simply, the structures of monolayers of the same substance are compared on different subphases or substrates. Monolayer structure data exist for comparison of eicosanoic acid between pure water and SiO_x; of heneicosanoic acid between calcium-containing subphases for differing pH; and of docosanoic acid between pure water, polyvinylformal, and polymethylmethacrylate.

17.3.1. Eicosanoic Acid between Water and SiO_x

The structure of an eicosanoic monolayer on the water surface is known from synchrotron diffraction studies on homologous fatty acids[10-12] using the known variation with chain length of corresponding phase regions.[13] The structure on a silicon monoxide substrate has been determined by electron diffraction.[27] Although the latter study was only carried out at 23°C, three distinct phases were observed. These were characterized by (1) pronounced molecular tilt in the direction of the nearest neighbor (NN) at 0 mN/m; (2) slight molecular tilt in the direction of the next nearest neighbor (NNN) at 7 mN/m; and (3) upright, hexagonally packed molecules at 30 mN/m. Phases L_2, L'_2, and LS with these characteristics respectively also exist on the water surface, and in this order of surface pressure.[10,13] A corner of the region of stability of the L'_2 phase with NNN tilt extends to 18°C, and the L_2-L'_2-LS triple point occurs at 25 mN/m. In spite of the 5°C discrepancy in the temperature at which the L'_2 phase can occur, this is in excellent agreement with the conjecture and implies a difference $\Delta\varphi$ in headgroup interfacial tensions of 18 mN/m.

17.3.2. Heneicosanoic Acid on Ca^{2+} Solutions

An x-ray diffraction study[33] has been carried out at 5°C over a range of pH. With increasing pH the surface pressure of the transition between tilted and upright molecules decreases. This is the well-known condensing effect of divalent metal ions, which is the opposite of the expansion observed with monovalent cations.[34] Now, electrostatic repulsion between charged headgroups leads to a negative contribution to the headgroup interfacial tension, which can be reduced by the effect of more highly charged counter ions. However, it cannot be eliminated completely or become positive. Goddard's work[34] indicated that in addition to electrostatic interactions, there must be specific interactions between the counter ions and headgroups. For example, the counter-ions may displace water molecules near the headgroups, and increase γ by reducing the amount of hydrogen bonding.

The low-pH high-surface-pressure monolayer phase of heneicosanoic acid stable at 5°C is the CS phase, the same as on a pure water surface. However, as the pH was increased beyond ~ 6, there was a transition to the S phase, which is only stable on a pure water subphase at temperatures above 6.5°C. Although the temperature discrepancy in this case is less than for eicosanoic on silicon monoxide, the experimental conditions were under much finer control, indicating that it is a real effect. This is evidence against the most simple formulation of the principle of corresponding states, and indicates that not only the surface pressure but also the temperature of a triple point can be affected by conditions on the lamellar boundaries.

At very high pH, Shih *et al.*[33] observed a phase, *X*, which appears to be crystalline rather than a mesophase. It may correspond to the phase reported by Stenhagen[7,8] on a pure water surface at surface pressures above those of the LS phase. At present, the author is involved in grazing-incidence x-ray diffraction investigations of octadecanol, which also displays such a phase.[24] While metastable, just as in the fatty acids, the collapse rate of the octadecanol phase is slow enough to allow structural determination.

17.3.3. Docosanoic Acid Between Water and Polymers

The eicosanoic acid data are less than ideal because silicon monoxide is not a well-defined chemical substance, and its surface properties are strongly affected by the density of the metastable silanol groups introduced during fabrication.[35] Recent attempts[36,37] have been made to repeat this experiment using polymer substrates in the hope that these would be more reproducible.

On polymethylmethacrylate (PMMA) at 24°C, smectic I, smectic L, smectic BH, and *X* were observed, roughly in that order of increasing surface pressure. Unfortunately no one-to-one correspondence was possible, because often more than one structure was observed for each pressure, perhaps due to polymer heterogeneity. The smectic L and smectic BH phases only occur on the water surface at temperatures higher than 24°C. Hence, as in the heneicosanoic case, a workable theory must explain changes of transition temperature as well as surface pressure.

On Formvar (polyvinylformal) only the *X* phase was observed, independent of deposition surface pressure. The difference between Formvar and PMMA, both hydrophilic polymers, can be understood in terms of Fowkes's theory of adhesion.[28-31] The ester sites in PMMA can receive hydrogen bonds initiated by the carboxylic headgroups. The hydrophilic sites in Formvar are mainly alcohols, which are much better at initiating hydrogen bonds than receiving them. The Formvar/carboxylic interface thus has higher tension, and the effective surface pressure in the monolayer is higher for a given deposition pressure.

One result of this investigation is of particular relevance to the principle of corresponding states. A plot was made of all pairs (d_{02}, d_{11}) of in-plane *d*-spacings of packings observed on different substrates and at different temperatures. There were three different loci, corresponding to different local packings of aliphatic chains. In general, the points would be expected to cover an area of the plane. In fact, within experimental error, each locus was one dimensional. There is some evidence that these loci are the same for bilayers. Hence, it may be possible to describe the state of a monolayer, independent of its environment, by the particular locus and the position of a point along it.[38]

17.3.4. Counterexamples

In the preceding examples, the conjecture compares favorably to a range of experimental results, although changing the monolayer environment is seen to change not only the surface pressure of phase transitions by a significant amount, but also their temperature by a few degrees. There are a number of other cases that are less favorable. First, the surface gas phase occurs for all materials and at all temperatures at infinitesimal surface pressure, independent of headgroup, chain length, and subphase. Second, although the temperatures for the corresponding condensed phase triple points of Fig. 17.4 agree to within 3° in the ester and acid of the same total chain length, there are large discrepancies for the L_1-L_2-LS triple point. Hence, although the principle of corresponding states was introduced to describe the L_1-L_2 transition, its experimental validity appears to be restricted to transitions between condensed phases.

17.4. Refinement of the Principle

In the original proposals[25,26] of the principle of equivalent states, it was assumed that the interfacial tensions γ_{AT} and γ_{HW} of Fig. 17.5 were characteristic for the monolayer interfaces. However, for the surface gas phase in the limit of zero density, their sum must be equal to γ_{AW}, the surface tension of a clean water surface. For ester molecules, the greater freedom for molecular deformation in the L_1 phase compared with that in the L_2 phase must lead to greater interaction of the ester groups with water and hence a greater reduction in interfacial energy than occurs for the same transition in the fatty acids. Hence, a plausible amendment to the preceding scheme is to preserve the picture of Fig. 17.5, but to allow the interfacial tensions to vary within and between phases.

However, if the interfacial tensions are variable, dependent on the particular phase of the monolayer, it is no longer possible to define a unique bulk pressure at which the transition occurs. Nevertheless the underlying idea can be subjected to more rigorous thermodynamic analysis. Define F^s to be the surface excess Helmholtz free energy for the whole system, n the number of monolayer molecules, μ their chemical potential, σ the area per molecule, γ the surface tension, and f^s the surface excess Helmholtz free energy per molecule. From the following definitions:

$$\mu \equiv \frac{\partial F^s}{\partial n}(A, n)$$
$$\gamma \equiv \frac{\partial F^s}{\partial A}(A, n) \qquad (17.3)$$
$$f^s(\sigma) \equiv F^s(n\sigma, n)/n$$

the following relationships can be deduced[38]:

$$f^s(\sigma) = \mu(\sigma) + \sigma \cdot \gamma(\sigma) \qquad (17.4a)$$

$$\gamma(\sigma) = \frac{\partial f^s}{\partial \sigma} \qquad (17.4b)$$

The thermodynamic parameters are in general also functions of temperature, but to simplify the equations, the temperature dependence is kept implicit.

If the total surface tension γ can be subdivided into an interfacial part and a bulk part proportional to molecular length λ, then f^s must be linear in λ:

$$f^s(\sigma) = I(\sigma) + \sigma \cdot \gamma_{AW} + \lambda \cdot B(\sigma) \qquad (17.5)$$

The term $\sigma \cdot \gamma_{AW}$ has been included to give a simpler mathematical expression for the surface pressure π, which is more commonly reported for insoluble monolayers than surface tension. Note that a similar linear dependence of free energy on chain length has been used successfully to describe bulk isotropic phases of long-chain compounds.[32]

Then, in conformity with Fig. 17.5, the total surface tension varies linearly with λ:

$$\gamma = f^{s'} = \frac{\mathrm{d}f^s}{\mathrm{d}\sigma} = \gamma_{AW} - \varphi - \lambda P \qquad (17.6)$$

where the symbols φ and P are used for the negatives of the derivatives of I and B, respectively:

$$\varphi = -\frac{\mathrm{d}I}{\mathrm{d}\sigma} \qquad (17.7)$$

$$P = -\frac{\mathrm{d}B}{\mathrm{d}\sigma} \qquad (17.8)$$

where φ and P will be called the interfacial and internal, pressures, respectively.

In a second-order transition, the phase remains uniform, and only its symmetry changes as the threshold is crossed. In this case, Eq. (17.6) indicates that the simple form of the conjecture is valid.

The first-order transition between two phases, 1 and 2, is more complicated. It is determined by the identity, not of the molecular Helmholtz free energies f_1^s and f_2^s for the same σ and T, but of the chemical potentials μ_1 and μ_2 for the same γ and T. For each phase, μ is given in terms of f^s from Eq. (17.4a) by

$$\mu = f^s - \gamma \cdot \sigma \tag{17.9}$$

Now it is possible to determine the conditions under which the conjecture is valid for a first-order transition. Expressed in the preceding terms, it states that the surface tension γ_{12} at which the two chemical potentials are equal,

$$\mu_1(\gamma_{12}) = \mu_2(\gamma_{12}) \tag{17.10}$$

has the following dependence on φ and λ:

$$\gamma_{12} = \gamma_{\mathrm{AW}} - \varphi - \lambda P_{12} \tag{17.11}$$

Clearly this is only possible or meaningful if the interfacial pressure φ is the same in the two phases.

If φ is assumed to be a constant, then the principle can be partly justified. Integrating Eq. (17.7) within phase 1 gives:

$$I_1(\sigma) = -\varphi\sigma + K_1 \tag{17.12}$$

while from Eq. (17.6):

$$-B_1'(\sigma_1) \equiv P = (\gamma_{\mathrm{AW}} - \varphi - \gamma)/\lambda \tag{17.13}$$

In Eq. (17.9), expressing f_1^s explicitly, we obtain

$$\begin{aligned} \mu_1 &= -\varphi\sigma_1 + K_1 + \gamma_{\mathrm{AW}}\sigma_1 + \lambda B_1(\sigma_1) - \gamma\sigma_1 \\ &= \lambda P\sigma_1 + \lambda B_1(\sigma_1) + K_1 \end{aligned} \tag{17.14}$$

Equation (17.8) defines σ_1 as a function of P, which through Eq. (17.13) is in turn a function of γ, so that Eq. (7.14) effectively defines μ_1 as a function of γ:

$$\mu_1(\gamma) = \lambda[P\sigma_1(P) + B_1(P)] + K_1 \tag{17.15}$$

Since σ_1 is defined in terms of B_1, the expression in brackets is independent of interfacial parameters. Denoting it by $\Gamma_1(P)$ and substituting P

from Eq. (17.13),

$$\mu_1(\gamma) = \lambda\Gamma_1[(\gamma_{AW} - \varphi - \gamma)/\lambda] + K_1 \qquad (17.16)$$

Hence, the constant term K of the interfacial Helmholtz free energy remains as a separate additive term in the chemical potential, but the term that is linear in σ has the effect of shifting the contribution from the bulk to different surface tensions. This latter effect is the objective basis of the conjecture. However, the K term adds slight complications. Equations (17.10) and (17.11) can be satisfied simultaneously, but only on the condition that $K_1 = K_2$. When this is the case, the pressure P_{12} of Eq. (17.11) is the root of

$$\Gamma_1(P_{12}) = \Gamma_2(P_{12}) \qquad (17.17)$$

In general, the two K's will not be equal, and the solution to (17.10) is then not given by Eq. (17.11). An approximate solution can be derived by taking the first two terms in the power series expansion for the two bulk Gibbs free energies $\Gamma(P)$. Both are differentiable, with derivatives equal to the functions $\sigma(P)$ introduced in Eq. (17.15). Hence, the variation of the phase transition surface tension is given to a good approximation by

$$\gamma_{12} \approx \gamma_{AW} - \varphi + (K_1 - K_2)/[\sigma_1(P_{12}) - \sigma_2(P_{12})] - \lambda P_{12} \qquad (17.18)$$

Hence, the conjecture is correct to the extent that the surface tension of a particular phase transition varies linearly with chain length and shows the expected gradient. In addition, the zero intercept depends on the interfacial tension φ. However, there is an additional term.

If now both φ and K differ between phases, Eq. (17.10) can be analyzed in a similar way, using linear approximations to describe the variation of the free energies. The resulting surface tension is given by:

$$\begin{aligned}\gamma_{12} &\approx \gamma_{AW} - \frac{\sigma_1\varphi_1 - \sigma_2\varphi_2}{\sigma_1 - \sigma_2} + \frac{K_1 - K_2}{\sigma_1 - \sigma_2} - \lambda P_{12} \\ &= \gamma_{AW} + \frac{\Delta I}{\Delta\sigma} - \lambda P_{12} \qquad (17.19)\end{aligned}$$

Once again, the surface pressure of the transition varies linearly with chain length with the expected gradient, but with a zero chain length intercept differing systematically from the value φ expected from Fig. 17.5.

Postscript. Since the submission of this chapter, the ideas presented in Section 17.4 have been compared with experimental data. It has now been shown[38] that they explain to within measurement accuracy the variation of the surface pressure with chain length of the two well-established second-order phase transitions L_2-LS and L'_2-S.

17.5. Conclusions

Monolayer systems of aliphatic chains can exist in many different states of order. While these include gaseous, liquid, and crystalline solid states, a very important area of the phase diagram is occupied by mesophase states, and a complete understanding of these lamellae must include them. Although they resemble both liquids and solids in specific respects, mesophases have many of their own distinct morphologies and characteristics.

In light of the new experimental evidence that has become available, it is no longer tenable to claim that the monolayer phases are those allowed by the headgroups, with the tails determining merely their regions of stability. Instead, evidence has been presented that the opposite is the case, summed up in the principle of corresponding states.

A new statement of this principle has been derived from plausible thermodynamic arguments, and it has been shown that the original conjecture is asymptotically true for second-order phase transitions in the long-chain limit. Such transitions in the lamellae occur at specific values of internal pressure, independent of headgroup, but the headgroups determine the offset of the corresponding surface pressure. The offset is not a constant characteristic of the headgroup/subphase interface but may vary with both surface pressure and temperature.

It has been shown that first-order transitions, unlike those of second order, need not, in general, occur at characteristic values of internal pressure. As a result, the region of stability of a phase bounded by at least one such transition can expand or contract when the monolayer is moved to a different environment. Hence, the formalism presented in this paper can explain the observed shifts in the temperature of triple points in terms of the differing interfacial Helmholtz free energies I of each phase and in each environment.

A lot of interest has been expressed recently in self-organizing monolayers of aliphatic chain compounds, for example, alkyltrihydroxysilanes on silica, and thiols on gold. The molecules in these films are sometimes claimed to be covalently bonded to the substrate by unbreakable bonds. If this were true, they would not be subject to the thermodynamic arguments described in this paper. However, in the former, the normal spacing of close-packed aliphatic chains, roughly 0.5 nm, is incompatible with extensive crosslinking of the headgroups, while their quality can be improved by refluxing under boiling toluene, indicating a degree of molecular mobility. For the latter, it should be noted that the interactions between sulphur and gold can be classified under Fowkes's electrophile-nucleophile scheme.[29,31] The diffraction peaks observed from these films indicate considerable molecular mobility, and an approach to thermodynamic equilibrium. The author believes that the present considerations may be the basis for understanding chain order in these systems as well as in the more weakly bonded monolayers studied thus far.

References

1. Devaux, H. (December 1903). *Proc. Verb. Soc. Sci. Phys. Nat. (Bordeaux).*
2. Adam, N. K. (1922). *Proc. Roy. Soc. (London)* **A101,** 452.
3. Adam, N. K. (1926). *Proc. Roy. Soc. (London)* **A112,** 362.
4. Fischer, A., and Sackmann, E. (1986). *J. Coll. Interface Sci.* **112,** 1.
5. Von Tscharner, V., and McConnell, H. M. (1981). *Biophys. J.* **36,** 421.
6. Ställberg-Stenhagen, S., and Stenhagen, E. (1945). *Nature* **156,** 239.
7. Stenhagen, E. (1955). In: *Determination of Organic Structures by Physical Methods* (E. A. Braude, and F. C. Nachod, eds.). Academic Press, New York.
8. Daguerre, A. (1957). *J. Rech. Centre Nat. Rech. Sci. Lab. Bellevue (Paris)* **8,** 273.
9. Lundquist, M. (1971). *Chem. Scripta* **1**(5), 197.
10. Kenn, R. M., Böhm, C., Bilbo, A. M., Peterson, I. R., and Möhwald, H. (1991). *J. Phys. Chem.* **95,** 2092.
11. Lin, B., Bohanon, T. M., Shih, M. C., and Dutta, P. (1990). *Phys. Rev. Lett.* **65,** 191.
12. Schlossman, M. L., Schwartz, D. K., Pershan, P. S., Kawamoto, E. H., Kellogg, G. J., and Lee, S. (1991). *Phys. Rev. Lett.* **66,** 1599.
13. Bibo, A. M., and Peterson, I. R. (1990). *Adv. Mater.* **2,** 309.
14. Bibo, A. M., Knobler, C. M., and Peterson, I. R. (1991). *J. Phys. Chem.* **95,** 5591.
15. Shih, M. C., Bohanon, T. M., Mikrut, J. M., Zschack, P., and Dutta, P. (1992). *Phys. Rev. A* **45,** 5734.
16. Abraham, B. M., Miyano, K., Ketterson, J. B., and Xu, S. Q. (1983). *Phys. Rev. Lett.* **51,** 1975.
17. Gallot, B., and Skoulios, A. E. (1962). *Acta Cryst.* **15,** 826.
18. Giroud-Godquin, A.-M., Latour, J.-M., and Marchon, J.-C. (1985). *Inorg. Chem.* **24,** 4452.
19. Chandrasekhar, S. (1977). *Liquid Crystals.* Cambridge University Press, Cambridge, pp. 120–122, 195–197.
20. Chandrasekhar, S. (1977). *Liquid Crystals.* Cambridge University Press, Cambridge, pp. 60–63, 108–112.
21. Nelson, D. R., and Halperin, B. I. (1979). *Phys. Rev. B* **19,** 2457.
22. Nelson, D. R., and Halperin, B. I. (1980). *Phys. Rev. B* **21,** 5312.
23. Peterson, I. R., Steitz, R., Krug, H., Voigt-Martin, I. (1990). *J. Phys. (France)* **51,** 1003.
24. Lawrie, G. A., and Barnes, G. T. (in press). *J. Coll. Interface Sci.*
25. Kimelberg, K., and Papahadjopoulos, D. (1971). *Biochim. Biophys. Acta* **233,** 805.
26. Albrecht, O., Gruler, H., and Sackmann, E. (1978). *J. Phys. (France)* **39,** 301.
27. Fischer, A., and Sackmann, E. (1986). *J. Coll. Interface Sci.* **112,** 1.

28. Fowkes, F. M., Tischler, D. O., Wolfe, J. A., Lannigan, L. A., and Ademu-John, C. M. (1984). *J. Polym. Sci. Polym. Chem. Ed.* **22,** 547.
29. Fowkes, F. M. (1987). *J. Adhesion Sci. Technol.* **1,** 7.
30. Fowkes, F. M. (1990). *J. Adhesion Sci. Technol.* **4,** 669.
31. Good, R. J., Srivatsa, N. R., Islam, M., Huang, H. T. L., and Van Oss, C. J. (1990). *J. Adhesion Sci. Technol.* **4,** 607.
32. Williamson, A. G., and Scott, R. L. (1970). *Trans. Faraday Soc.* **66,** 335.
33. Shih, M. C., Bohanon, T. M., Mikrut, J. M., Zschack, P., and Dutta, P. (1992). *J. Chem. Phys.* **96,** 1556.
34. Goddard, E. D., Kao, O., and Kung, C. (1967). *J. Coll. Interface Sci.* **24,** 297.
35. Tijburg, I., Jagiello, J., Vidal, A., and Papirer, E. (1991). *Langmuir* **7,** 2243.
36. Steitz, R., Mitchell, E. M., and Peterson, I. R. (1991). *Thin Solid Films* **205,** 124.
37. Engel, M., Merle, H. J., Peterson, I. R., Riegler, H., and Steitz, R. (1991). *Ber. Bun.-Ges. Phys. Chem.* **95,** 1514.
38. Peterson, I. R., Brzezinski, V., Kenn, R. M., and Steitz, R. (1992). *Langmuir* **8,** 2995.

Subject Index

A

AA, *see* Arachidic acid
Absorption band profile, porphyrin-doped Langmuir-Blodgett films, 301
Adhesion, cancer cells to silane films, 136
Alcohol monolayer, headgroup orientation, 169–170
Aliphatic alcohol monolayer, x-ray structure, 149–153
Alkyl-lithium reagents, surface modification, 96–99
α-Amino acids
  crystal structure, 176
  hydrogen bonding, 178
  interaction energy, 179
p-Aminobenzoic acid
  orientation in films, 52
  intercalation of Cu^{2+}, 44
Amphiphile
  electron acceptor, 244, 251–253
  electron donor, 244, 251–253
  ferrocenes as, 244, 247, 249–252
  in molecular diode, 244–245
Amphiphilic alcohol films, 145–181
Amphiphilic dyes, 296
Amphiphilic mesophases, 368–369
Amphiphilic monolayer
  nucleation template, 146
  phases, 367–371
Amphiphilic porphyrin
  dimer formation in monolayer, 71
  molecular area, 71
  molecular area with axial ligand, 73
  oxo-form of porphyrin as axial ligand, 73
Amphiphilic porphyrin films, 69–82; *see also* Polymer networks; Porphyrin networks
  formulas, 70
  spectra, 80
Amphiphilic receptors, 225
Amphiphilic triads, 246–249
Antenna molecules, 254–256
  and energy transfer, 254
  spectra, 255
  structure, 247
Arachidic acid
  latice constant, 31
  monolayers, 16–18
Area, molecular, *see* Molecular area

B

Bacteriorhodopsin
  photocell, 287–290
  photocycle, 280–281
  photoelectric behavior, 279–292
    at interface, 283–287
Bandgap energy
  cadmium sulfide film, 24–26
  zinc sulfide film, 25
Bandgap engineering, 11–12, 21–23, 29
Binding modulation, spatial, 224
Binding site, depth versus binding character, 213–214
Biomimetic systems, specificity, 210
Biomineralization, 11, 29
Biomolecular interaction
  association rate, 217–218
  bond strength, 219
  concentration effect, 214–215
  force measurement, 207–222
  ligand effect, 214–215
  membrane fluidity, 212–213
  receptor blocking, 215–219
  specificity, 210
  steric inhibition, 213–214
  steric interaction, 210–212
  surface mobility, 210–212
  temperature effects, 212–213
Biosensors, 129, 135–137, 207

Biotin-streptavidin film
bond strength, 219
layer failure scheme, 219
receptor concentration, 214
Biotin surface, description, 208–209
Blocking agent, 215–219
addition procedure, 190
n-octadecyl mercaptan, 184, 189–202
n-octadecylsilane, 190–202

C

Cadmium arachidate films, 296–312
absorption threshold, 16
doped with porphin, 302–304
Cadmium stearate films
lattice constant, 31
transfer ratios, 322–323
water evaporation rate, 320
x-ray diffraction, 323–325
Cadmium sulfide films
character, 19
electrochemical apparatus, 17
exciton diameter, 21
particle size, 16, 21
Cancer cells, adhesion to silane film, 136
Carbon monoxide, reduction in Langmuir-Blodgett films, 270–271
Catalysts
for electroless metallization, 118
ice nucleation, 145
Charge separation units, 243–246
Condensation, substrate-mediated, 356–358
Conductivity, 26–27
Contact angle
organic solvents on *p*-aminobenzoic acid, 47
organic solvents–gold, 192
water–*p*-aminobenzoic acid, 59
water on α-amino octanoic acid, 177
water–diamond, 124–125
water–ethylene diamine surface, 132
water–gold, 192
water–ligand-coated gold, 189
water–phenyltrichlorosilane, 114
water–polypropylene, 92
Contact potential difference, on multilayer films, 260–265
Copper ion complex, formation, 183–205
Corresponding state theory, 371–383
experimental verification, 374–378
refinement, 378–382
Crystal morphology
lead selenide, 32
lead sulfide–arachidic acid film, 30
Crystal structure
alkyl alcohols, 151
alcohol monolayers, 148, 152
α-amino acids, 176
Crystallinity, interaction energy, 155

D

Dark resistance, cadmium sulfide particulate film, 26–27
Defect, *see* Pinhole
Deuterium oxide, nucleation, 159
Devices, molecular, 256, 265–267
Diamond, self-assembled films on, 124–125
Dielectric friction, 343
Diffusion coefficient
fab′-lipids, 228
lateral versus chain entanglement, 338–339
Director field, 369
Docosanoic acid films
phase diagram, 368
on water and polymers, 377
Doping, nanocrystalline particulate films, 22

E

Eicosanoic acid film, on water and SiO, 375–376
Electrical double layer, in lipid membrane, 249–253
Electrochemical generation, silver film, 15
Electrode
associated character of film-covered gold, 191–193
line, 332–333
Electroless metallization, 117–123
by ligand complexation, 118–123
Electron acceptor amphiphiles, 244, 251–253
Electron donor amphiphiles, 244, 251–253
Electron hopping
in mixed monolayers, 344–346
osmium *tris*-diphenylphenanthroline film, 340–346
solvent effects, 343–346
Electron transfer
free energy difference, 244–245
photoinduced
in Langmuir-Blodgett film, 249
rate, 243–245
rate in monolayer, 244–245
Energy diagram
amphiphilic triad film, 248
lipid membrane, 266
molecular photodiode, 244–245

photocurrent via triad molecule, 266
photodiode in lipid film, 244–245
Energy transfer
antenna molecules and, 254
hole-burning spectroscopy, 299–300
Epitaxy
aliphatic alcohols on ice, 152–155
lead selenide, 33–37
lead sulfide, 31–32
water on α-amino acids, 175
Ester-alcohol monolayer, structure, 169–170
Ethyl eicosanoate film, phase diagram, 373

F

Fab′-lipid, diffusion coefficient, 228
Fab′-lipid films
diffusion, 228–230
gas-fluid transition model, 230–233
preparation, 226
Fatty acids, phase diagrams, 372–372
Ferrocene amphiphilics, 244, 247, 249–252
Film-forming chemicals
N-(2-aminoethyl)(3-aminopropyl)-trimethoxysilane, 131
amphiphilic porphyrin, 70
arachidic acid, 30
behenic acid, 263
benzyltrichlorosilane, 115
cadmium arachidate, 296
cadmium stearate, 319–323
p-(chloromethyl)phenyltrichlorosilane, 123
cholesterol, 324
L-α-dimyristophosphatidylethanolamine, 355–361
dioctadecyldimethylammomium bromide, 20
dioctadecylferrocene, 339
dipalmitoylphosphatidylcholine, 322
docosanoic acid, 368, 372
ethyl eicosanoate, 372
ethyl stearate, 325
heneicosanoic acid film, 376–377
lanthanum stearate, 325–326
(3-mercaptopropyl)trimethoxysilane, 129
nickel hexadecylcyclams, 267–269
18-nonadecenyltrichlorosilane, 45
N-octadecylferrocenecarboxamide, 336
n-octadecyl mercaptan, 185
n-octadecyltrichlorosilane, 46
oleic acid, 324
osmium *tris*-diphenylphenanthroline film, 340–346
partially fluorinated undecyloxypropionic acid, 258
perfluorolauric acid, 263
phenethyltrichlorosilane, 115
phosphatidylethanolamine, 324
poly(1-naphthol), 185
stearyl alcohol, 324
2,2′-thio*bis*ethyl acetoacetate, 185
3,3′-thio*bis*propyl acetoacetate, 186
2-(trimethoxysilyl)ethyl-2-pyridine, 120
trimethoxysilylpropyldiethylenetriamine, 136
Film lifting, comparison of methods, 321–328
Fluoropolymers
alkylation, 96–97
fluorine removal, 99–100
indirect synthesis modification, 100
oxidation, 100–102
surface modification, 95–104
Force measurement
biomolecular interaction, 207–222
in monolayer, 208
technique, 208
Force profile
and association rate, 217–218
biotin-inactive streptavidin, 216
biotin-streptavidin, 210
temperature effect on biotin-streptavidin attraction, 212
Free energy difference, in lipid membranes, 249–252
Freezing temperature
alcohol-covered water, 149–153, 171
versus aliphatic chain characteristics, 149
carboxylic acid-covered water, 149–150, 171
water under amide alcohols, 165
water on α-aminooctanoic acid, 177
water under ester alcohols, 165
water under mixed alcohol layers, 160–163, 165–168, 172–174
Friction, dielectric, *see* Dielectric friction

G

Gold electrode
blocking component, 190–192
surface treatment, 188–189

H

Headgroup
effect on quenching rate, 252–253
interfacial tension difference, 373–375
Hole-burning spectroscopy
apparatus, 296–299
energy transfer, 299–300

Hole-burning spectroscopy (*continued*)
frequency shift from environment, 306–308
hole shape in magnetic field, 308–310
hole width, 299–306
hole width versus number of layers, 302–305
hole width versus temperature, 303–305
Langmuir-Blodgett films, 293–314
shifts in magnetic field, 308–311
Stark effect, 308–311
Hydrogen bonded films
p-aminobenzoic acid intercalation, 48–62
defects, 47–48
infrared characterization, 46–48
matrices, 42
Hydrogen bonds, 45–64
in α-amino acids, 178
in biotin-streptavidin surface, 210
in film matrices, 42–68
in ice nucleation, 164
in monolayer, 164
substrate/film interaction, 377
templates, 42
Hydrophobic interaction, in biotin-blocked streptavidin attraction, 216–217

I

Ice, structure, 147–148
Ice nucleation, 149–158
alcohol chain length, 155–157
on amino acid, 174–177
catalysts, 145
effect of film solvent, 172–174
factors, 149
headgroup orientation, 164
by organic solvents, 170–174
templates, 145–181
under immiscible alcohols, 162–163
under miscible alcohols, 160–162
under modified amide-alcohol layers, 164–169
Image sensor, 288–290
Infrared layer, thickness measurement, 46
Interaction energy
crystallinity, 155
water–α-amino acids, 179
Intercalated films, preparation, 48–49
Intercalation
p-aminobenzoic acid, 44
copper, 44
lateral transport in, 49–50
matrix stability, 51
p-nitroaniline, 54
reversal conditions, 49
Interphase, 89–91
diffuse, production, 100
Ionic binding constant, monolayers, 197
Ionic networks, 72–76
structure, 74–76
Ionic replacement, reversible, 195
Ionic size, monolayer binding, 198–199

L

Langmuir wetting configuration, 350
Langmuir-Blodgett films
adhesion promoters, 109
amide alcohols, 164–169
amphiphilic alcohols, 145–181
artificial photosynthesis, 267–271
cadmium arachidate, 296–312
catalytic amphiphilic nickel cyclams, 267–271
chain length, 155–157
controllable factors, 351–355
dyes for, 296
effect on spectral broadening, 302
electrical field control, 358–361
electron hopping in, 339–346
ester alcohols, 164–169
horizontal lifting mechanisms, 317–18, 326–328
and ice nucleation, 145–181
ionic networks, 72–78
lead sulfide crystals in, 30
low temperature dynamics, 300–306
multilayered, 267–271, 349
odd/even carbon chain, 157–158
polymeric, 69–82
purple membrane, 283–284
dye-doped, spectroscopy, 293–314
Stark effect, 297
stripe deposition, 356–362
structure control, 349–364
structure variety, 350–351
substrate/monolayer interactions, 355–361
transfer ratios, 322–323
water evaporation rate from cadmium stearate, 320
X and Y lifting methods, 317–318
Langmuir-Blodgett techniques, recognition patterns, 223–236
Lattice constant
arachidic acid, 31
cadmium stearate, 31
Layer thickness, infrared measurement, 46
LB, *see* Langmuir-Blodgett films

Lead selenide
crystal morphology, 32
epitaxy, 33–37
particulate film, 22–23, 33–37
Lead sulfide
crystal structure, 31–32
epitaxy under films, 31
Lead sulfide–arachidic acid film, crystal morphology, 30
Ligand
organosilane, 121
octadecylimidazole, 73
planar alignment by, 72–76
Ligating films, 120–123
Light sensor, sandwich structure, 287–290
Line electrode, 332–333
Lipid complex films
biotin-streptavidin, 225–234
crystallization in, 233
diffusion coefficient, 228
fab′-lipid, 226–234
Lipid membranes
contact potential difference, 260–265
charge separation units in, 242–249
electrical double layer, 249–253
electron transfer kinetics, 249–254
energy diagram, 266
ferrocene derivatives in, 244, 247, 249–253
free energy difference, 249–252
by Langmuir-Blodgett technique, 241–249
mixed monolayer, 247–249
molecular photodiode, 243
multilayered systems, 243–246
nanodomains by phase separation, 256–267
phase separation by atomic force microscopy, 265–267
photodiode energy diagram, 244–245
reaction centers, 257
scanning surface-potential microscopy, 260–265
simulation of transfer process, 254–256
spacer molecules, 243
triad systems, 246–249
Lithium, polymer modifying agents, 96
Lithography
film requirements, 110–112
nanometer, 267–268
patterned films, 109–141

M

Mass spectrometry, differencial electrochemical, 270–271
Matrices
hydrogen-bonded multilayers, 42
hydrogen-bonded silane, 42–68
monolayer, 12
multilayer, 45–48
sealing components, 184
stability to intercalation, 51
Membranes, *see* Lipid membranes; Purple membrane
Mercaptosilane films, 128–130
Mesophase
amphiphilic, 368–390
definition, 369
Metal films, adhesion, 123–124
Metal oxide semiconductors, 112
capacitor, 127
transitor, 112, 128
Microscope, scanning tunnel, 16–18
Mobility, polymer surface chains, 85–87
Molecular area
alkyl alcohols, 150, 164
p-aminobenzoic acid, 56
amphiphilic porphyrin, 71
arachidic acid, 31
ester alcohols, 169
in mixed monolayers, 75
N-octadecylferrocenecarboxamide, 336
osmium *tris*-diphenylphenanthroline, 340, 342
Molecular devices, 256, 265–267
Molecular photodiode
energy diagram, 244–245
multilayered stacks, 245
Molecular rearrangement
in biotin film, 211–212
versus temperature, 212
Monolayer assemblies
electron transfer, 239–241
reaction centers, 239–241
Monolayers, *see also* Amphiphilic monolayers
aliphatic alcohol, x-ray structure, 149–153
character, 12–14
biotin-streptavidin, 209–221
chelation in, 184
corresponding state theory, 371–383
electrochemistry, 331–346
force measurement, 208
formation, 12–14
hydrocarbon/fluorohydrocarbon, 258–261
ionic binding constants, 197
ionic size effect on binding, 198–199
matrices, 12
mixed-charge surfactants, 22

Monolayers (*continued*)
phase diagrams of aliphatic chains, 367–384
tetradentate ligand, 183–205
translational diffusion, 333–339
Motion sensor, 288–290

N

Nanocrystalline particulate films
absorption coefficient, 23
absorption spectra, 23–25
bandgap energy
cadmium sulfide, 24–26
zinc sulfide, 25
cadmium sulfide
particle size, 25
resistivity, 25–26
characterization, 16–17
chemical generation, 14
copper sulfide doping, 22
crystal structure of silver, 28
electrochemical generation, 15
growth mechanism, 19–20
lead selenide, 22–23, 33–37
lead sulfide, 23
microscopic characterization, 18
mixed sulfides, 22
semiconductor, 23–26
silver, 15, 26–29
silver resistivity, 29
size quantization, 21–23
template, 12
zinc sulfide, 23
Nanodomains
analysis by microscopy, 257–265
by phase separation, 256–267
Nickel cyclams, as amphiphilic catalysts, 268–271
Nucleation, *see also* Ice nucleation
alcohol chain length effect on ice, 157–158
on crystalline amino acids, 174–177
deuterium oxide under alcohols, 159
headgroup orientation, 169–170
hydrogen bonding in, 164
induced by organic solvents, 170–174

O

N-Octadecylferrocenecarboxamide
apparent molar mass, 336–338
electrochemistry, 333–339
lateral diffusion coefficient, 333–337
monolayer diffusion, 333–339
Organosilane films
monomers, 113
photochemical properties, 115–117
two-step patterning, 122–123
ultraviolet photochemistry, 112–117
Organosilane ligands, 121
Organosilanes
structure of aromatic hydrocarbon, 113
structure of ligands with nitrogen, 121
Osmium *tris*-diphenylphenanthroline film
electron exchange rate, 342–344
electron exchange rate versus solvation, 343
Oxidation, double bonds to acid groups, 45
Oxidative coupling
by diffusion, 78–79
by reactants from substrate, 79–81
polymer networks, 76

P

PABA, *see p*-Aminobenzoic acid
p-aminobenzoic acid films
contact angle for organic solvents, 47
defect ratio, 47–48
Paint, 221
Particle size, cadmium sulfide, 25
Pattern formation methods, 224–225
PCTFE, see Poly(chlorotrifluoroethylene)
Phase diagram
aliphatic chain molecules, 367–383
docosanoic acid, 368
ethyl eicosanoate, 373
fatty acid films versus chain length, 372–373
fatty acid films versus temperature, 372–373
Phase separation, by scanning surface-potential microscopy, 260–265
Phase transfer catalyst
1,8-diazabicyclo[5.4.0]undec-7-ene, 100
tetrabutylammonium ion, 100
Photochemistry, (3-mercaptopropyl)trimethoxysilane, 129
Photocurrent
energy diagram for triad molecule, 266
function of purple membrane layers, 287
generation mechanism, 285–287
rectification by biorhodopsin films, 284–286
rectification mechanism, 285
versus amphiphilic triad shape and orientation, 247–249
Photoelectric conversion efficiency, 249–254
Photoelectrochemical cell, purple membrane, 283–285

Photosynthesis
artificial
Langmuir-Blodgett films for, 271
reaction centers in, 231–243
multielectron transfer reaction, 269
Pinhole, determination, 189
PM, *see* Purple membrane
Poly(chlorotrifluoroethylene)
kinetics of modification, 97–99
mechanism of alkylation, 96–99
surface modification, 95–100
Polyethylene, chlorination types, 102
Poly(vinylidene fluoride)
removal of fluorine, 99–100
surface modification, 101–103
Polymer networks, 69–82
coupling reaction, 76
forced planar alignment, 72–76
formation requirements, 69–70
Langmuir-Blodgett technique, 69–82
oxidative coupling, 76
perpendicular alignment, 71–72
Polymer surface
alkyl-lithium reagents, 96–99
carbonyl modification, 103–104
chain mobility, 85–87
chemical modification, 89–104
types, 91
chlorination, 102
definition, 85–87
environmental effect, 86–87
interphase modification, 89–90
modification methods, 87–88
modification of fluoropolymer, 100–102
modification of poly(chlorotrifluoro-ethane), 95–100
modified, characteristics, 90–91
polypropylene, hydroxylation, 91–95
types, 87
variation, 85–87
Polypropylene hydroxylation, 91–95
characterization of surface, 92–95
hydroboration, 94
rate, 91–93
side reactions, 94–95
Porphyrin films, 70–82
coupling prior to formation, 79–81
covalent linking, 76–81
diffusion polymerized, 78–79
electron spin resonance, 71
ionic linking, 72–76
linear dichroism, 71
mixed porphyrins, 71
structure of Langmuir-Blodgett films, 77–79
Position sensitive detector, diagram, 260, 265
Proton pumping, 280, 286
PTCS, *see* Phenyltrichlorosilane
Phenyltrichlorosilane film, mass spectra, 114
photochemistry mechanism, 114–117
ultraviolet spectra, 113–114
Photodiode, molecular, 244–245
Purple membrane, 280–290
on SnO_2, 283–285
photoelectrochemical cell, 283–285
quantum efficiency, 284–285

Q

Q state, 21
Quantum effect, purple membrane, 283–285
Quantum wires, 22
Quartz microbalance, 77
equation, 318

R

Rayleigh criterion, 111
Reaction center
in artificial photosynthesis, 231–243
in lipid membranes, 257
photosynthetic, 239–240, 280
single molecule, 265–267
Reactivity template, 133
Rearrangement, molecular, 211–212
Receptor, amphiphilic, 225
Recognition patterns, Langmuir-Blodgett films, 223–236
Resistivity, silver in particulate film, 29

S

SA films, *see* Self-assembled films
Scanning tunneling microscope, arachidic acid films, 16–18
Second harmonic analysis, silane films, 57
Self-assembled films
amphiphilic alcohols, 146–149
biological patterning, 128–130
cell adhesion, 135–138
chemically controlled, 42–45
complexation, 196–199
coordination geometry, 197–199
defect removal, 193–194
on diamond, 124–125
electroless metallization, 117–123
on functionalized diamond, 124–125
ionic binding selectivity, 197–200
ionic competition, 194–196
on gold, 183–205

Self-assembled films (*continued*)
ligand complexation, 119–123
ligation reversal, 195–196
organosilane, 112–141
patterned, 109–141
pattern transfer, 125–128
pinhole effect, 200–202
preparation method for blocking agent, 190
protein, 129–130
third level, 41–68
wetting control, 130–133
Sensors, 288–290
Silane films
cooperative effect of intercalants, 57
copper ion intercalation, 48–62
infrared spectroscopy, 50–53
intercalation reversibility, 51–54
molecular alignment, 50–59
p-nitroaniline intercalation, 54
preparation, 45–47
second harmonic analysis, 57
silane orientation, 57–58
thiol-terminated, 128–130
ultraviolet spectroscopy, 50–53
Silane matrices, 41–68
Silane-*p*-aminobenzoic acid film
contact angle, 59
second harmonic generation, 57–59
stoichiometry, 54–57
wettability, 59
Silver films
crystal structure, 28
electrochemical generation, 15
reflectivity, 16
resistivity, 29
transfer efficiency, 16
Silver ions, defect removal in mercaptan film, 193–194
Spacer molecules
effect on quenching rate, 252–253
lipid membrane, 243,246
polyimide Langmuir-Blodgett film, 246
Spatial modulation, in molecular binding, 224
Spectral hole
as label, 306–312
width, model for variation, 305–306
Spectroscopy, hole-burning, *see* Hole-burning spectroscopy
Stark effect, 297, 308–311
STM, *see* Scanning tunneling microscope
Streptavidin surface, description, 208–209
Supramolecular assemblies, 1–6, 42
Surface character, voltammogramic analysis, 188, 191, 194–196, 199, 201
Surface density, intercalants, 56

T

TBEA, *see* 2,2′-thio*bis*ethyl acetoacetate
TBPA, *see* 3,3′-thio*bis*propyl acetoacetate
Template
amphiphilic film nucleation, 146
hydrogen bonded, 42
ice nucleation, 145–181
nanocrystalline particulate films, 12
reactivity, 133
2,2′-Thio*bis*ethyl acetoacetate
as receptor, 184
preparation, 186
3,3′-Thio*bis*propyl acetoacetate
as receptor, 184
preparation, 186
Transfer process
efficiency in silver films, 16
vertical dip film, 319–321
Y-horizontal dip film, 321–326
Transfer rates
versus lifting method, 323–326
versus measurement method, 324–325
Transfer ratios, cadmium stearate, 322–323
Two-dimensional polymers, *see* Polymer networks

V

Vertical dip Langmuir-Blodgett film, transfer process, 319–321

W

Water, *see* Ice; Ice nucleation
Wettability, silane-*p*-aminobenzoic acid film, 59
Wetting configuration, Langmuir, 350
Wetting control, 130–133
Wires, quantum, 22

X

X-ray diffraction
alcohol-covered water ice, 154–155
cadmium stearate films, 323–325

Y

Y-horizontal dip Langmuir-Blodgett films, transfer process, 321–326

Z

Zinc sulfide, bandgap energy, 25

ISBN 0-12-523485-6

Nitric Oxide in Plant Resilience to Abiotic Stress